Probability at Saint-Flour

For further volumes:
http://www.springer.com/series/10212

Saint-Flour Probability Summer School

Founded in 1971, the Saint-Flour Probability Summer School is organised every year by the mathematics department of the Université Blaise Pascal at Clermont-Ferrand, France, and held in the pleasant surroundings of an 18th century seminary building in the city of Saint-Flour, located in the French Massif Central, at an altitude of 900 m.

It attracts a mixed audience of up to 70 PhD students, instructors and researchers interested in probability theory, statistics, and their applications, and lasts 2 weeks. Each summer it provides, in three high-level courses presented by international specialists, a comprehensive study of some subfields in probability theory or statistics. The participants thus have the opportunity to interact with these specialists and also to present their own research work in short lectures.

The lecture courses are written up by their authors for publication in the LNM series.

The Saint-Flour Probability Summer School is supported by:

- Université Blaise Pascal
- Centre National de la Recherche Scientifique (C.N.R.S.)
- Ministère délégué à l'Enseignement supérieur et à la Recherche

For more information, see back pages of the book and
http://math.univ-bpclermont.fr/stflour/

Jean Picard
Summer School Chairman
Laboratoire de Mathématiques
Université Blaise Pascal
63177 Aubière Cedex
France

Rick Durrett • Thomas M. Liggett
Frank Spitzer • Alain-Sol Sznitman

Interacting Particle Systems at Saint-Flour

Rick Durrett
Mathematics Department
Duke University
Durham, NC, USA

Thomas M. Liggett
Department of Mathematics
University of California
Los Angeles, CA, USA

Frank Spitzer (1926–1992)

Alain-Sol Sznitman
Department of Mathematics
Swiss Federal Institute of Technology
Zürich, Switzerland

This book contains reprints of lectures originally published in the Lecture Notes in Mathematics volumes 1608 (1995), 1464 (1991), 598 (1977) and 390 (1974).

ISBN 978-3-642-25297-6
Springer Heidelberg Dordrecht London New York

Library of Congress Control Number: 2011943168

Mathematics Subject Classification (2010): 60K25; 60Jxx; 60-02; 60K35; 60G20; 81S25

Printed on acid-free paper

Springer is part of Springer Science+Business Media (www.springer.com)

Preface

The *École d'Été de Saint-Flour*, founded in 1971 is organised every year by the *Laboratoire de Mathématiques* of the *Université Blaise Pascal* (Clermont-Ferrand II) and the *CNRS*. It is intended for PhD students, teachers and researchers who are interested in probability theory, statistics, and in applications of stochastic techniques. The summer school has been so successful in its 40 years of existence that it has long since become one of the institutions of probability as a field of scholarship.

The school has always had three main simultaneous goals:
1. to provide, in three high-level courses, a comprehensive study of 3 fields of probability theory or statistics;
2. to facilitate exchange and interaction between junior and senior participants;
3. to enable the participants to explain their own work in lectures.

The lecturers and topics of each year are chosen by the Scientific Board of the school. Further information may be found at http://math.univ-bpclermont.fr/stflour/

The published courses of Saint-Flour have, since the school's beginnings, been published in the *Lecture Notes in Mathematics* series, originally and for many years in a single annual volume, collecting 3 courses. More recently, as lecturers chose to write up their courses at greater length, they were published as individual, single-author volumes. See www.springer.com/series/7098. These books have become standard references in many subjects and are cited frequently in the literature.
As probability and statistics evolve over time, and as generations of mathematicians succeed each other, some important subtopics have been revisited more than once at Saint-Flour, at intervals of 10 years or so .

On the occasion of the 40th anniversary of the *École d'Été de Saint-Flour,* a small ad hoc committee was formed to create selections of some courses on related topics from different decades of the school's existence that would seem interesting viewed and read together. As a result Springer is releasing a number of such theme volumes under the collective name "Probability at Saint-Flour".

Jean Bertoin, Erwin Bolthausen and K. David Elworthy

Jean Picard, Pierre Bernard, Paul-Louis Hennequin
(current and past Directors of the *École d'Été de Saint-Flour*)

September 2011

Table of Contents

INTRODUCTION AUX PROCESSUS DE MARKOV

A PARAMETRE DANS Z_ν

par Frank L. SPITZER

Je suis très reconnaissant à Messieurs AMARA, VILLARD, MONTADOR, DEMONGEOT, LEDRAPPIER, HENION qui ont rédigé les conférences et qui y ont apporté beaucoup d'améliorations.

Originally published in: *Ecole d'Eté de Probabilités de Saint-Flour III – 1973*, Lecture Notes in Mathematics, Vol. **390**, 114–188, DOI: 10.1007/BFb0082435, © Springer-Verlag Berlin Heidelberg 1974, Reprint by Springer-Verlag Berlin Heidelberg 2012

CHAPITRE I

CHAMPS ALEATOIRES ET LIMITES THERMODYNAMIQUES

Soit $\Omega = \{0, 1\}^{\mathbb{Z}_\nu}$, où $\mathbb{Z}_\nu$ désigne le produit cartésien de ν ensembles d'entiers $\mathbb{Z}$. On va étudier une classe $\mathcal{M}_\nu$ de mesures de probabilités sur $(\Omega, \mathcal{F})$, où $\mathcal{F}$ est la σ-algèbre produit sur Ω . On peut voir Ω comme l'ensemble des configurations de particules sur $\mathbb{Z}_\nu$, considéré comme ensemble de sites : pour $x \in \mathbb{Z}_\nu$, $\omega(x) = 1$ si le site x est occupé, et $\omega(x) = 0$ sinon.

Pour $\nu = 1, \mathcal{M}_\nu = \mathcal{M}_1 = \mathcal{M}$ sera la classe de mesures μ qui correspondent aux chaînes de Markov stationnaires, à matrice de transition strictement positive :

Définition 1

Une mesure de probabilité μ sur $(\Omega, \mathcal{F})$ appartient à $\mathcal{M}$, s'il existe une matrice stochastique $M = [M(i, j)]$, $i, j = 0, 1$; $M(i, j) > 0$, avec $\pi = \pi M$ son (unique) mesure invariante, de sorte que

$$(1)\ \mu\left[\omega : \omega_k = x_o, \omega_{k+1} = x_1, \dots \omega_{k+n} = x_n\right] = \pi(x_o)\, M(x_o, x_1) \dots M(x_{n-1}, x_n),$$

pour tout $k \in \mathbb{Z}$ et toute suite $x_o, x_1, \dots, x_n$ à valeurs dans $\{0, 1\}$.

Le but principal de ce cours est de généraliser cette définition d'une façon naturelle de une dimension à plusieurs. Donc on va étudier la notion de chaînes stationnaires de Markov quand le paramètre t (le temps) appartient à l'espace $\mathbb{Z}_\nu$ de dimension $\nu \geqslant 1$. Pour introduire les idées on va caractériser la classe $\mathcal{M} = \mathcal{M}_1$ d'une nouvelle manière, qui se prête mieux à la généralisation. J'appelle cela méthode A, ou méthode de Gibbs. Une autre méthode, plus récente, sera traitée dans le chapitre III, dans les définitions 1 et 4. Elle s'appuie sur les probabilités conditionnelles.

Méthode A

Soit $Q = [Q(i, j)]$, $i, j \in \{0, 1\}$ une matrice positive, $Q(i, j) > 0$. Soit $\Omega_n = \{0, 1\}^{[-n,n]}$ pour tout entier $n \geqslant 1$, et soit φ une application de la frontière $\{-n-1, n+1\}$ de l'intervalle $[-n, n]$ dans $\{0, 1\}$. Disons que $\varphi(-n-1) = a$, $\varphi(n+1) = b$. Soit μ_n^{φ} la densité de probabilité définie sur Ω_n par

$$(2) \quad \mu_n(\omega) = Z_n^{-1}(\varphi)\, Q(a, \omega_{-n})\, Q(\omega_{-n}, \omega_{-n+1}) \ldots Q(\omega_n, b), \qquad \omega \in \Omega_n.$$

Ici $Z_n(\varphi)$ est une constante de normalisation, telle que

$$\sum_{\omega \in \Omega_n} \mu_n^{\varphi}(\omega) = 1.$$

En effet on voit que

$$(3) \qquad Z_n(\varphi) = Q^{2n+2}(a, b).$$

On peut considérer μ_n^{φ} comme mesure de probabilité sur $(\Omega, \mathfrak{F})$ d'une façon arbitraire compatible avec la condition de frontière, par exemple en supposant que $\omega_k = 0$ pour $k \leqslant -n-2$ et $k \geqslant n+2$.

Question : quelles sont toutes les limites vagues μ, qu'on peut obtenir par le passage à la limite (limite thermodynamique)

$$(4) \qquad \lim_{n \to \infty} \mu_n^{\varphi_n} = \mu$$

en utilisant des suites arbitraires de fonctions φ_n qui spécifient les valeurs à la frontière ? Appelons cette classe de mesures $\mathcal{G}$.

Théorème 1

$\mathcal{G} = \mathcal{M}$. *De plus, chaque suite* $\mu_n^{\varphi_n}$ *a une limite vague* μ *, qui est indépendante de la suite* φ_n *, et qui dépend de Q de la façon suivante :* μ *est la mesure de la chaîne de Markov stationnaire avec*

$$(5) \qquad M(i, j) = \frac{Q(i,j)\, r(j)}{\lambda\, r(i)}, \quad i, j = 1, 2,$$

où λ *est la plus grande valeur propre de M, et* $Mr = \lambda r$.

Démonstration

C'est une application du théorème de Frobenius sur les matrices positives. On sait que

$$(6) \qquad \lim_{n\to\infty} \frac{Q^n(i,j)}{\lambda^n} = \ell(j)\, r(i),$$

si $\ell Q = \lambda \ell$, $Qr = \lambda r$, et $\sum_i \ell(i)\, r(i) = 1$. Supposons donc que $\mu \in \mathcal{G}$, et que μ est une limite vague dans le sens de (4), en utilisant la matrice Q et les fonctions φ_n telles que $\varphi_n(n+1) = b_n$, $\varphi_n(-n-1) = a_n$. Pour illustrer l'idée générale on regarde l'évènement que $\omega_0 = \alpha$, $\omega_1 = \beta$. Alors, selon (3)

$$\mu_n^{\varphi_n}[\omega_0 = \alpha,\ \omega_1 = \beta] = \frac{Q^{n+1}(a_n, \alpha)\, Q(\alpha,\beta)\, Q^n(\beta, b_n)}{Q^{2n+2}(a_n, b_n)}.$$

Utilisant (6) et (5)

$$\lim_{n\to\infty} \mu_n^{\varphi_n}[\omega_0 = \alpha,\ \omega_1 = \beta] = \frac{1}{\lambda}\, \ell(\alpha)\, Q(\alpha,\beta)\, r(\beta) = \ell(\alpha)\, r(\alpha)\, M(\alpha,\beta);$$

Mais $\pi(\alpha) = \ell(\alpha)\, r(\alpha)$ satisfait à l'équation $\pi M = \pi$, donc on a démontré que $\mu \in \mathcal{M}$, avec la matrice M donnée par (5). Pour démontrer que $\mu \in \mathcal{M} \Longrightarrow \mu \in \mathcal{G}$, on vérifie aisément qu'on peut prendre $Q = M$.

Remarque : il est clair que tout $\mu \in \mathcal{M}$ peut s'obtenir à l'aide d'une grande famille de matrices Q différentes. La nature de cette ambiguité sera expliquée dans le chapitre II.

Le théorème 1 suggère qu'on doit définir $\mathfrak{M}_\nu$, pour $\nu \geq 1$, de la même façon que la classe $\mathcal{G}$, c'est-à-dire en utilisant comme blocs de construction les éléments d'une matrice positive Q. D'abord quelques notations. Soit

$$\Lambda_n = [-n, n]^\nu \quad , \Omega_n = \{0, 1\}^{\Lambda_n} , Q = Q(i, j) > 0 ; i, j \in \{0, 1\},$$

$$\partial \Lambda_n = \{x : x \in \mathbb{Z}_\nu \setminus \Lambda_n ; |y - x| = 1 \text{ pour un } y \in \Lambda_n\}, \bar{\Lambda}_n = \Lambda_n \cup \partial \Lambda_n ,$$

$$\varphi : \partial \Lambda_n \to \{0, 1\}.$$

Les fonctions φ sont les valeurs à la frontière de Λ_n.

Posons $\bar{\omega}(x) = \begin{cases} \omega(x) \text{ pour } x \in \Lambda_n \\ \varphi(x) \text{ pour } x \in \partial \Lambda_n \end{cases}$

On définit les densités de probabilités $\mu^\varphi_{\Lambda_n}$ sur Ω_n par

$$\text{(7)} \qquad \mu^\varphi_{\Lambda_n}(\omega) = Z_n^{-1}(\varphi) \prod_{\substack{\{x,y\}:x,y \in \bar{\Lambda}_n \\ |x-y|=1}} Q(\bar{\omega}_x, \bar{\omega}_y).$$

Evidemment (7) est l'analogue de (2). Donc on espère définir la classe $\mathfrak{M}_\nu$ comme la totalité des limites vagues des mesures données par (7). Ce n'est pas tout à fait satisfaisant, comme nous le verrons dans le chapitre III, théorème 2 (i). On doit en effet considérer aussi les limites vagues de combinaisons convexes de densités $\mu^\varphi_{\Lambda_n}$, selon φ .

Définition 2

$\mathcal{G}_Q$ est l'ensemble de toutes les mesures de probabilités μ sur $(\Omega, \mathcal{F})$ qui sont limites faibles de la forme

$$\text{(8)} \qquad \mu = \lim_{n\to\infty} \sum_{\varphi:\partial \Lambda_n \to \{0,1\}} c_n(\varphi) \mu^\varphi_{\Lambda_n} ;$$

où $c_n(\varphi) \geq 0$, $\sum_{\varphi:\partial \Lambda_n \to \{0,1\}} c_n(\varphi) = 1$. Finalement $\mathfrak{M}_\nu = \bigcup_Q \mathcal{G}_Q$.

<u>Remarque</u> : dans le chapitre III, cette définition sera remplacée par les définitions 1 et 4 (équivalentes)

En dimension un, nous avons démontré (théorème 1) que chaque classe $\mathcal{G}_Q$ consiste en une seule mesure μ, parce que la limite vague de $\mu_{\Lambda_n}^{\varphi_n}$ existe et est indépendante de $\{\varphi_n\}$. L'intérêt principal de la théorie en dimension $\nu \geqslant 2$, et de ses applications à la mécanique statistique, est du au fait que <u>la limite peut maintenant dépendre de la suite</u> φ_n des conditions à la frontière. Nous commençons par étudier ce phénomène dans un cas un peu artificiel, mais en revanche très simple. Nous démontrons ensuite le théorème remarquable, que pour $\Omega = \{0,1\}^{\mathbb{Z}^2}$, $\mathcal{G}_Q$ contient plusieurs éléments, si

$$Q = \begin{pmatrix} e^{\frac{J}{2}} & e^{-\frac{J}{2}} \\ e^{-\frac{J}{2}} & e^{\frac{J}{2}} \end{pmatrix}, \text{ et } J \text{ suffisamment grand.}$$

<u>L'arbre infini</u> : Construction : on part d'une origine O ; on obtient Λ_1 en construisant 3 branches partant de O. On obtient Λ_n, par récurrence, en construisant, à partir de chaque bout de Λ_{n-1}, 2 branches. On suppose que toutes les branches ainsi obtenues ne se coupent qu'à leurs extrémités.

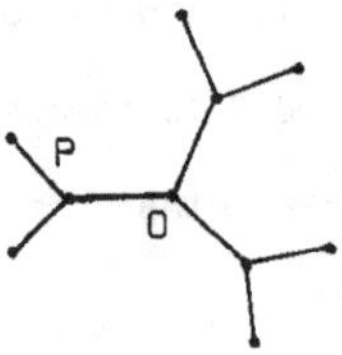

L'arbre infini est $\Lambda_\infty = \bigcup_n \Lambda_n$. Soit $\varphi : \partial \Lambda_n \to \{0, 1\}$, et soit

$$Q = \begin{pmatrix} a & 1-a \\ 1-a & a \end{pmatrix}, \quad \frac{1}{2} \leqslant a < 1.$$

On définit $\mu_{\Lambda_n}^{\varphi}$ sur $\Omega_n = \{0, 1\}^{\Lambda_n}$ suivant (7).

<u>Proposition</u>

$$\text{Pour } \varphi \equiv 1, \quad \lim_{n \to \infty} \mu_{\Lambda_n}^{\varphi} [\omega_0 = 1] \begin{cases} > \frac{1}{2} & \text{si } a > \frac{3}{4} \\ = \frac{1}{2} & \text{si } a \leqslant \frac{3}{4} \end{cases}$$

<u>Remarque</u> : pour $a > \frac{3}{4}$ nous voyons que $\mathcal{G}_Q$ contient au moins deux éléments. En effet, en partant de $\varphi_n \equiv 0$, et de $\varphi_n \equiv 1$

$$\mu^\circ = \lim_{n_k \to \infty} \mu^\circ_{n_k} \quad \text{et} \quad \mu^1 = \lim_{n_m} \mu^1_{n_m}$$

sont différents (on a choisi des sous-suites pour assurer la convergence), car par symétrie

$$\mu^\circ \ [\omega_o = 0] = \mu^1 \ [\omega_o = 1], \quad \mu^\circ \ [\omega_o = 1] = \mu^1 \ [\omega_o = 0] .$$

Donc

$$\mu^1 \ [\omega_o = 1] > \frac{1}{2} \Longrightarrow \mu^\circ \ [\omega_o = 1] = 1 - \mu^1 \ [\omega_o = 1] < \frac{1}{2} .$$

<u>Démonstration</u>

Λ_n est formé de 3 grandes branches identiques de longueur n : soit B_n l'une d'elles.

Soit : $R_n\ (1, 1) = \sum_\omega \prod_{\substack{x,y \in \overline{B}_n \\ |x-y|=1}} Q\ (\bar\omega_x, \bar\omega_y)$ la sommation sur les $\omega \in \{0, 1\}^{B_n}$ tels que $\omega_o = i$ et $\bar\omega \equiv 1$, sur ∂B_n $i = 0, 1$

$$\mu^1_n \ [\omega_o = 1] = Z_n^{-1} \ R_n^3\ (1,1) = \frac{R_n^3\ (1, 1)}{R_n^3\ (0,1) + R_n^3\ (1,1)}$$

Posons : $X_n = R_n\ (0, 1)$, $Y_n = R_n\ (1, 1)$, $T_n = \dfrac{X_n}{Y_n}$

On peut obtenir B_{n+1} en collant en P, 2 branches B_n ayant même valeur en P, d'où

$$X_{n+1} = Q\ (0, 1)\ Y_n^2 + Q\ (0, 0)\ X_n^2$$

$$Y_{n+1} = Q\ (1, 1)\ Y_n^2 + Q\ (1, 0)\ X_n^2$$

et $T_{n+1} = \dfrac{Q(0,1) + Q\ (0,0)\ T_n^2}{Q(1,1) + Q(1,0)\ T_n^2} = f(T_n)$ où $f(x) = \dfrac{a}{1-a} - \dfrac{2a-1}{(1-a)^2} \dfrac{1}{\frac{a}{1-a} + x^2}$

Suivant les valeurs de a, f(x) a pour graphe :

$a \leqslant 3/4$

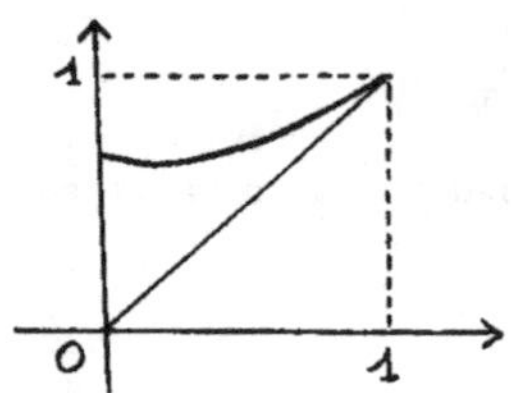

$a > 3/4$

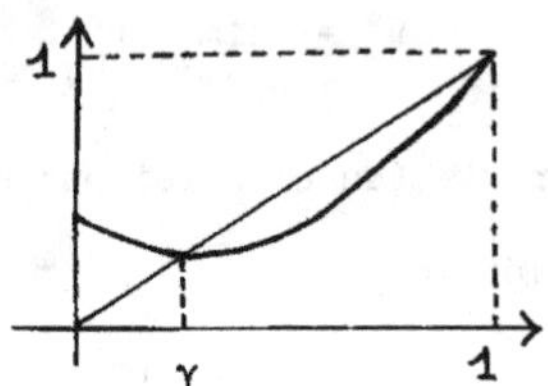

d'où pour $a \leqslant 3/4$ $\lim_{n \to +\infty} T_n = 1$

pour $a > 3/4$ $\lim_{n \to +\infty} T_n = \gamma < 1$

ainsi pour $a > 3/4$

$$\mu_n^1 \left[\omega_o = 1\right] = \frac{1}{1 + T_n^3} \to \frac{1}{1 + \gamma^3} > 1/2$$

Le Modèle d'Ising

C'est un modèle simple pour étudier la transition de phase dans un gaz ou pour le magnétisme. Il y a un grand nombre d'articles récents donnant un aperçu des progrès dans ce domaine [1, 2, 3] . Ce qui importe pour nous c'est que le modèle d'Ising en dimension $\nu \geqslant 1$ (cas symétrique ou à champ magnétique extérieur zéro) équivaut à l'étude de la classe $\mathcal{G}_Q$ de la définition 2, avec la matrice

$$\mathcal{G}_Q = \begin{pmatrix} e^{\frac{J}{2}} & e^{-\frac{J}{2}} \\ e^{-\frac{J}{2}} & e^{\frac{J}{2}} \end{pmatrix}, \quad J \geqslant 0.$$

Si $\nu = 1$, le théorème 1 entraîne que $\mathcal{G}_Q$ contient un seul élément.

Pour $\nu = 2$ (et aussi pour $\nu \geqslant 2$ par des méthodes analogues) on a le résultat suivant.

Théorème 2 *(Dobrushin et Griffiths [4], [5])*

Si $\nu = 2$, et si J est suffisamment grand, $\mathcal{G}_Q$ contient plusieurs éléments différents.

Remarque : les travaux récents [2] ont montré que, pour $J \geqslant 0$, $\nu = 2$ $\mathcal{G}_Q$ contient un seul élément, si et seulement si $0 \leqslant J \leqslant J_c$, où $J_c = \log(\sqrt{2}+1)$ (racine de sh $J = 1$). Pour $\nu \geqslant 3$ la valeur critique J_c n'est pas connue.

Démonstration

Pour mettre mieux en évidence la symétrie de Q on va construire les mesures de $\mathcal{G}_Q$ sur l'espace $\Sigma = \{+1, -1\}^{\mathbb{Z}_2}$. Alors le produit de l'équation (7) devient :

$$\mu_\Lambda^\varphi(\sigma) = Z_\Lambda^{-1}(\varphi) \prod_{\substack{\{x,y\}:x,y\in\bar\Lambda \\ |x-y|=1}} Q(\bar\sigma_x, \bar\sigma_y) \tag{9}$$

$$= Z_\Lambda^{-1}(\varphi) \exp\left[\frac{J}{2} \sum_{\substack{\{x,y\}:x,y\in\bar\Lambda \\ |x-y|=1}} \bar\sigma(x)\,\bar\sigma(y)\right], \quad \Lambda\subset\mathbb{Z}_2, \sigma\in\{+1,-1\}^\Lambda$$

Evidemment les densités μ_Λ^φ sont invariantes par rapport à la transformation $+ \to -$ et $- \to +$. Donc la démonstration sera achevée si on prend comme condition de frontière $\varphi \equiv 1$ sur $\partial\Lambda$, pour tout Λ (on notera $\mu_\Lambda^\varphi = \mu_\Lambda^+$), et si on montre que, pour tout $\varepsilon > 0$

$$\limsup_{\Lambda\nearrow\mathbb{Z}_2} \mu_\Lambda^+[\sigma(x) = -1] \leqslant \varepsilon, \quad \text{pour tout } x \in \mathbb{Z}_2, \tag{10}$$

si J est suffisamment grand.

Pour tout $\sigma\in\{+1,-1\}^\Lambda$ (dite configuration dans Λ) on définit le <u>contour</u> de σ comme la ligne brisée fermée qu'on obtient en séparant par des segments médians les points voisins x, y de $\overline{\Lambda}$ pour lesquels $\sigma_x \neq \sigma_y$.

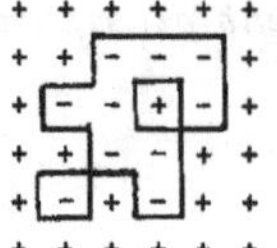

La longueur du contour sera $|\sigma|$. Dans l'exemple ci-contre $|\sigma| = 24$.

La formule (9) peut s'écrire

$$(10) \qquad \mu_\Lambda^+ (\sigma) = \tilde{Z}_\Lambda^+ \; e^{-|\sigma| J} \quad ,$$

où $\tilde{Z}_\Lambda^+$ est une constante de normalisation. Donc les configurations σ les plus probables ont les contours les plus courts. Pour expliciter cette idée on regarde les boucles d'un contour, c'est-à-dire les différentes courbes simples fermées qui sont présentes dans un contour. Dans l'exemple ci-dessus on a 3 boucles, de longueur 4, 4 et 16. Si β est une boucle, et σ une configuration, soit

$$I_\beta (\sigma) = \begin{cases} 1 \text{ si la boucle } \beta \text{ est présente dans le contour de } \sigma \\ 0 \text{ sinon} \end{cases}$$

La longueur d'une boucle β sera dénotée $|\beta|$. Soit $E_\Lambda^+ [.]$ l'espérance par rapport à μ_Λ^+ .

Lemme de Peierls

Soit β une boucle, et $|\beta| = b$. Alors

$$E_\Lambda^+ (I_\beta) \leq e^{-Jb} \text{ , indépendamment de } \Lambda \text{ .}$$

Pour la démonstration on fixe β . Pour chaque configuration σ dont le contour contient β , soit σ^* la configuration modifiée telle que

$$\sigma^*_x = \begin{cases} \sigma_x & \text{si } x \text{ est à l'extérieur de } \beta \\ -\sigma_x & \text{si } x \text{ est à l'intérieur de } \beta \end{cases}$$

Il est clair que $|\sigma^*| = |\sigma| - b$. Maintenant soit $\sum_\beta = \{\sigma$: le contour de σ contient $\beta\}$. On trouve

$$E^+_\Lambda\ [I_\beta] = \frac{\sum_{\sigma \in \Sigma_\beta} e^{-J|\sigma|}}{\sum_{\sigma \in \Sigma} e^{-J|\sigma|}} = e^{-Jb} \frac{\sum_{\sigma \in \Sigma_\beta} e^{-J|\sigma^*|}}{\sum_{\sigma \in \Sigma} e^{-J|\sigma|}} \leq e^{-Jb}$$

La dernière inégalité est due au fait que chaque terme du numérateur est présent dans le dénominateur.

Maintenant on trouve, pour Λ suffisamment grand pour que $x \in \Lambda$, que $\mu^+_\Lambda\ [\sigma(x)=-1]$

$= \mu^+_\Lambda\ [\ \exists$ une boucle β dans le contour de σ , avec x à l'intérieur]

$\leq \sum_{b=4}^{\infty} \mu^+_\Lambda\ [\ \exists$ une boucle β, avec $|\beta| = b$, avec x à l'intérieur]

$\leq \sum_{b=4}^{\infty} e^{-bJ}\ N_b$,

où N_b est le nombre de boucles de longueur $|\beta| = b$, ayant x à l'intérieur. Une telle boucle se trouve dans un carré de côté b, d'où $N_b \leq b^2\ 3^b$. Donc

$$\mu^+_\Lambda\ [\sigma\ (x) = -1] < \sum_{b=4}^{\infty} b^2\ (3\ e^{-J})^b = f\ (J).$$

et comme $f\ (J) \to 0$ quand $J \to \infty$, la démonstration de (10) est achevée.

<u>Remarque</u> : un calcul profond et remarquable de L. Onsager [6] (dont les détails n'ont été complètement justifiés que récemment [31]) donne le résultat

$$\lim_{\Lambda \nearrow \mathbb{Z}_2} \mu^+_\Lambda\ [\sigma\ (x) = +1] = \frac{1}{2} + \begin{cases} \left[1 - \frac{1}{(\text{sh } J)^4}\right]^{1/8} & \text{si sh } J \geq 1 \\ 0 & \text{si sh } J \leq 1. \end{cases}$$

C'est la célèbre formule pour la magnétisation spontanée.

CHAPITRE II

ETATS DE MARKOV ET DE GIBBS FINIS

Λ est un graphe fini sans direction et sans boucles et $\Omega = 2^{\Lambda}$.

On note : $x \sim y$ si $x = y$ ou x et y voisins

$$\partial x = \{y \in \Lambda \; : y \sim x \text{ et } y \neq x\}$$

$$\sum = \{S \subset \Lambda \; : S \neq \emptyset \text{ et } \forall (x, y) \in S \times S \quad x \sim y\}$$

Les éléments de $\sum$ s'appellent les simplexes du graphe Λ .

Définition 1

Une densité de probabilité μ sur Ω est un état de Markov si et seulement si

(i) $\forall A \in \Omega$, $\mu(A) > 0$

(ii) $\forall A \in \Omega$, $\forall x \in \Lambda \setminus A$ $\dfrac{\mu[A \cup \{x\}]}{\mu(A)} = \dfrac{\mu[(A \cap \partial x) \cup \{x\}]}{\mu(A \cap \partial x)}$

Remarque : $\dfrac{\mu(A \cup \{x\})}{\mu(A \cup \{x\}) + \mu(A)} = \dfrac{1}{1 + \dfrac{\mu(A)}{\mu(A \cup \{x\})}}$ ne dépend que de $A \cap \partial x$ d'après (ii) pour $x \in \Lambda \setminus A$; (ii) exprime donc que la probabilité que le point x soit occupé, sachant l'état d'occupation de $\Lambda \setminus \{x\}$, ne dépend que de l'occupation des voisins de x.

On appelle potentiel de Grimmett une application V de $\sum$ dans $\mathbb{R}$.

Définition 2

Une densité de probabilité μ sur Ω est un état de Gibbs si et seulement s'il existe un potentiel de Grimmett V tel que :

$$\mu(A) = Z^{-1} \exp \sum_{\substack{B \subset A \\ B \in \Sigma}} V(B) \qquad \forall A \in 2^{\Lambda} \; (Z^{-1} = \mu(\emptyset)) .$$

On a alors le :

Théorème [7]

La densité μ est un état de Markov si et seulement si c'est un état de Gibbs et le potentiel de Grimmett V est déterminé de manière unique :

$$V(A) = \sum_{B \subset A} (-1)^{|A \setminus B|} \operatorname{Log} \mu(B) \quad , \qquad A \in \Sigma .$$

Démonstration

Elle est basée sur une formule combinatoire de Moebius : si Λ est un ensemble fini quelconque et f et g deux applications de 2^{Λ} dans $\mathbb{R}$

$$g(A) = \sum_{B \subset A} f(B) , \forall A \in 2^{\Lambda} \iff f(A) = \sum_{B \subset A} (-1)^{|A \setminus B|} g(B) , \forall A \in 2^{\Lambda} .$$

1°) Soit μ un état de Markov sur Ω , définissons,pour tout $A \in \Omega$, V (A) par :

$$V(A) = \sum_{B \subset A} (-1)^{(A \setminus B)} \operatorname{Log} \mu(B)$$

Si $A \neq \emptyset$ n'est pas un simplexe,V (A) = 0 ; en effet il existe dans A deux points x et y distincts et non voisins et,en remarquant que tout $B \subset A$ est de l'une des formes C, $C \cup \{x\}$, $C \cup \{y\}$ ou $C \cup \{x, y\}$ avec $C \subset A \setminus \{x,y\}$,il vient :

$$V(A) = \sum_{C \subset A \setminus \{x,y\}} (-1)^{(A \setminus C)} \operatorname{Log} \frac{\mu(C)\, \mu(C \cup \{x,y\})}{\mu(C \cup \{x\})\, \mu(C \cup \{y\})} = 0$$

puisque d'après (ii) l'argument du logarithme vaut 1 pour tout $C \subset A \setminus \{x,y\}$.

Alors par l'inversion de Moebius :

$$\operatorname{Log} \mu(A) = \sum_{B \subset A} V(B) = V(\emptyset) + \sum_{\substack{B \subset A \\ B \in \Sigma}} V(B)$$

si bien que $\mu(A) = \mu(\emptyset) \exp \sum_{\substack{B \subset A \\ B \in \Sigma}} V(B), \forall A \in \Omega$; μ est un état de Gibbs admettant comme potentiel la restriction de V à $\sum$.

L'unicité du potentiel de Grimmettd'un état de Gibbs découle de la formule de Moebius (on peut définir V et V' sur Ω tout entier en posant $V'(B) = V(B)$ si $B \notin \sum$)

2°) Inversement soit μ un état de Gibbs de potentiel V,montrons que c'est un état de Markov. La condition (i) est trivialement vérifiée.

D'autre part : $$\frac{\mu(A \cup \{x\})}{\mu(A)} = \exp\left[\sum_{\substack{S \in \Sigma \\ S \subset A \cup \{x\}}} V(S) - \sum_{\substack{S \in \Sigma \\ S \subset A}} V(S)\right]$$

donc $$\frac{\mu(A \cup \{x\})}{\mu(A)} = \exp \sum_{\substack{S \in \Sigma \\ x \in S \\ S \subset A \cup \{x\}}} V(S)$$ ne dépend que de $A \cap \partial x$ car un simplexe contenant x ne peut contenir aucun point hors de ∂x.

CHAPITRE III

LES ETATS DE MARKOV ET DE GIBBS SUR $\mathbb{Z}_\nu$

Dans ce chapitre nous allons montrer l'équivalence entre les états de Markov (la généralisation multidimensionnelle des processus stationnaires de Markov) et les états de Gibbs locaux (l'analogue des états de Gibbs définis au chapitre précédent). Dans un premier temps nous donnerons les définitions nécessaires et nous prouverons un lemme technique, tout en énonçant les trois théorèmes principaux. Ensuite nous ferons la démonstration de ces théorèmes.

Notation

Soit $(\Omega, \mathcal{F})$ l'espace défini par $\Omega = \{0, 1\}^{\mathbb{Z}_\nu}$ et $\mathcal{F}$ la tribu engendrée par les cylindres finis.

Définition 1. [8]

Une mesure de probabilité μ sur $(\Omega, \mathcal{F})$ est un état de Markov ($\mu \in \mathfrak{M}_\nu$) si

a) $\mu\,(C) > 0$ pour tout C cylindre fini,

b) $\mu\,[\omega\,(x) = 1 \mid \omega\,(y) = f\,(y),\ y \in A]$

$= \mu\,[\omega\,(x) = 1 \mid \omega\,(y) = f\,(y),\ y \in A \cap \partial x]$,

pour tout ensemble fini A contenu dans $\mathbb{Z}_\nu$ tel que $\partial x \subset A$ et $x \notin A$, et pour toute fonction $f : A \to \{0, 1\}$.

c) Pour tout $a \in \mathbb{Z}_\nu$

$\mu\,[\omega\,(x + a) = 1 \mid \omega\,(y + a) = f\,(y),\ y \in x]$

$= \mu\,[\omega\,(x) = 1 \mid \omega\,(y) = f\,(y),\ y \in \partial x]$.

Théorème 1

Lorsque $\nu = 1$ *les états de Markov sont exactement les processus stationnaires de Markov ; c'est-à-dire que* $\mathcal{M} = \mathcal{M}_1$ *(voir chapitre I, définition 1 pour la définition de* $\mathcal{M}$ *).*

Remarque : Ce résultat justifie notre description des états de Markov comme une généralisation multidimensionnelle des processus stationnaires de Markov.

Définition 2

Une application $U : \mathbb{Z}_\nu \times \mathbb{Z}_\nu \to \mathbb{R}$ est dite un potentiel local si

a) $U(x, y) = 0$ si $|x - y| > 1$

b) $U(x, x) = U(y, y)$ pour tout $x, y \in \mathbb{Z}_\nu$

c) $U(x, y) = U(y, x)$ pour tout $x, y \in \mathbb{Z}_\nu$

d) $U(x + a, y + a) = U(x, y)$ pour tout $x, y, a \in \mathbb{Z}_\nu$.

Nous noterons par u_0 la valeur commune des $U(x, x)$.

Notation

Si B et A sont des sous-ensembles finis de $\mathbb{Z}_\nu$ on écrit

$$U(A, B) = \sum_{x \in A} \sum_{y \in B} U(x, y)$$

et

$$U(A) = U(A, A).$$

Définition 3

Soient U un potentiel local et $\Lambda \subset \mathbb{Z}_\nu$ un ensemble fini. Si $Y \subset \partial \Lambda$, l'état de Gibbs fini Π_Λ^Y sur Λ avec potentiel U et à valeurs de frontière Y est la mesure de probabilité discrète sur $\mathcal{P}(\Lambda)$ $(= 2^\Lambda)$ dont la probabilité en un point est

$$\Pi_\Lambda^Y(A) = Z_\Lambda^{-1}(Y)\exp\left[-\frac{1}{2}U(A\cup Y)\right] \text{ pour tout } A\in P(\Lambda)$$

où

$$Z_\Lambda(Y) = \sum_{B\subset\Lambda} \exp\left[-\frac{1}{2}U(B\cup Y)\right]$$

Remarque : $\Pi_\Lambda^Y(A)$ est interprétée comme la probabilité que la partie occupée de Λ soit exactement A, étant donné que la partie occupée de $\partial\Lambda$ est Y.

Nous démontrons maintenant un lemme technique qui fait voir la relation entre des états de Gibbs finis ayant le même potentiel. Ce résultat nous sera utile dans la démonstration du théorème 2.

Notation

Si μ est une mesure de probabilité sur $(\Omega, \mathfrak{F})$ et si $A\subset\Lambda$ où Λ est un sous-ensemble fini de $\mathbb{Z}_\nu$

$$\mu_\Lambda(A) = \mu\{\omega : \omega(x) = 1, x\in A \,;\, \omega(x) = 0, x\in\Lambda\setminus A\}.$$

Lemme

Soient Λ et Λ' deux ensembles finis (dans $\mathbb{Z}_\nu$) tels que $\Lambda\subset\Lambda'$. Si $Y'\subset\partial\Lambda'$ nous avons la relation suivante :

$$(1)\qquad \sum_{B\subset\Lambda'\setminus\Lambda} \Pi_{\Lambda'}^{Y'}(A\cup B) = \sum_{\substack{Y\subset\partial\Lambda \\ Y',Y \text{ compatibles}}} \alpha_{Y'}(Y)\,\Pi_\Lambda^Y(A) \quad \text{pour tout } A\subset\Lambda,$$

où

$$\alpha_{Y'}(Y) = \sum_{C\subset\Lambda'\setminus(\partial\Lambda\cap\Lambda')} \Pi_\Lambda^{Y'}(C\cup\tilde{Y}) \quad \text{où } \tilde{Y} = Y\cap\Lambda'.$$

On dit que Y' et Y sont compatibles si $Y\cap\partial\Lambda' = Y'\cap\partial\Lambda$.

Remarque. Si $\Lambda' \supset \bar{\Lambda}$ la relation devient

$$\sum_{B \subset \Lambda' \setminus \Lambda} \Pi^{Y'}_{\Lambda'} (A \cup B) = \sum_{Y \subset \partial\Lambda} \alpha_{Y'}(Y)\, \Pi^{Y}_{\Lambda} (A)$$

où

$$\alpha_{Y'}(Y) = \sum_{C \subset \Lambda' \setminus \partial\Lambda} \Pi^{Y'}_{\Lambda'} (C \cup Y).$$

Démonstration

Si l'on admet au départ que $\partial\Lambda'$ est occupé en Y', le coefficient $\alpha_{Y'}(Y)$ est la probabilité que $\partial\Lambda$ soit occupé en Y, et le membre de gauche de la relation (1) est la probabilité que Λ soit occupé en A. Par conséquent, il suffit de montrer que $\Pi^{Y}_{\Lambda}(A)$ est la probabilité conditionnelle que Λ soit occupé en A, étant donné que $\partial\Lambda$ est occupé en Y, où Y et Y' sont compatibles. Si l'on écrit $\tilde{Y}' = Y' \setminus (Y \cap Y')$, nous avons que

$$\frac{\text{Prob}\{A, Y \text{ et } Y' \text{ soient les parties occupées de } \Lambda, \partial\Lambda \text{ et } \partial\Lambda' \text{ resp.}\}}{\text{Prob}\{Y \text{ et } Y' \text{ soient les parties occupées de } \partial\Lambda \text{ et } \partial\Lambda' \text{ resp.}\}}$$

$$= \frac{\displaystyle\sum_{C \subset \Lambda' \setminus (\bar{\Lambda} \cap \Lambda')} \Pi^{Y'}_{\Lambda'} (A \cup \tilde{Y} \cup C)}{\displaystyle\sum_{C \subset \Lambda' \setminus (\bar{\Lambda} \cap \Lambda')} \sum_{B \subset \Lambda} \Pi^{Y'}_{\Lambda'} (B \cup \tilde{Y} \cup C)}$$

$$= \frac{Z^{-1}_{\Lambda'}(Y')}{Z^{-1}_{\Lambda'}(Y')} \quad \frac{\displaystyle\sum_{C} \exp\left(-\frac{1}{2} U (A \cup Y \cup C \cup \tilde{Y}')\right)}{\displaystyle\sum_{C}\sum_{B} \exp\left(-\frac{1}{2} U (B \cup Y \cup C \cup \tilde{Y}')\right)}$$

car $Y \cup \tilde{Y}' = \tilde{Y} \cup Y'$. Puisque $U(B \cup Y \cup C \cup \tilde{Y}') = U(B \cup Y) + U(Y \cup C \cup \tilde{Y}') - U(Y)$ pour tout B dans Λ (B et $C \cup \tilde{Y}'$ étant trop éloignés l'un de l'autre pour pouvoir réagir l'un sur l'autre) ce dernier quotient devient

$$\frac{\left[\sum_{C} \exp\left(-\frac{1}{2} U (Y \cup C \cup Y') + \frac{1}{2} U (Y)\right)\right] \exp\left(-\frac{1}{2} U (A \cup Y)\right)}{\left[\sum_{C} \exp\left(-\frac{1}{2} U (Y \cup C \cup \tilde{Y}') + \frac{1}{2} U (Y)\right)\right] \sum_{B \subset \Lambda} \exp\left(-\frac{1}{2} U (B \cup Y)\right)}$$

$$= \frac{\exp\left(-\frac{1}{2} U (A \cup Y)\right)}{Z_{\Lambda} (Y)} = \pi_{\Lambda}^{Y} (A)$$

et la démonstration est complète.

Maintenant nous abordons la méthode B (méthode des probabilités conditionnelles) pour définir les états de Gibbs. Cette méthode est l'invention de Dobrushin [9] et de Lanford et Ruelle [10]. La définition 4 va remplacer la définition 2 du chapitre I.

Définition 4

Si U est un potentiel local sur $\mathbb{Z}_{\nu}$, une mesure de probabilité μ sur $(\Omega, \mathfrak{J})$ est un état de Gibbs local à potentiel U ($\mu \in \mathcal{G}_U$) si

a) $\mu_{\Lambda} (A) > 0$ pour tout $A \subset \Lambda \subset Z_{\nu}$, Λ fini,

et

b) $\dfrac{\mu_{\bar{\Lambda}}(A \cup Y)}{\mu_{\partial\Lambda} (Y)} = \pi_{\Lambda}^{Y} (A)$ pour tout $A \subset \Lambda$, $Y \subset \partial\Lambda$, Λ fini, si

$\bar{\Lambda} = \Lambda \cup \partial\Lambda$

Théorème 2

Pour tout potentiel local U, $\mathcal{G}_U$ *est*

i) exactement l'ensemble (non vide) des limites de la forme (par rapport à la topologie vague) :

$$\lim_{n} \sum_{Y \subset \partial \Lambda_n} C_n (Y) \, \pi_{\Lambda_n}^{Y} (.),$$

où les Λ_n *croissent vers* $\mathbb{Z}_{\nu}$ *et où* $C_n (Y) \geq 0$ *pour tout n et tout* $Y \subset \partial\Lambda_n$ *, avec* $\sum_{Y \subset \partial\Lambda_n} C_n (Y) = 1.$

ii) convexe

et

iii) compact (dans la topologie vague).

Théorème 3

i) Si U et U' sont des potentiels locaux distincts sur $\mathbb{Z}_\nu$

$$\mathcal{G}_U \cap \mathcal{G}_{U'} = \emptyset$$

ii) $\mathcal{M}_\nu = \bigcup_U \mathcal{G}_U$.

Démonstration des théorèmes

Nous allons démontrer les théorèmes dans l'ordre suivant : 2, 3, et 1.

Démonstration (Théorème 2)

Considérons une suite $\{V_n\}$ de sous-ensembles finis de $\mathbb{Z}_\nu$ telle que $V_n \uparrow \mathbb{Z}_\nu$. Pour chaque n, et chaque choix de $Y \subset \partial V_n$ nous pouvons regarder l'état de Gibbs fini $\Pi_n^Y(.) = \Pi_{V_n}^Y(.)$ comme une mesure de probabilité sur $(\Omega, \mathfrak{F})$; si $A \subset \Lambda$, ensemble fini de $\mathbb{Z}_\nu$,

$$\Pi_n^Y(E_{\Lambda,A}) = \sum_{B \subset V_n \setminus \Lambda} \Pi_n^Y(A \cup B) \quad \text{si} \quad \Lambda \subset V_n$$

$$= 0 \qquad\qquad \text{sinon.}$$

où $E_{\Lambda,A} = \{\omega \in \Omega \ : \omega(x) = 1,\ x \in A\ ;\ \omega(x) = 0,\ x \in \Lambda \setminus A\}$.

$(\Omega, \mathfrak{F})$ pouvant s'interpréter comme étant $([0, 1]^{\mathbb{Z}_\nu}, \mathfrak{B})$, toute suite de mesures bornées possède une sous-suite convergeant vaguement. Par conséquent l'ensemble des limites vagues des combinaisons convexes des états de Gibbs finis est non vide.

Soit μ une telle limite vague, c'est-à-dire :

$$\mu(E_{\Lambda,A}) = \lim_n \sum_{Y \subset \partial V_n} C_n(Y)\, \Pi_n^Y(E_{\Lambda,A})$$

où $E_{\Lambda,A}$ est un cylindre fini et $\sum_{Y \subset \partial V_n} C_n(Y) = 1$, avec $C_n(Y) \geq 0$. Il faut montrer que $\mu \in \mathcal{G}_U$. Supposons que $\mu(E_{\Lambda,A}) > 0$ pour tout Λ, A et prenons $\Lambda \subset Z_\nu$ fini, $A \subset \Lambda$, et $B \subset \partial\Lambda$. Si n est suffisamment grand pour que $\overline{\Lambda} \subset V_n$, nous avons, selon le lemme précédent, que

$$\Pi_n^Y(E_{\overline{\Lambda},A\cup B}) = \sum_{Y' \subset \partial\overline{\Lambda}} C'_n(Y')\, \Pi_{\overline{\Lambda}}^{Y'}(E_{\overline{\Lambda},A\cup B}),$$

$$\Pi_n^Y(E_{\partial\Lambda,B}) = \sum_{Y' \subset \partial\overline{\Lambda}} C'_n(Y')\, \Pi_{\overline{\Lambda}}^{Y'}(E_{\partial\Lambda,B}).$$

Une autre application du lemme montre que

$$\Pi_{\overline{\Lambda}}^{Y'}(E_{\overline{\Lambda},A\cup B}) = \Pi_\Lambda^B(A)\quad \Pi_{\overline{\Lambda}}^{Y'}(E_{\partial\Lambda,B})$$

pour tout $Y' \subset \partial\overline{\Lambda}$. Il s'ensuit que

$$\frac{\mu_{\overline{\Lambda}}(A\cup B)}{\mu_{\partial\Lambda}(B)} = \frac{\mu(E_{\overline{\Lambda},A\cup B})}{\mu(E_{\partial\Lambda,B})} = \Pi_\Lambda^B(A),\ A \subset \Lambda,\ B \subset \partial\Lambda .$$

Il suffit donc de montrer que $\mu(E_{\Lambda,A}) = \mu_\Lambda(A) > 0$ pour tout couple Λ, A, ou de montrer qu'il existe un $M > 0$ tel que pour chaque n et pour chaque $Y \subset \partial V_n$: $M < \Pi_n^Y(E_{\Lambda,A})$. Selon le lemme $\Pi_n^Y(E_{\Lambda,A})$ est une combinaison convexe de la forme

$$\sum_{B \subset \partial\Lambda} \alpha_n(B)\, \Pi_\Lambda^B(A) \geq M = \min_{B \subset \partial\Lambda} \Pi_\Lambda^B(A) > 0,$$

et il s'ensuit que μ_{Λ} (A) est positive; par conséquent $\mathcal{G}_U$ contient toutes les limites vagues de mesures de la forme

$$\sum_{Y \subset V} C(Y) \ \pi_V^Y (.).$$

D'autre part, si $\mu \in \mathcal{G}_U$, pour tout $A \subset \Lambda$ et $Y \subset \partial\Lambda$

$$\mu_{\bar{\Lambda}} (A \cup Y) = \mu_{\partial\Lambda} (Y) \ \pi_{\Lambda}^Y (A),$$

d'où

$$\mu \ (E_{\Lambda,A}) = \sum_{Y \subset \partial\Lambda} \mu_{\partial\Lambda} (Y) \ \pi_{\Lambda}^Y (A).$$

En utilisant le lemme démontré précédemment, nous savons que, si $V \supset \bar{\Lambda}$ et si $C(Y) = \mu_{\partial V} (Y)$ pour $Y \subset \partial V$,

$$\begin{aligned}
\sum_{Y \subset \partial V} C(Y) \ \pi_V^Y (E_{\Lambda,A}) &= \sum_{Y \subset \partial V} \mu_{\partial V} (Y) \sum_{B \subset V \setminus \Lambda} \pi_V^Y (B \cup A) \\
&= \sum_{Y \subset \partial V} \mu_{\partial V} (Y) \ (\sum_{Y' \subset \partial\Lambda} \alpha_{Y'} (Y) \ \pi_{\Lambda}^{Y'} (A)) \\
&= \sum_{Y' \subset \partial\Lambda} \left[\sum_{Y \subset \partial V} \mu_{\partial V} (Y) \ \alpha_{Y'} (Y) \right] \pi_{\Lambda}^{Y'} (A) \\
&= \sum_{Y' \subset \partial\Lambda} \mu_{\partial\Lambda} (Y') \ \pi_{\Lambda}^{Y'} (A) \\
&= \mu \ (E_{\Lambda,A}).
\end{aligned}$$

Il s'ensuit que toute $\mu \in \mathcal{G}_U$ est de la forme voulue.

$\mathcal{G}_U$ est convexe : si μ, $\mu' \in \mathcal{G}_U$ et si $\lambda \in (0, 1)$ il est très facile de voir que $\lambda\mu + (1 - \lambda) \mu'$ est positive et que

$$(\lambda\mu + (1 - \lambda) \ \mu')_{\bar{\Lambda}} (A \cup Y) = (\lambda\mu + (1 - \lambda) \ \mu')_{\partial\Lambda} (Y) \ \pi_{\Lambda}^Y (A)$$

pour tout $A \subset \Lambda$, $Y \subset \partial\Lambda$.

Enfin $\mathcal{G}_U$ est compact ; si $\{\mu_\tau\}_{\tau \in T}$ est un ensemble filtrant à droite dans $\mathcal{G}_U$ qui converge dans la topologie vague vers μ, mesure de probabilité, et si Λ, A sont donnés, $\mu_\tau\,(E_{\Lambda,A}) \geq \frac{M_A}{M} > 0$ pour tout $\tau \in T$ (car μ_τ est la limite de mesures avec cette propriété). Par conséquent, μ est aussi positive sur chaque cylindre fini $E_{\Lambda,A}$.

D'autre part, pour chaque $\tau \in T$, μ_τ satisfait à

$$\frac{\mu_\tau\,(E_{\Lambda,A\cup Y})}{\mu_\tau\,(E_{\partial\Lambda,Y})} = \Pi^Y_\Lambda\,(A) \quad \text{pour tout } A,\ Y, \text{ et } \Lambda$$

d'où nous avons que

$$\frac{\mu\,(E_{\Lambda,A\cup Y})}{\mu\,(E_{\partial\Lambda,Y})} = \Pi^Y_\Lambda\,(A),$$

et nous voyons que $\mu \in \mathcal{G}_U$. C.Q.F.D.

<u>Démonstration</u> (Théorème 3)

i) Supposons que $\mu \in \mathcal{G}_U \cap \mathcal{G}_{U'}$, pour deux potentiels locaux U, U'. Nous allons montrer que U = U'. Prenons d'abord un ensemble $\Lambda \subset \mathbb{Z}_\nu$, fini, et posons $Y = \emptyset$ $A = \emptyset$.

$$Z^{-1}_{\Lambda,U}(\emptyset) = \Pi^{\emptyset}_{\Lambda,U}(\emptyset) = \frac{\mu_{\overline{\Lambda}}(\emptyset)}{\mu_{\partial\Lambda}(\emptyset)} = \Pi^{\emptyset}_{\Lambda,U'}(\emptyset) = Z^{-1}_{\Lambda,U'}(\emptyset)$$

Maintenant si $A = \{x\} \subset \Lambda$,

$$Z^{-1}_{\Lambda,U}(\emptyset) \exp\left(-\frac{1}{2} U(x,x)\right) = \frac{\mu_{\overline{\Lambda}}(\{x\})}{\mu_{\partial\Lambda}(\emptyset)} = Z^{-1}_{\Lambda,U'}(\emptyset) \exp\left(-\frac{1}{2} U'(x,x)\right)$$

d'où $u_0 = U(x,x) = U'(x,x) = u'_0$. Si $A = \{x,y\}$ où $|x-y| = 1$, choisissons $\Lambda \supset \{x, y\}$. Alors

$$\exp(-u_0 - U(x,y)) = \exp(-u'_0 - U'(x,y))$$

d'où

$$U(x,y) = U'(x,y) ;$$

Il s'ensuit que U = U'.

ii) Supposons que $\mu \in \mathcal{G}_U$ pour un potentiel local U. Choisissons $\Lambda \subset \mathbb{Z}_\nu$ fini et $Y \subset \partial\Lambda$. $\mu_{\Lambda,Y}(.) = \dfrac{\mu_{\overline{\Lambda}}(Y \cup .)}{\mu_{\partial\Lambda}(Y)}$ est un état de Gibbs (au sens du chapitre II) sur 2^{Λ} avec potentiel V :

$$V(\{x\}) = -\frac{1}{2} u_0 - \sum_{y \in Y} U(x,y)$$

$$V(\{x,y\}) = -U(x,y).$$

Alors $\mu_{\Lambda,Y}$ est un état de Markov sur 2^{Λ} (d'après Grimmett). Prenons x, Λ tels que $\partial x \subset \Lambda$ ensemble fini, et $x \notin \Lambda$. Utilisons la notation $E_{\Lambda,A} = \{\omega \in \Omega : \omega(y) = 1, y \in A ; \omega(y) = 0, y \in \Lambda \setminus A\}$. Si nous démontrons que, pour tout $A \subset \Lambda$, $\mu\{\omega(x) = 1 \mid E_{\Lambda,A}\} = \mu\{\omega(x) = 1 | E_{\Lambda,A\cap\partial x}\}$ nous pourrons en conclure que $\mu\{\omega(x) = 1 \mid E_{\Lambda,A}\} = \mu\{\omega(x) = 1 | E_{\partial x,A\cap\partial x}\}$ en vertu d'égalités sur des probabilités conditionnelles. Or en utilisant la propriété markovienne de $\mu_{\Lambda,Y}$, nous avons

$$\mu\left\{\omega(x)=1 \mid E_{\Lambda,A}\right\} = \sum_{Y\subset\partial\Lambda} \mu\left\{\omega(x)=1 \mid E_{\Lambda,A\cup Y}\right\}\ \mu\left(E_{\partial\Lambda,Y}\right)$$

$$= \sum_{Y} \mu\left(E_{\partial\Lambda,Y}\right)\left[\frac{\mu_{\Lambda,Y}\left(A\cup\{x\}\right)}{\mu_{\Lambda,Y}\left(A\cup\{x\}\right)+\mu_{\Lambda,Y}(A)}\right]$$

$$= \sum_{Y} \mu\left(E_{\partial\Lambda,Y}\right)\left[\frac{\mu_{\Lambda,Y}\left((A\cap\partial x)\cup\{x\}\right)}{\mu_{\Lambda,Y}\left((A\cap\partial x)\cup\{x\}\right)+\mu_{\Lambda,Y}(A\cap\partial x)}\right]$$

$$= \mu\left\{\omega(x)=1 \mid E_{\Lambda,A\cap\partial x}\right\} \quad .$$

La probabilité conditionnelle $\mu\left\{\omega(x)=1 \mid E_{\partial x,A\cap\partial x}\right\}$ est invariante par translation, car U l'est, et par conséquent $\mu_{\Lambda,Y}$ aussi. Puisque la condition de positivité pour un état de Gibbs est la même que celle d'un état de Markov, il s'ensuit que $\mu\in\mathfrak{M}_\nu$.

Maintenant supposons que $\mu\in\mathfrak{M}_\nu$. Pour tout ensemble fini $\Lambda\subset\mathbb{Z}_\nu$ et $Y\subset\partial\Lambda$, la mesure

$$\nu_\Lambda^Y(A) = \mu\left\{E_{\Lambda,A} \mid E_{\partial\Lambda,Y}\right\}$$

est un état de Markov fini sur 2^Λ. (Il suffit de voir que

$$\frac{\nu_\Lambda^Y\left(A\cup\{x\}\right)}{\nu_\Lambda^Y(A)+\nu_\Lambda^Y\left(A\cup\{x\}\right)} = \frac{\nu_\Lambda^Y\left((A\cap\partial x)\cup\{x\}\right)}{\nu_\Lambda^Y(A\cap\partial x)+\nu_\Lambda^Y\left((A\cap\partial x)\cup\{x\}\right)}$$

en vertu de la propriété markovienne de μ).

En appliquant le théorème de Grimmett à ν_Λ^Y , nous voyons que c'est un état de Gibbs fini sur 2^Λ et que $\nu_\Lambda^Y(A) = \nu_\Lambda^Y(\emptyset)\exp\left(\sum_{\substack{B\in\Sigma\\ B\subset A}} V_\Lambda^Y(B)\right)$, où σ est l'ensemble des simplexes de Λ et où V_Λ^Y est le potentiel donné par $V_\Lambda^Y(A) = \sum_{B\subset A}(-1)^{|A\setminus B|}\log\ \nu_\Lambda^Y(B)$ pour $A\in\Sigma$. Les seuls simplexes de

Λ sont les singletons et les couples $\{x, y\}$ où $|x - y| = 1$. Pour le moment nous supposons que $Y = \phi$, et écrivons $\nu_\Lambda^\phi = \nu_\Lambda$, $V_\Lambda^\phi = V_\Lambda$, $\{x\} = x$, $\{x\} \cup \{y\} = x \cup y$.

La formule de Grimett donne

$$V_\Lambda (x) = \log \frac{\nu_\Lambda (x)}{\nu_\Lambda (\phi)} , \quad x \in \Lambda$$

$$V_\Lambda (x \cup y) = \log \frac{\nu_\Lambda (x \cup y)\, \nu_\Lambda(\phi)}{\nu_\Lambda (x)\, \nu_\Lambda (y)} , \quad |x - y| = 1, \quad x, y \in \Lambda .$$

On voit facilement que le potentiel V_Λ est déterminé par les probabilités conditionnelles de μ, i.e.,

$$\mu \left[\omega_x = 1 \mid \omega = 0 \text{ sur } \partial x\right] = \frac{\nu_\Lambda(x)}{\nu_\Lambda(\phi) + \nu_\Lambda(x)} = \frac{1}{1 + \left(\frac{\nu_\Lambda (x)}{\nu_\Lambda (\phi)}\right)^{-1}} = \frac{1}{1 + \exp\left[-V_\Lambda(x)\right]}$$

$$\mu \left[\omega_x = 1 \;\middle|\; \begin{matrix} \omega_y = 1 \\ \omega = 0 \text{ sur } \partial x \backslash y \end{matrix}\right] = \frac{\nu_\Lambda (x \cup y)}{\nu_\Lambda (x \cup y) + \nu_\Lambda(x)} = \frac{1}{1 + \exp\left[-V_\Lambda(x \cup y) - V_\Lambda(y)\right]} .$$

Mais $\mu \in \mathfrak{M}_\nu$ et Def. 1, (c), entraîne que les probabilités conditionnelles sont invariantes par translation. Donc le potentiel V_Λ est invariant et indépendant de Λ. Cela nous permet de définir le potentiel local

$$U (x, x) = - 2 V_\Lambda (x) , \quad x \in \mathbb{Z}_\nu$$

$$U (x, y) = - V_\Lambda (x, y) , \quad |x - y| = 1, \quad x, y \in \mathbb{Z}_\nu$$

$$= 0 \qquad \text{si} \quad |x - y| > 1,$$

pour tout Λ qui contient x, y. Il s'ensuit que ν_Λ a la représentation

$$\nu_\Lambda (A) = Z_\Lambda^{-1} \exp \left[- \frac{1}{2} U (A)\right] , \qquad A \subset \Lambda .$$

ν_Λ est un état de Gibbs fini, dans le sens de Déf.3, en effet $\nu_\Lambda (A) = \pi_\Lambda^\phi (A)$.

Si la frontière $\partial\Lambda$ de Λ est occupée dans $Y \subset \partial\Lambda$, il faut modifier le potentiel V_Λ près de la frontière $\partial\Lambda$ pour obtenir V_Λ^Y. Exactement comme dans le cas $Y = \phi$, on voit que les probabilités conditionnelles de μ déterminent le potentiel V_Λ^Y. Donc ils déterminent ν_Λ^Y comme état de Gibbs sur 2^Λ. Mais il est évident que Π_Λ^Y est aussi un état de Gibbs sur 2^Λ. Si son potentiel $U(x,y)$ est défini comme plus haut,il aura les probabilités conditionnelles désirées. Par conséquent

$$\Pi_\Lambda^Y(A) = \nu_\Lambda^Y(A) = \frac{\mu_{\bar{\Lambda}}(A \cup Y)}{\mu_{\partial\Lambda}(Y)}, \quad A \subset \Lambda, \; Y \subset \partial\Lambda .$$

Il s'ensuit que $\mu \in \mathcal{G}_U$, et la preuve du théorème 3 est complète.

Remarques :

Le théorème 3 montre qu'on peut substituer pour la condition (b) dans la définition 4, la condition

$$\text{(b)'} \qquad \frac{\mu_{\bar{\Lambda}}(A \cup Y)}{\mu_{\partial\Lambda}(Y)} = \Pi_\Lambda^Y(A), \text{ pour } A \subset \Lambda, \quad |\Lambda| = 1.$$

(les conditions (a) et (b)' entraînent que μ est un état de Markov, donc un état de Gibbs local)

Voir [19, 34] pour les résultats analogues concernant les états de Gibbs sur $\{0, 1\}^S$, S un ensemble dénombrable quelconque.

Démonstration (Théorème 1).

i) $\mathcal{M} \subset \mathcal{M}_1$: soit μ un processus stationnaire de Markov. La positivité et l'invariance par translation des probabilités conditionnelles de μ sont des conséquences directes de la positivité de M et l'invariance par translation de μ.

Il faut montrer que pour tout $A \subset Z$ fini, et toute $f : A \to \{0, 1\}$ où $x \notin A$ et $\partial x \subset A$,

$$\mu\{\omega_x = 1 \mid \omega_y = f_y,\ y \in A\} = \mu\{\omega_x = 1 \mid \omega_y = f_y,\ y \in A \cap \partial x\}. \quad (*)$$

Supposons d'abord que A soit consécutif : c'est-à-dire que

$A = \{y, y+1, \ldots, x-1, x+1, \ldots z-1, z\}$.

Le membre de gauche de $(*)$ devient

$$\frac{\mu\{\omega_x = 1 ; \omega_t = f_t,\ t \in A\}}{\mu\{\omega_x = 0 ; \omega_t = f_t,\ t \in A\} + \mu\{\omega_x = 1 ; \omega_t = f_t,\ t \in A\}}$$

$$= \frac{M(f_{x-1}, 1)\, M(1, f_x)}{M(f_{x-1}, 0)\, M(0, f_{x+1}) + M(f_{x-1}, 1)\, M(1, f_{x+1})} \cdot \frac{\pi(v_{x-1})}{\pi(v_{x-1})}$$

et ceci est exactement le membre de droite de $(*)$. Lorsque A n'est pas consécutif le principe de la démonstration reste le même. On fait la somme des probabilités des différentes possibilités sur les "trous" de A, et après l'élimination des facteurs communs.le nouveau quotient est exactement la probabilité que $\omega_x = 1$, conditionnée par les valeurs que prend ω sur $A \cap \partial x$. Par conséquent μ est un état de Markov de dimension 1.

ii) $\mathfrak{M}_1 \subset \mathfrak{M}$. Si μ est dans $\mathfrak{M}_1$, il est un état de Gibbs local pour un potentiel local U (théorème 3). D'après le théorème 2, il est donc la limite vague, pour une suite d'intervalles I_n qui croissent vers $\mathbb{Z}$, de mesures de la forme $\sum_{Y \subset \partial I_n} C(Y)\ \Pi_{I_n}^Y(.)$ où $C(Y) \geq 0$, $\sum C(Y) = 1$.

$$\text{Or } \Pi_{I_n}^Y (A) = Z_{I_n}^{-1}(Y) \exp\left(-\frac{1}{2}\ U(A \cup Y)\right)$$

$$= \mu_n(A) \text{ (définition)}$$

$$= \tilde{Z}_{I_n}^{-1}(Y)\ Q(V_{a-1}, V_a) \ldots Q(V_b, V_{b+1})$$

où $I_n = [a, b]$ et où $V_t = 1$ si $t \in A \cup Y$, et $= 0$ sinon.

En effet, il s'agit d'écrire

$$\log Q(1, 0) = \log Q(0, 1) - \frac{1}{4} u_0$$

$$\log Q(0, 0) = 0$$

$$\text{et} \quad \log Q(1, 1) = - U(0, 1) - \frac{1}{2} u_0 .$$

Le théorème 1 du chapitre I nous assure que μ_n^Y converge vers une mesure $\bar{\mu}$ dans $\mathfrak{M}$ qui est indépendante du choix de Y. Il s'ensuit que la limite commune des $\{\Pi_{I_n}^Y\}$ s'identifie nécessairement à la mesure $\mu \in \mathfrak{M}_1$ qui est la limite de combinaisons convexes des $\Pi_{I_n}^Y$.

(N.B. $|\partial I_n| = 2$ pour tout n et par conséquent les combinaisons sont toujours de quatre termes).

Nous avons donc $\mathfrak{M}_1 \subset \mathfrak{M}$.

Dans le chapitre qui suit on étudiera pour quels potentiels locaux on a transition de phase, c'est-à-dire pour quels U on a $|\ \ _U| > 1$. On sait déjà (théorème 1) que c'est impossible si la dimension $\nu = 1$, tandis que c'est possible (théorème 2, chapitre I) si $\nu > 2$).

CHAPITRE IV

TRANSITION DE PHASE POUR LE MODELE D'ISING D'UN GAZ

Nous supposerons dans ce chapitre que le potentiel local U est isotrope, de la forme :

$$U(x, y) = \begin{cases} u_0 & , \text{ si } x = y \\ u_1 & , \text{ si } |x-y| = 1 \\ 0 & , \text{ si } |x-y| > 1 \end{cases}$$

<u>Question</u> : Dans quelle partie du plan (u_0, u_1) la classe $\mathcal{G}_U$ est-elle réduite à un seul élément ?

Nous démontrerons l'unicité dans le demi-plan $u_1 \leq 0$, lorsque $u_0 + 2\nu u_1 \neq 0$, c'est-à-dire en dehors d'une demi-droite d'origine O ; ce sera l'objet du théorème de Ruelle ; puis nous donnerons des indications sur la manière de résoudre le cas $u_1 > 0$.

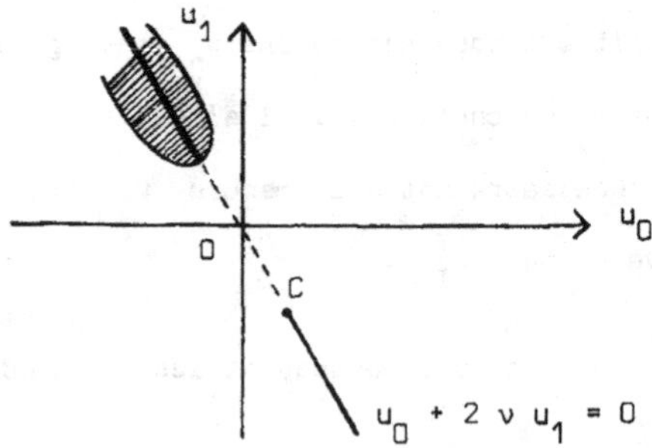

Un calcul simple montre que le modèle d'Ising du chapitre I, correspond au modèle présent avec la condition que $u_0 + 2\nu u_1 = 0$, et $u_1 = -2J$.

Donc nous savons aussi (théorème 2, chapitre I) l'existence d'un point C sur la demi-droite, tel que $|\mathcal{G}_U| > 1$ si le point (u_0, u_1) est en dessous de C. Le théorème 1 ci-dessous montre que, inversement $|\mathcal{G}_U| = 1$, si (u_0, u_1) est suffisamment proche de l'origine.

Proposition 1

Un état de Markov μ appartient à $\mathcal{G}_U$ si et seulement si

$$\mu\left(\{\omega(x) = 1\} \mid \mathcal{A}_k\right) = \frac{1}{1 + e^{\frac{u_0}{2} + ku_1}}, \forall k \in [0, 2\nu], \forall x \in \mathbb{Z}_\nu, \text{ où}$$

$\mathcal{A}_k$ désigne l'évènement $\{\omega \in \{0, 1\}^{\mathbb{Z}_\nu} = \Omega \;;\; \omega(y_1) = 1$, pour k voisins y_1 de x ; $\omega(y_2) = 0$ pour les $2\nu - k$ voisins restants y_2 de x$\}$.

Démonstration

Condition nécessaire :

$$\mu\left(\{\omega(x) = 1\} \mid \mathcal{A}_k\right) = \frac{\mu\left(\{\omega(x) = 1\} \cap \mathcal{A}_k\right)}{\mu(\mathcal{A}_k)}.$$

Soit Y l'ensemble des k voisins de x occupés ; nous avons :

$$\mu\left(\{\omega(x) = 1\} \cap \mathcal{A}_k\right) = \mu_{x \cup \partial x}(Y \cup x) \text{ et } \mu(\mathcal{A}_k) = \mu_{\partial x}(Y)$$

d'où
$$\mu\left(\{\omega(x) = 1\} \mid \mathcal{A}_k\right) = \Pi_x^Y(x) = Z_x^{-1}(Y)\, e^{-\frac{1}{2} U(x \cup Y)} = \frac{1}{1 + e^{\frac{u_0}{2} + k u_1}},$$

car $Z_x(Y)\left(\Pi_x^Y(x) + \Pi_x^Y(\emptyset)\right) = e^{-\frac{1}{2} U(x \cup Y)} + e^{-\frac{1}{2} U(Y)}$

et $U(x \cup Y) - U(Y) = U(x) + 2\, U(x, Y) = u_0 + 2\, k\, u_1$

Condition suffisante :

$\forall A \subset \Lambda$ fini $\subset \mathbb{Z}_\nu$, nous avons (d'après le théorème III 3 b)

(1) $\mu_\Lambda(A) > 0$, puisque μ est un état de Markov

et

$$(2) \quad \frac{1}{1 + e^{\frac{1}{2}(U(x \cup Y) - U(Y))}} = \frac{\mu_{x \cup \partial x}(X \cup Y)}{\mu_{\partial x}(Y)} = \mu(\{\omega(x) = 1\} \mid \mathcal{A}_k)$$

$$= \frac{1}{1 + e^{\frac{u_0}{2} + |Y|\, u_1}}, \quad \forall Y \subset \partial x$$

Or, comme μ est un état de Markov, c'est un état de Gibbs appartenant à une classe $\mathcal{G}_{U_1}$ d'où, comme les relations ci-dessus déterminent entièrement U (cf démonstration du théorème 3 du chapitre III), et comme les classes sont disjointes, $U_1 = U$.

Remarque : la proposition 1 est évidemment valable pour le modèle d'Ising ; on remarque que :

$$u_0 + 2\nu u_1 = 0 \Longleftrightarrow \mu(\{\omega(x) = 1\} \mid \mathcal{A}_k) = \mu(\{\omega(x) = 0\} \mid \mathcal{A}_{2\nu-k})$$

On retrouve donc la symétrie de μ pour les configurations obtenues en échangeant les + et les -, vue dans le modèle avec champ magnétique nul étudié dans le chapitre I.

Théorème 1

$\mathcal{G}_U$ possède un seul élément si :

(i) pour u_0 fixé, u_1 est suffisamment petit

(interaction faible)

ou (ii) pour u_1 fixé, u_0 est positif, suffisamment grand

(densité basse)

Démonstration

Démontrons (i) et (ii) en utilisant une méthode fondée sur l'équation de Kirkwood-Salsburg.

Définition

On appelle fonction de corrélation de la mesure μ sur $\Omega = \{0, 1\}^{\mathbb{Z}_\nu}$ la fonction ρ définie par :

$$\rho\ (A) = \mu\ (\{\omega\ ;\ \omega\ (x) = 1,\ \forall x \in A\}),\ \forall A \subset \mathbb{Z}_\nu$$

Lemme 1

ρ détermine entièrement μ_Λ , pour toute partie finie Λ de $\mathbb{Z}_\nu$, donc détermine entièrement μ .

Démonstration

Nous avons, par définition de ρ et de μ_Λ :

$$\rho\ (A) = \sum_{A \subset B \subset \Lambda} \mu_\Lambda(B)$$

d'où le résultat, en utilisant la formule d'inversion de Möbius :

$$\mu_\Lambda\ (A) = \sum_{C \subset \Lambda \setminus A} (-1)^{|C|}\ \rho\ (A \cup C)$$

Montrons maintenant que ρ satisfait à l'équation de Kirkwood-Salsburg ; nous prouverons ensuite que cette équation a une solution unique sous les conditions (i) ou (ii) ; l'application du lemme 1 achèvera alors la démonstration.

Lemme 2

Soit ρ la fonction de corrélation d'un état μ de $\mathcal{G}_U$; soit x un point quelconque de $\mathbb{Z}_\nu$ et A une partie finie de $\mathbb{Z}_\nu$ possédant x ; alors, si $A'_x = A^c \cap \partial x$ et si on pose :

$$g_x(A, B) = \left(1 + e^{\left(\frac{U(x,x)}{2} + \sum_{y \in (A \setminus x) \cup B} U(x,y)\right)}\right)^{-1}, \quad \forall B \subset A'_x,$$

on a :

$$\rho(A) = \sum_{C \subset A'_x} \rho((A \setminus x) \cup C)(-1)^{|C|} \sum_{B \subset C} g_x(A, B)(-1)^{|B|}$$

<u>Démonstration</u>

Nous avons, d'après la proposition 1 :

$$\mu(\{\omega \quad (x) = 1\} | \mathcal{A}_{|(A \cap \partial x) \cup B|}) = g_x(A, B)$$

d'où :

$$\rho(A) = \sum_{B \subset A'_x} \mu(\{\omega \ ; \omega(y) = 1, \forall y \in A \cup B \ ; \omega(y) = 0, \forall y \in A'_x \setminus B\})$$

$$= \sum_{B \subset A'_x} g_x(A,B)\, \mu(\{\omega \ ; \omega(y) = 1, \forall y \in (A \setminus x) \cup B \ ; \omega(y) = 0, \forall y \in A'_x \setminus B\})$$

(en appliquant la formule de Bayes et la propriété (2) de la définition d'un état de Markov)

$$= \sum_{B \subset A'_x} g_x(A, B)\, \mu_{(A \setminus x) \cup A'_x}((A \setminus x) \cup B)$$

$$= \sum_{B \subset A'_x} g_x(A, B) \sum_{E \subset A'_x \setminus B} (-1)^{|E|} \rho((A \setminus x) \cup B \cup E),$$

d'où la relation cherchée, en posant $C = B \cup E$.

L'équation de Kirkwood-Salsburg s'écrit alors :

$$\rho(A) = \sum_{D \subset \mathbb{Z}_\nu} K_x(A, D)\, \rho(D), \ \forall x \in A \text{ tel que } |A| < +\infty,$$

où le noyau $K_x(A, D)$ est sommable, valant 0 pour tout D sauf pour un nombre fini.

Par la méthode habituelle des opérateurs de contraction, nous obtenons :

Lemme 3

Soit ρ la fonction de corrélation de l'état μ de $\mathcal{G}_U$; si K est une contraction, c'est-à-dire si :

$$\sum_{D \in \mathbb{Z}_\nu} |K_x (A, D)| \leqslant k < 1, \forall x \in A \text{ tel que } |A| < +\infty, \text{ alors } \mathcal{G}_U \text{ a}$$

un seul élément.

Démonstration

Soit μ et $\tilde{\mu}$ dans $\mathcal{G}_U$, de fonctions de corrélation ρ et $\tilde{\rho}$; alors pour tout x de A tel que $|A| < +\infty$, nous avons, d'après le lemme 2 :

$$\rho (A) - \tilde{\rho} (A) = \sum_{D \in \mathbb{Z}_\nu} K_x (A, D) (\rho (D) - \tilde{\rho} (D))$$

d'où $|\rho (A) - \tilde{\rho} (A)| \leqslant \sum_{D \subset \mathbb{Z}_\nu} |K_x (A, D)| \; |\rho (D) - \tilde{\rho} (D)|$

$$\leqslant k \sup_{D \in \mathbb{Z}_\nu} (|\rho (D) - \tilde{\rho} (D)|), \text{ avec } k < 1.$$

Comme le sup est atteint pour D fini, alors $\rho (A) = \tilde{\rho} (A)$, d'où, d'après le lemme 1, $\mu = \tilde{\mu}$.

Il est aisé de voir que les conditions (i) ou (ii) nous placent dans les hypothèses du lemme 3 pour le noyau K, d'où la conclusion de la preuve du théorème 1.

Nous allons maintenant démontrer le théorème de Ruelle [11] ; pour cela, nous utiliserons deux théorèmes auxiliaires :

Inégalité de Griffiths-Holley (corollaire du théorème 5, chapitre VI)

Soit Λ un ensemble fini arbitraire et soit μ_1 et μ_2 deux densités de probabilité sur 2^Λ telles que :

$$\mu_1 (A \cup B) \; \mu_2 (A \cap B) \geqslant \mu_1 (A) \mu_2 (B), \forall A, B \in 2^\Lambda ,$$

alors, pour toute fonction f à valeurs réelles croissante sur 2^Λ (partiellement ordonné par l'inclusion, la croissance étant large), on a :

$$\sum_{A \subset \Lambda} \mu_1 (A)\, f (A) \geqslant \sum_{A \subset \Lambda} \mu_2 (A)\, f (A)$$

(ce qui équivaut à : il existe une probabilité ν sur $2^\Lambda \times 2^\Lambda$ de marginales μ_1 et μ_2 telles que :

$$\nu (A, B) > 0 \longrightarrow B \subset A)$$

Exemple

Si μ_1 et μ_2 ont pour densités $\Pi_\Lambda^{Y_1}$ et $\Pi_\Lambda^{Y_2}$ respectivement, c'est-à-dire sont deux états de Gibbs correspondant à U tel que $u_1 \leqslant 0$, alors l'hypothèse de ce théorème est satisfaite si $Y_1 \supset Y_2$.

Théorème de Lee-Yang (cf démonstration dans [1] p. 108)

Soit Λ un ensemble fini arbitraire et soit un noyau $a (x, y)$ symétrique sur Λ tel que :

$$- 1 \leqslant a (x, y) = a (y, x) \leqslant 1, \qquad \forall x, y \in \Lambda$$

Alors les zéros de $\mathcal{G} (z) = \sum_{A \subset \Lambda} z^{|A|} \prod_{x \in A} \prod_{y \in \Lambda \setminus A} a (x, y)$

se trouvent sur le cercle $|z| = 1$

Définition

Un potentiel local U est dit attractif, si $U (x, y) \leqslant 0$, lorsque $x \neq y$.

Théorème de Ruelle

Si U est attractif ($u_1 \leqslant 0$), alors il ne peut y avoir transition de phase que sur la droite d'équation $u_0 + 2\, \nu\, u_1 = 0$.

Démonstration

Elle comporte 11 étapes.

(1) a) Pour toute partie finie Λ de $\mathbb{Z}_\nu$ et $Y \subset \partial\Lambda$, nous désignerons par ρ_Λ^Y la fonction de corrélation de l'état correspondant à Π_Λ^Y ; nous avons :

$$\rho_\Lambda^Y (A) = \sum_{\Lambda \supset B \supset A} \Pi_\Lambda^Y (B)$$

Posons : $\rho_\Lambda^+ = \rho_\Lambda^{\partial\Lambda}$ et $\rho_\Lambda^- = \rho_\Lambda^{\emptyset}$

Alors, si $Y_1 \supset Y_2$, comme les états correspondant à $\Pi_\Lambda^{Y_1}$ et $\Pi_\Lambda^{Y_2}$ satisfont à l'hypothèse du théorème de Griffiths-Holley, en prenant pour fonction croissante sur 2^Λ la fonction χ_A définie par :

$$\chi_A (B) = \begin{cases} 1, \text{ si } B \supset A \\ 0 \text{ sinon} \end{cases}, \text{ cela pour tout } A \subset \Lambda,$$

nous avons :

$$\sum_{B \subset \Lambda} \chi_A (B) \quad \Pi_\Lambda^{Y_1} (B) \geq \sum_{B \subset \Lambda} \chi_A (B) \, \Pi_\Lambda^{Y_2} (B)$$

d'où $\rho_\Lambda^{Y_1} (A) \geq \rho_\Lambda^{Y_2} (A)$, $\forall A \subset \Lambda$,

puisque $\rho_\Lambda^Y (A) = \sum_{A \subset B \subset \Lambda} \Pi_\Lambda^Y (B)$

b) Soit $\Lambda' \supset \Lambda$; alors, pour $A \subset \Lambda$,

$$\rho_{\Lambda'}^+ (A) \leq \rho_\Lambda^+ (A), \quad \rho_{\Lambda'}^- (A) \geq \rho_\Lambda^- (A).$$

Montrons la première inégalité ; en utilisant la projectivité des Π_Λ^Y démontrée au chapitre III (lemme), nous avons :

$$\rho_{\Lambda'}^+ (A) = \sum_{Y \subset \partial\Lambda} C (Y) \, \rho_\Lambda^Y (A),$$

où $\sum_{Y \subset \partial\Lambda} C (Y) = 1$

(on applique la formule de Bayes).

Or $\rho_\Lambda^Y (A) \leq \rho_\Lambda^{\partial\Lambda} (A)$, $\forall Y \subset \partial\Lambda$, d'après a), d'où le résultat.

(2) Pour toute suite $\{\Lambda_n\}$ de parties finies de $\mathbb{Z}_\nu$ croissant vers $\mathbb{Z}_\nu$, la suite $\{\rho^+_{\Lambda_n}(A)\}$ (resp. $\{\rho^-_{\Lambda_n}(A)\}$) tend donc vers une limite $\rho^+(A)$ (resp. $\rho^-(A)$), indépendante de la suite $\{\Lambda_n\}$, fonction de corrélation définissant une probabilité μ^+ sur Ω(resp. μ^-) ; nous avons alors :

$$\mu^+ = \mu^- \longrightarrow |\mathcal{G}_U| = 1$$

(En effet, d'après le théorème 2 du chapitre III, tout état μ de Gibbs de $\mathcal{G}_U$ est limite étroite de combinaisons convexes de mesures à support l'ensemble des cylindres à base partie d'un ensemble fini $\Lambda_n \subset Z_\nu$, où $\{\Lambda_n\}$ est une suite croissant vers $\mathbb{Z}_\nu$; or, pour tout cylindre à base $A \subset \Lambda_n$, la combinaison convexe au rang n a une valeur comprise entre $\rho^-_{\Lambda_n}(A)$ et $\rho^+_{\Lambda_n}(A)$, d'où le résultat).

(3) Remarquons que μ^+ et μ^- sont invariantes par translation ; c'est-à-dire que nous avons :

$$\forall A \subset \mathbb{Z}_\nu \ , \ \forall x \in \mathbb{Z}_\nu \ , \ \rho^+_-(A) = \rho^+_-(A+x),$$

car $\rho^+_\Lambda(A) = \rho^+_{\Lambda+x}(A+x)$, et $\rho^+_\Lambda(A) \searrow \rho^+(A)$ et de même $\rho^+_{\Lambda+x}(A+x) \searrow \rho^+(A+x)$. Même raisonnement pour $\bar{\rho}$.

(4) Montrons que $\mu^+ = \mu^-$, si et seulement si $\rho^+(o) = \rho^-(o)$

<u>Condition nécessaire</u> : évidente

<u>Condition suffisante</u> : par translation, $\rho^+(x) = \rho^-(x)$, $\forall x \in \mathbb{Z}_\nu$, On achève la démonstration par induction sur la cardinalité de A, en utilisant le théorème de Griffiths-Holley appliqué à la fonction monotone $f = \chi_A + \chi_x - \chi_{A\cup x}$ (car alors $\rho^+_\Lambda(A\cup x) \leqslant \rho^-_\Lambda(A\cup x)$, si $\rho^+_\Lambda(A) = \rho^-_\Lambda(A)$)

(5) Comme $\rho^\pm(o) = \rho^\pm(x)$, $\forall x \in \mathbb{Z}_\nu$, on a :

$$\rho^{\pm}(o) = \lim_{\Lambda_n \uparrow \mathbb{Z}_\nu} \frac{1}{|\Lambda_n|} \sum_{x \in \Lambda_n} \rho^{\pm}_{\Lambda_n}(x)$$

(6) Soit $Z^{-}_{\Lambda}(\lambda) = \sum_{A \subset \Lambda} e^{-\frac{1}{2}U(A) - \lambda |A|}$

et $Z^{+}_{\Lambda}(\lambda) = \sum_{A \subset \Lambda} e^{-\frac{1}{2}U(A) - \lambda |A| - U(A, \partial\Lambda)}$,

où Λ est une partie finie quelconque de $\mathbb{Z}_\nu$.

Remarquons que $Z^{-}_{\Lambda}(-\operatorname{Log} t) = \varphi(t)$ est la fonction génératrice de la variable aléatoire $|A|$ égale au nombre de points occupés d'une configuration de Λ , pour la probabilité de densité Π^{-}_{Λ} (au coefficient $Z^{-1}_{\Lambda}(\phi)$ près)

Soit $P^{\pm}_{\Lambda}(\lambda) = \frac{1}{|\Lambda|} \operatorname{Log} Z^{\pm}_{\Lambda}(\lambda)$

Alors $P(\lambda) = \lim_{\Lambda_n \uparrow \mathbb{Z}_\nu} P^{\pm}_{\Lambda_n}(\lambda)$, pour toute suite adéquate $\{\Lambda_n\}$ de parties finies de $\mathbb{Z}_\nu$ croissant vers $\mathbb{Z}_\nu$.

La démonstration utilise une version affaiblie du théorème de Frobenius sur les matrices positives pour le démontrer, voir [1] p. 22. L'égalité des limites de $P^{-}_{\Lambda}(\lambda)$ et $P^{+}_{\Lambda}(\lambda)$ revient au fait que

$$\frac{|\partial\Lambda_n|}{|\Lambda_n|} \to 0 \quad \text{lorsque } n \to \infty .$$

(7) Nous avons, en posant $S^{\pm}_{\Lambda} = \frac{1}{|\Lambda_n|} \sum_{x \in \Lambda_n} \rho^{\pm}_{\Lambda_n}(x)$:

$$S^{\pm}_{\Lambda_n} = -\frac{\partial}{\partial\lambda} (P^{\pm}_{\Lambda_n}(\lambda))\big|_{\lambda=0}$$

En effet $S^{\pm}_{\Lambda_n}$ est l'espérance de la variable aléatoire $\frac{|A|}{|\Lambda_n|}$ pour la probabilité de densité $\Pi^{\pm}_{\Lambda_n}$

Nous en déduisons :

$$\rho^{\pm}(o) = -\lim_{\Lambda_n \uparrow \mathbb{Z}_\nu} \left(\frac{\partial}{\partial\lambda}(P^{\pm}_{\Lambda_n}(\lambda))\Big|_{\lambda=0}\right)$$

(8) $P(\lambda)$ est une fonction convexe sur $\mathbb{R}$; en effet les $P_{\Lambda_n}(\lambda)$ sont des fonctions convexes sur $\mathbb{R}$, pour tout n ; car $Z^-_{\Lambda_n}(\lambda) = Z^-_{\Lambda_n}(o)\, E^-_{\Lambda_n}\left[e^{-\lambda|A|}\right]$

d'où, en appliquant l'inégalité de Schwarz :

$$Z^-_{\Lambda_n}\left(\frac{\lambda_1+\lambda_2}{2}\right) \leqslant \left(Z^-_{\Lambda_n}(\lambda_1)\, Z^-_{\Lambda_n}(\lambda_2)\right)^{\frac{1}{2}} \quad , \forall (\lambda_1, \lambda_2) \in \mathbb{R}^2$$

$$\text{d'où } P^-_{\Lambda_n}\left(\frac{\lambda_1+\lambda_2}{2}\right) \leqslant \frac{P^-_{\Lambda_n}(\lambda_1) + P^-_{\Lambda_n}(\lambda_2)}{2} \quad , \forall (\lambda_1, \lambda_2) \in \mathbb{R}^2,$$

puisque la fonction Log est croissante, ce qui établit la convexité de $P^-_{\Lambda_n}(\lambda)$, pour tout n, donc celle de $P(\lambda)$ par passage à la limite.

(9) <u>Lemme</u> :

Soit $\{f_n\}$ une suite de fonctions convexes sur $\mathbb{R}$ de limite f, lorsque n augmente indéfiniment, telle que les f_n et f soient différentiables à l'origine, alors :

$$\lim_{n\to+\infty} f'_n(o) = f'(o) .$$

Si nous posons $f_n = P^-_{\Lambda_n}$, alors, si la fonction P est différentiable en 0, nous aurons :

$$\rho^+(o) = \rho^-(o),$$

d'où la solution du théorème de Ruelle, d'après (4).

(10) Comme il est plus facile d'établir l'analyticité de P, nous supposerons λ complexe. Montrons que, s'il existe un disque $|\lambda| \leqslant \delta$, $\delta > 0$ tel que Z_{Λ_n} ne s'annule pas à l'intérieur, et si δ est indépendant de n, alors P est analytique dans le disque : or cela résulte du fait que P est limite de polynômes en $e^{-\lambda}$, en utilisant les résultats sur les familles normales.

(11) Considérons une quantité intermédiaire entre $Z^-_{\Lambda_n}$ et $Z^+_{\Lambda_n}$:

$$Z_{\Lambda_n}(\lambda) = \sum_{A\subset\Lambda_n} e^{-\frac{1}{4}U(A) - \frac{1}{4}U(A\cup\partial\Lambda_n) - \lambda|A|}$$

Etudions les zéros de $Z^{\pm}_{\Lambda_n}$: si $u_1 \leqslant 0$ et si $u_0 + 2\nu u_1 \neq 0$, alors il existe un δ tel que $Z_{\Lambda_n}(\lambda) \neq 0$ pour $|\lambda| \leqslant \delta$.

Comme $U(A \cup \partial\Lambda_n) = U(A) + 2U(A, \partial\Lambda_n) + U(\partial\Lambda_n)$:

$$Z_{\Lambda_n}(\lambda) = e^{-\frac{1}{4}U(\partial\Lambda_n)} \sum_{A\subset\Lambda_n} e^{-\frac{1}{2}U(A) - \frac{1}{2}U(A,\partial\Lambda_n) - \lambda|A|}$$

$$= k \sum_{A\subset\Lambda_n} e^{-(\lambda+\frac{u_0}{2})|A| - \frac{1}{2}u_1\sum_{x\in A}(2\nu - \#_x)}$$

où $\#_x$ désigne le nombre de voisins de x dans $\Lambda_n \setminus A$

$$= k \sum_{A\subset\Lambda_n} e^{-(\lambda+\frac{u_0+2\nu u_1}{2})|A|}\, e^{\frac{u_1}{2}\sum_{x\in A}\sum_{y\in\Lambda_n\setminus A} I(x,y)}$$

($I(x, y) = 0$, si $x = y$ ou x non voisin de y et $I(x, y) = 1$, si x voisin de y)

$$= k \sum_{A\subset\Lambda_n} z^{|A|} \prod_{x\in A}\ \prod_{y\in\Lambda_n\setminus A} a(x, y),$$

où $z = e^{-(\lambda+\frac{u_0+2\nu u_1}{2})}$ et $a(x, y) = \begin{cases} 1, & \text{si } |x-y| \neq 1 \\ e^{\frac{u_1}{2}}, & \text{si } |x-y| = 1 \end{cases}$

Nous avons :

$-1 \leqslant a(x, y) = a(y, x) \leqslant 1$ pour $u_1 \leqslant 0$,

d'où, comme z n'est pas sur le cercle $|z| = 1$, si $|\lambda|$ est suffisamment petit et si $u_0 + 2\nu u_1 \neq 0$, nous pouvons, en appliquant le théorème de Lee-Yang, affirmer qu'il existe un δ tel que $Z_{\Lambda_n} \neq 0$ pour $|\lambda| \leqslant \delta$. La démonstration du théorème de Ruelle est ainsi achevée.

Remarques

(1) Exemple : $\nu = 2$

D'après le théorème 1, chapitre I, il existe une demi-droite d'équation $\begin{cases} u_0 + 4\,u_1 = 0 \\ u_1 \leqslant k < 0 \end{cases}$ correspondant à la transition de phase ; un point $\begin{pmatrix} u_0 \\ u_1 \end{pmatrix}$ de cette demi-droite correspond à un potentiel local attractif U, tel que $|\mathcal{G}_U| \neq 1$; soit $\mu \in \mathcal{G}_U$, nous avons :

k	$\mu\,(\{\omega\,(x) = 1\} \mid \mathcal{A}_k)$
0	$(1 + e^{-2\,u_1})^{-1} = \alpha$
1	$(1 + e^{-u_1})^{-1} = \beta$
2	$\frac{1}{2}$
3	$(1 + e^{u_1})^{-1} = 1 - \beta$
4	$1 - \alpha$

α et β sont voisins de 0, si $-u_1$ est grand (cas d'une grande attraction) ; alors, il existe deux états μ^+ et μ^- distincts dans $\mathcal{G}_U$. μ^+ correspond à une occupation presque totale, μ^- à une occupation presque nulle. On peut montrer qu'ils sont ergodiques, donc points extrêmaux de $\mathcal{G}_U$.

(2) Dans le modèle d'Ising étudié au chapitre I, le cas $J \geqslant 0$ (ferromagnétisme) correspond au cas attractif et le cas $J \leqslant 0$ (antiferromagnétique) correspond au cas répulsif ; dans le cas ferromagnétique, pour J suffisamment grand, il y a transition de phase et les deux états extrêmaux μ^+ et μ^- sont tels que :

- il n'existe presque que des + dans l'état μ^+
- il n'existe presque que des - dans l'état μ^-

(3) Que se passe-t-il dans le cas répulsif ?

(a) Dobrushin [12] a montré qu'il existe un voisinage de la demi-droite d'équation

$$\begin{cases} u_0 + 2\nu u_1 = 0 \\ u_1 \geq k' > 0 \end{cases}$$

pour lequel il y avait transition de phase.

(b) On peut montrer un résultat moins précis, dans le cas du modèle d'Ising.

Théorème

Pour J négatif tel que $|J|$ soit suffisamment grand, $|\mathcal{G}_U| \neq 1$.

Démonstration ($\nu = 2$)

Décomposons $\mathbb{Z}_\nu$ en $P \cup I$, où $P = \{(m, n) ; m + n \text{ pair}\}$;

Soit $\Sigma = \{+1, -1\}^{\mathbb{Z}_\nu}$; considérons l'application $*$ de Σ vers Σ qui à un élément σ associé σ^* par :

$$\sigma^*(x) = \begin{cases} \sigma(x), \text{ si } x \in P \\ -\sigma(x), \text{ sinon} \end{cases}$$

Nous avons : $|x-y| = 1 \longrightarrow \sigma^*(x)\sigma^*(y) = -\sigma(x)\sigma(y)$

L'application $*$ est donc telle qu'elle transforme le potentiel U correspondant à $J < 0$ en un potentiel V correspondant à $-J > 0$, d'où le résultat, comme

$|\mathcal{G}_U| = |\mathcal{G}_V| \neq 1$, pour $|J|$ suffisamment grand.

(Ce résultat est connu sous l'appellation de "symmetry breakdown").

(4) Question : quels sont les points extrêmaux invariants par translation de $\mathcal{G}_U$?

Dans le cas attractif :

. on pense qu'il n'y a que μ^+ et μ^-

. si on s'intéresse à tous les points extrêmaux, invariants ou non, on croit que μ^+ et μ^- sont encore les seuls, si $\nu = 2$. Mais récemment Dobrushin [13] a démontré qu'il existe une infinité dénombrable si $\nu \geqslant 3$.

Voici la raison intuitive de ce phénomène. Imposons sur une suite croissante de cubes Λ_n la condition de frontière, que la demi-frontière supérieure soit occupée (+) tandis que la demi-frontière inférieure soit vide (-). Dobrushin a montré que la limite μ de ces états finis de Gibbs existe toujours ($\nu \geqslant 2$). Si $\nu = 2$ il trouve que

$$\mu = \frac{1}{2} \mu^+ + \frac{1}{2} \mu^- ,$$

donc le mélange (non-ergodique) des états à haute et à basse densité. Mais si $\nu \geqslant 3$ l'influence de la frontière est plus forte, et μ est un état non-invariant par translation. Sa densité $\mu\left[\omega(x) = 1\right]$ est une fonction monotone croissante en x quand x se déplace vers le haut dans la direction verticale.

CHAPITRE V

CARACTERISATION VARIATIONNELLE DES ETATS DE GIBBS

1 - Caractérisation variationnelle d'un état fini de Gibbs

Soient Λ fini, μ une probabilité sur 2^Λ, U une fonction réelle sur 2^Λ.

On définit les quantités suivantes :

- l'entropie spécifique $S_\Lambda(\mu) = -\frac{1}{|\Lambda|} \sum_{A \in 2^\Lambda} \mu(A) \log \mu(A)$

  (en posant $0 \log 0 = 0$)

- l'énergie moyenne $E_\Lambda^U(\mu) = \frac{1}{|\Lambda|} \sum_{A \in 2^\Lambda} \mu(A) \frac{U(A)}{2}$

- l'énergie libre $F_\Lambda^U(\mu) = E_\Lambda^U(\mu) - S_\Lambda(\mu)$.

- la fonction de répartition $Z_\Lambda(U) = \sum_{A \in 2^\Lambda} e^{-\frac{1}{2}U(A)}$

- la pression $P_\Lambda(U) = \frac{1}{|\Lambda|} \log Z_\Lambda(U)$

On note ν_Λ la mesure de Gibbs sur 2^Λ

$$\nu_\Lambda(A) = (Z_\Lambda(U))^{-1} e^{-\frac{1}{2}U(A)} .$$

On a la caractérisation suivante :

Théorème 1

i) Pour toute probabilité μ, $F^U(\mu) \geq -P_\Lambda(U)$

ii) On a $F_\Lambda^U(\mu) = -P_\Lambda(U)$ *si et seulement si* μ *est la mesure de Gibbs* ν_Λ

Démonstration

On calcule $F_\Lambda^U(\mu) + P_\Lambda(U)$.

$$F_\Lambda^U(\mu) + P_\Lambda(U) = \frac{1}{|\Lambda|} \sum_{A \in 2^\Lambda} \mu(A) \left[\log \frac{\mu(A)}{e^{-\frac{1}{2}U(A)}} + \log Z_\Lambda(U) \right]$$

$$= \frac{1}{|\Lambda|} \sum_{A \in 2^\Lambda} \nu_\Lambda (A) \left(\frac{\mu (A)}{\nu_\Lambda (A)} \log \frac{\mu (A)}{\nu_\Lambda (A)} \right)$$

il suffit de remarquer que pour tout t réel positif ou nul, $t \log t \geq t-1$ avec égalité seulement au point $t = 1$.

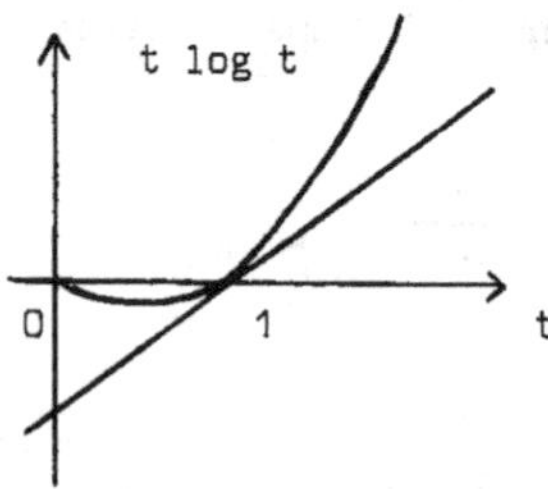

Alors $F^U_\Lambda (\mu) + P_\Lambda (U) \geq \frac{1}{|\Lambda|} \sum_{A \in 2^\Lambda} \nu_\Lambda (A) \left(\frac{\mu (A)}{\nu_\Lambda (A)} - 1\right) = 0$

ce qui montre i), et il n'y a égalité que si il y a égalité pour chaque terme de la somme. Il y a donc égalité si et seulement si μ est la mesure de Gibbs, ce qui montre ii).

2 - Le théorème de Lanford et Ruelle

On va énoncer une généralisation du théorème précédent.

Soit $\mathfrak{J}$ l'ensemble des probabilités sur $\Omega = \{0, 1\}^{\mathbb{Z}_\nu}$ invariantes par les opérateurs τ_x de translation des coordonnées par x.

Soit U une énergie définie par un potentiel invariant (cf Ruelle p. 20) ; U est une fonction réelle des parties finies de $\mathbb{Z}_\nu$, invariante par les translations de $\mathbb{Z}_\nu$.

Par exemple $U(A) = \sum_{x \in A} \sum_{y \in A} U(x, y)$, si U est un potentiel local sur $\mathbb{Z}_\nu$.

On peut étendre les définitions précédentes.

Si Λ est un ensemble fini, soit μ_Λ la mesure image de $\mu \in \mathfrak{F}$ par la projection de $\{0, 1\}^{\mathbb{Z}_\nu}$ sur $\{0, 1\}^\Lambda$. Dans les propositions suivantes Λ tend vers l'infini au sens de Van Hove (cf Ruelle [1] p. 14) ; on peut prendre seulement les limites de long de la suite Λ_n des cubes de côté 2n+1 centrés en O.

Proposition (Ruelle [1] p. 180)

$S_\Lambda(\mu_\Lambda)$ converge vers une fonction $s(\mu)$ affine semi-continue supérieurement sur $\mathfrak{J}$; $s(\mu)$ s'appelle l'entropie de μ.

Proposition

$E^U_\Lambda(\mu_\nu)$ converge vers une fonction affine continue $e_U(\mu)$ sur $\mathfrak{J}$.
Pour un potentiel local, un calcul simple donne

$$e_U(\mu) = \frac{1}{2}\left\{U(0,0)\ \mu\left[\omega_0 = 1\right] + \sum_{|x|=1} U(0,x)\ \mu\left[\omega_0 = \omega_x = 1\right]\right\}$$

Soit $f_U(\mu) = e_U(\mu) - s(\mu)$ l'énergie libre de μ par rapport à U.

Proposition (Ruelle [1] p. 22)

$P_\Lambda(U)$ converge vers une limite $P(U)$, pour tout potentiel local U.
Toutes ces propositions permettent d'écrire, en passant à la limite

$$f_U(\mu) \geqslant -P(U) \text{ pour tout } \mu \text{ de } \mathfrak{J}$$

Enfin $\mathcal{G}_U$ est convexe compact et invariant par les translations. $\mathcal{G}_U \cap \mathfrak{J}$ est alors non vide car tout point d'accumulation de $\frac{1}{|\Lambda|}\sum_{x\in\Lambda}\tau_x\circ\mu(.)$, $\mu\in\mathcal{G}_U$ est une mesure de Gibbs invariante. Inversement on a le résultat beaucoup plus profond, que chaque mesure $\mu\in\mathfrak{J}$ qui minimise $f_U(\mu)$, est forcément un élément de $\mathcal{G}_U$.

Théorème 2 (Lanford-Ruelle) [10]

On a $f_U(\mu) = -P(U)$ si et seulement si μ appartient à $\mathcal{G}_U \cap \mathfrak{J}$.

3 - Equivalence des ensembles

On veut pouvoir introduire la mesure de Gibbs de manière "naturelle" à partir des principes de la thermodynamique.

Plus précisément si Λ est une région finie, U une énergie, "l'ensemble microcanonique" est la densité de probabilité uniforme sur 2^{Λ}.

On peut alors calculer les énergies des configurations et se restreindre à celles qui donnent

$$|\Lambda| \quad t \leqslant U\ (A) \leqslant \quad |\Lambda| \quad t + \delta$$

On obtient "un ensemble grand canonique".

Les théorèmes "d'équivalence des ensembles" [1, 3] montrent que pour des valeurs convenables de t et λ, les ensembles canoniques convergent, quand Λ tend vers l'infini, vers une probabilité μ de $\mathcal{G}_{\lambda U}$. La constante λ doit être telle que l'énergie moyenne de μ, par rapport à U est égale à t. Cela s'explique comme suit : le conditionnement nous donne l'état d'entropie maximale parmi tous les états qui sont compatibles (dans le sens d'énergie moyenne) avec le conditionnement [3, 14, 15].

On peut montrer un tel théorème dans le cas élémentaire suivant :

Soit S fini, $\Omega = S^{\mathbb{N}}$, φ une application de S dans $\mathbb{Z}$. Soit p une probabilité sur S et P la probabilité sur Ω produit des lois égales à p. Soit enfin

$$\inf_{x \in S} \varphi < \rho < \sup_{x \in S} \varphi .$$

Soient $P_{n,k}$ pour n fixé et k une valeur possible de $\sum_{i=1}^{n} \varphi(x_i)$ la probabilité sur $\{0, 1\}^n$ définie par $P_{n,k}(A) = P(A \mid \sum_{i=1}^{n} \varphi(\omega_i) = k)$. On note encore $P_{n,k}$ un prolongement de $P_{n,k}$ à $\{0, 1\}^{\mathbb{N}}$.

On a le théorème suivant :

Théorème 3

Quand k et n tendent vers l'infini, avec $\frac{k}{n} \to \rho$, *la famille* $P_{n,k}$ *n'a qu'un point d'accumulation :*

$$\mu\left[\omega_1 = x_1, \ldots, \omega_m = x_m\right] = \prod_{j=1}^{m} f_\alpha(x_j) \quad \text{où } f_\alpha \text{ est uniquement}$$

déterminée par :

$$f_\alpha(x) = \frac{p(x)\, e^{-\alpha\, \varphi(x)}}{Z(\alpha)}, \quad \sum_{x\in S} f_\alpha(x) = 1, \quad \sum_{x\in S} \varphi(x)\, f_\alpha(x) = \rho$$

Démonstration

On remarque d'abord que les $P_{n,k}$ ont la propriété de symétrie suivante, dès que n est assez grand pour que la formule ait un sens :

$$P_{n,k}\left(\omega_1 = x_1 \ldots \omega_m = x_m \,\middle|\, \sum_{j=1}^{m} \varphi(\omega_j) = s\right) = Z_m^{-1}(\varphi, s) \prod_{1}^{m} p(x_j)$$

pour tout m et tous $x_1 \ldots x_m$ tels que $P_{n,k}\left[\sum_{j=1}^{m} \varphi(\omega_j) = s\right]$ est non nul.

($Z_n(\varphi, s)$ est la normalisation).

Tout point d'accumulation a donc la même propriété et on peut appliquer le lemme suivant :

Lemme

Soit μ une probabilité sur Ω telle que pour tout m et tous $x_1 \ldots x_m$ tels que $\sum_{j=1}^{m} \varphi(\omega_j) = s$,

$$\mu\left(\omega_1 = x_1 \ldots \omega_m = x_m \,\middle|\, \sum_{j=1}^{m} \varphi(\omega_j) = s\right) = Z_m^{-1}(\varphi, s) \prod_{j=1}^{m} p(x_j)$$

Alors $\mu = \int \nu_\alpha\, dF(\alpha)$, combinaison convexe arbitraire des mesures ν_α qui sont les probabilités sur Ω produit de lois égales à f_α ;

$$f_\alpha(x) = \frac{p(x)\, e^{-\alpha\, \varphi(x)}}{Z(\alpha)}, \quad x \in S, \quad \sum_{x\in S} f_\alpha(x) = 1.$$

D'après le lemme, tout point d'accumulation de la famille des $P_{n,k}$ est de la forme $\int \nu_\alpha dF(\alpha)$.

De plus μ est obtenue quand $\frac{k}{n} \to \rho$, et donc on doit avoir :

$$\int \varphi(\omega_1)\, d\mu = \rho, \int \varphi(\omega_1)\ \varphi(\omega_2)\, d\mu = \rho^2 .$$

Donc si $h(\alpha) = \sum_x \nu_\alpha(x)\ \varphi(x) = Z^{-1}(\alpha) \sum_{x \in S} p(x)\ \varphi(x)\ e^{-\alpha\varphi(x)}$

dF doit vérifier :

$$\rho = \int h(\alpha)\, dF(\alpha) \qquad \rho^2 = \int h^2(\alpha)\, dF(\alpha).$$

La mesure dF est donc concentrée sur l'ensemble des α tel que $h(\alpha) = \rho$

Le théorème est démontré si on vérifie qu'il n'y a qu'un α tel que :

$$h(\alpha) = \frac{\sum_x p(x)\ \varphi(x)\ e^{-\alpha\varphi(x)}}{\sum_x p(x)\ e^{-\alpha\varphi(x)}} = \rho$$

Or $h(\alpha)$ est décroissante ($h'(\alpha) < 0$) de $\sup_{x \in S} \varphi(x)$ à $\inf_{x \in S} \varphi(x)$ et ρ est entre ces deux limites.

<u>Démonstration du lemme</u>

Si on pose $F_n(t) = \mu(\sum_{k=1}^{n} \varphi(\omega_k) = t)$, on a :

$$F_n(t) = \sum_{s \in S} \mu(\sum_{k=1}^{n+1} \varphi(\omega_k) = s,\ \varphi(\omega_{n+1}) = s-t)$$

$$= \sum_s \mu\left[\varphi(\omega_{n+1}) = s-t \mid \sum_{k=1}^{n+1} \varphi(\omega_k) = s\right] F_{n+1}(s)$$

Or d'après l'hypothèse sur μ :

$$\mu(\varphi(\omega_{n+1}) = s-t \mid \sum_{k=1}^{n+1} \varphi(\omega_k) = s) = Z_{n+1}^{-1}(\varphi, s) \sum_{x_1 \ldots x_n \in A_{s,t}} (\prod_{i=1}^{n+1} p(x_i))$$

où $A_{s,t} = \left[x_1, \ldots x_n ; \sum_{j=1}^{n} \varphi(x_j) = t , \varphi(x_{n+1}) = s-t\right]$.

Par suite on a

$$\mu\left[\varphi(\omega_{n+1}) = s-t \mid \sum_{k=1}^{n+1} \varphi(\omega_k) = s\right] = \frac{Z_n(\varphi, t)}{Z_{n+1}(\varphi, s)} \; q(s-t),$$

si $q(s-t) = \sum_{x:\varphi(x)=s-t} p(x)$.

Donc $F_n(t) = \sum_{s \in S} q(s-t) \dfrac{Z_n(\varphi, t)}{Z_{n+1}(\varphi, s)}$.

En posant $G_n(t) = \dfrac{F_n(t)}{Z_n(\varphi, t)}$, $G_n(t)$ vérifie donc

$$G_n(t) = \sum_s q(s-t)\, G_{n+1}(s)$$

Les solutions de l'équation $G_n(t) = \sum_s q(s-t)\, G_{n+1}(s)$ sont des combinaisons convexes de solutions extrêmales $a^n e^{bt}$ avec $a = (\sum_s q(s)\, e^{bs})^{-1}$.

D'où $G_n(t) = \dfrac{e^{bt}}{(\sum_s q(s)\, e^{bs})^n}\, dM(b)$.

$$\text{Mais} \quad \mu\left[\omega_1 = x_1, \ldots, \omega_n = x_n\right] = \frac{F_n(\sum_{j=1}^{n} \varphi(\omega_j))}{Z_n(\varphi, s)} \prod_{k=1}^{n} p(x_k)$$

$$= G_n(\sum_j \varphi(\omega_j)) \prod_{k=1}^{n} p(x_k)$$

$$= \int dM(b) \frac{\prod_{k=1}^{n} e^{b\varphi(x_k)}\, p(x_k)}{\left[\sum_x p(x)\, e^{b\varphi(x)}\right]^n}$$

$$= \int dM(b)\, \nu_b\left[\omega_1 = x_1, \ldots, \omega_n = x_n\right] \quad \text{c.q.f.d.}$$

Donc nous avons vu que le théorème 3 est lié à la frontière de Martin d'un processus de Markov associé. Un travail récent de P. Martin Löf [32] donne l'espoir que c'est ainsi aussi dans le cas des théorèmes généraux qui expriment l'équivalence des ensembles.

CHAPITRE VI

EVOLUTIONS TEMPORELLES

Dans les chapitres précédents nous avons étudié des mesures décrivant la configuration à l'équilibre d'un système de particules. Nous considérons maintenant des évolutions markoviennes de ces systèmes. Nous nous intéresserons plus particulièrement à celles pour lesquelles les états de Gibbs sont des états d'équilibre i.e. les mesures de Gibbs sont invariantes.

Nous introduisons deux types d'évolution dans les deux cas d'un espace de phase fini (VI-2) puis dénombrable (VI-3).

1 - Rappels sur les processus de Markov à valeurs dans un ensemble fini

Soit Γ un ensemble fini dont les éléments sont notés A, B, C. Un semi-groupe fortement continu de noyaux markoviens sur Γ est défini par une famille $(P_t)_{t\geq 0}$ de matrices telle que :

$$P_t \times P_s = P_{t+s} \qquad\qquad P_t\,(A, B) \geq 0$$

$$P_0\,(A, B) = \delta_{A,B} \qquad\qquad \sum_{B \in \Gamma} P_t\,(A, B) = 1$$

l'application $t \rightsquigarrow P_t\,(A, B)$ est continue en 0.

Dans la suite tous les semi-groupes considérés seront de ce type. Nous avons la caractérisation suivante :

Théorème 1

Pour qu'une famille $(P_t)_{t\geq 0}$ *de matrices soit un semi-groupe il faut et il suffit que pour tout* $t \geq 0$

$$P^t = \exp t\,G = \sum_{n=0}^{\infty} \frac{t^n}{n!} G^n$$

où $G = [G\,(A,\,B)]$ *est une matrice satisfaisant à :*

$$G\,(A,\,B) \geq 0 \text{ si } A \neq B \text{ et } \sum_B G\,(A,\,B) = 0 \text{ pour tout } A \in \Gamma\ .$$

G est appelé ***générateur*** *du semi-groupe* (P_t).

Remarquons que, puisque

$$\lim_{t \downarrow 0} \frac{1}{t} P_t\,(A,\,B) = G\,(A,\,B) \text{ pour } A \neq B\ ,$$

G $(A,\,B)\,dt$ *décrit l'évolution du processus dans l'intervalle de temps* $(0,dt)$.

Définition

Soit (P_t) un semi-groupe de générateur G, on dit que (P_t) ou G est **irréductible** si :

pour tout $A, B \in \Gamma$ il existe une suite $(A_k)_{k=1}^n$, $A_k \in \Gamma$ telle que

$$G\,(A,\,A_1)\,.\,G\,(A_1,\,A_2)\,\ldots\,G\,(A_{n-1},\,A_n)\,.\,G\,(A_n,\,B) \neq 0.$$

Concernant l'invariance des probabilités pour le semi-groupe (P_t) de générateur G nous avons le

Théorème 2

(a) Pour que la probabilité μ *sur* Γ *soit invariante pour* (P_t) *il faut et il suffit que* $\mu G = 0$

(b) Supposons G irréductible alors

i) pour tout $t > 0$ *et tout* $A, B \in \Gamma$, $P_t(A, B) > 0$,

ii) il existe une et une seule probabilité invariante μ *et, pour tout* $A, B \in \Gamma$,

$$\lim_{t \to +\infty} P_t(A, B) = \mu(B) > 0$$

Démonstration

Nous prouvons seulement que pour toute probabilité ν sur Γ

$$\lim_{t \to +\infty} \nu P_t = \mu$$

Pour toute probabilité λ sur Γ on pose

$$F(\lambda) = \sum_{A \in \Gamma} \lambda(A) \operatorname{Log} \frac{\lambda(A)}{\mu(A)}$$

avec la convention $0 \operatorname{Log} 0 = 0$. Si $\nu_t = \nu P_t$, $F(\nu_t)$ est dérivable et

$$\frac{dF(\nu_t)}{dt} = \sum_{A \in \Gamma} \frac{d\nu_t(A)}{dt} \operatorname{Log} \frac{\nu_t(A)}{\mu(A)}$$

$$= \sum_{A,B \in \Gamma} \nu_t(B)\, G(B,A) \operatorname{Log} \frac{\nu_t(A)}{\mu(A)},$$

puisque $\frac{d\nu_t}{dt} = \nu_t G$. Utilisant $G1 = 0$ et $\mu G = 0$, il vient :

$$\frac{dF(\nu_t)}{dt} = \sum_{A \in \Gamma} \left[\sum_{B \in \Gamma} \{z(A,B) - z(A,B) \operatorname{Log} z(A,B) - 1\}\, G(B,A) \frac{\mu(B)}{\mu(A)} \nu_t(A) \right]$$

en posant :

$$z(A, B) = \frac{\nu_t(B)\, \mu(A)}{\nu_t(A)\, \mu(B)} .$$

D'après l'inégalité $x-1 \leqslant x \log x$ si $x \geqslant 0$, nous avons $\frac{dF(\nu_t)}{dt} \leqslant 0$, avec égalité si et seulement si $z(A, B) = 1$ pour tout couple (A, B), $A \neq B$ pour lequel $G(B, A) \neq 0$. Supposons $\nu_{t_o} \neq \mu$, $\frac{\nu_{t_o}(A)}{\mu(A)}$ prend au moins deux valeurs distinctes ; posons $U_k = \left\{A : \frac{\nu_{t_o}}{\mu(A)} = a_k\right\}$. Puisque G est irréductible il existe i et j, $A_i \in U_i$ et $A_j \in U_j$ tels que $G(A_i, A_j) > 0$, par suite $\frac{dF(\nu_t)}{dt}(t_o) \neq 0$.

Nous avons prouvé que $\frac{dF(\nu_t)}{dt} \leqslant 0$ avec égalité seulement aux points où $\nu_t = \mu$. Pour établir la convergence des ν_t vers μ, il suffit de prouver que, si λ est la limite d'une suite ν_{t_n}, $\lambda = \mu$. Soit $c = \inf_t F(\nu_t) > -\infty$, d'après la décroissance et la continuité de $F(\nu_t)$ on peut écrire pour tout $t > 0$, $\lim_n F(\nu_{t_n+t}) = F(\lambda P_t) = 0$; par suite $\frac{dF(\lambda P_t)}{dt} = 0$ et $\lambda = \mu$.

Rappelons enfin la notion de réversibilité :

<u>Définition</u>

Un processus stationnaire $(X_t)_{t \in \mathbb{R}}$ est dit <u>réversible</u> si les processus $(X_t)_{t \in \mathbb{R}}$ et $(X_{-t})_{t \in \mathbb{R}}$ ont les mêmes lois.

<u>Théorème 3</u>

Un processus de Markov stationnaire défini par le sous-groupe (P_t) et la mesure invariante μ est réversible si :

(R) pour tout $A, B \in \Gamma$, $\mu(A) \,.\, G(A, B) = \mu(B) \,.\, G(B, A)$.

Réciproquement si la condition (R) est satisfaite μ est invariante pour (P_t) et le processus stationnaire correspondant est réversible.

2 - Cas d'un espace de phase fini

Λ désigne un ensemble fini et l'on pose $\Gamma = 2^{\Lambda}$

Chaque point de Λ est occupé par au plus une particule de sorte que la situation du système est décrite par un élément de Γ.

Intéraction (I) - Spin flip ou processus de naissance et mort

On donne pour tout $A \in \Gamma$ et tout $x \notin A$ deux réels > 0, $\beta(x, A)$, $\delta(x,A)$

Définition

On appelle processus de naissance et mort un processus de Markov sur Γ associé au générateur G défini par :

$G(A, A \cup x) = \beta(x, A)$, si $x \notin A$,

$G(A \cup x, A) = \delta(x, A)$, si $x \notin A$,

$G(A, B) = 0$ dans tous les autres cas, lorsque $A \neq B$,

$$G(A, A) = -\sum_{B \neq A} G(A, B) .$$

Cette évolution répond à la description intuitive suivante :

(i) Si A est la configuration du système à l'instant t, $\beta(x, A)\,dt$ est la probabilité de naissance d'une particule au point $x \notin A$ entre t et t + dt, $\delta(x, A \setminus x)\,dt$ est la probabilité de mort entre t et t + dt de la particule qui se trouve en $x \in A$.

(ii) on a au plus une naissance ou une mort entre t et t + dt.

Interaction (II) - Processus de saut avec exclusion

On donne une matrice stochastique irréductible $[p(x, y)]_{x,y \in \Lambda}$, et pour tout $A \in \Gamma$ et $x \notin A$, un réel strictement positif $c(x, A)$.

Définition

On appelle processus de saut avec exclusion un processus de Markov sur Γ associé au générateur G défini par :

$G(A \cup x, A \cup y) = c(x, A)\, p(x, y)$, si $x, y \notin A$,

$G(A, B) = 0$ dans tous les autres cas, lorsque $A \neq B$,

$$G(A, A) = - \sum_{B \neq A} G(A, B).$$

La description intuitive de cette évolution est la suivante :

(i) si A est la configuration à l'instant t, dans l'intervalle de temps t, t+dt la particule située en $x \in A$ saute avec la probabilité $c(x, A \setminus x)\, dt$, la distribution des sauts étant $p(x, y)$, $y \notin A$.

(ii) il y a au plus un saut dans l'intervalle t, t + dt.

Etude de la réversibilité de (I)

Nous supposons maintenant que Λ est un graphe fini satisfaisant aux hypothèses de Grimmett dont nous reprenons les notations.

On considère des processus de naissance et mort associés à des fonctions $\beta(x, A)$ et $\delta(x, A)$ ayant la propriété :

$$(L) \qquad \begin{aligned} \beta(x, A) &= \beta(x, A \cap \partial x) \\ \delta(x, A) &= \delta(x, A \cap \partial x) \end{aligned}$$

Théorème 4

Pour un processus de naissance et mort satisfaisant à (L), les conditions suivantes sont équivalentes :

(i) μ *est un équilibre réversible*

(ii) μ *est l'état de Gibbs associé au potentiel V et*

$$\frac{\beta(x,A)}{\delta(x,A)} = \exp \sum_{\substack{S \in \Sigma \\ x \in S \subset A \cup x}} V(S)$$

Démonstration

Remarquons que la condition (R) du théorème 3 s'écrit ici : pour tout $A \in \Gamma$ et $x \notin A$, $\dfrac{\beta(x, A)}{\delta(x, A)} = \dfrac{\mu(A \cup x)}{\mu(A)}$.

Supposons (i) vérifié ; d'après l'hypothèse (L) et la relation précédente, $\dfrac{\mu(A \cup x)}{\mu(A)}$ ne dépend que de $A \cap \partial x$, donc μ est un état de Gibbs au sens de Grimmett et il existe un potentiel V tel que

$$A \in \Gamma, \ \mu(A) = \mu(\emptyset) \exp \sum_{S \in \Sigma, S \subset A} V(S)$$

il vient alors

$$\frac{\beta(x,A)}{\delta(x,A)} = \frac{\mu(A \cup x)}{\mu(A)} = \exp \left\{ \sum_{S \subset A \cup x} V(S) - \sum_{S \subset A} V(S) = \exp \sum_{\substack{S \in \Sigma \\ x \in S \subset A \cup x}} V(S) \right.$$

La réciproque est immédiate.

Nous montrons maintenant sur un exemple comment l'hypothèse de réversibilité permet de réduire les paramètres de l'interaction (Interaction (I)). Soit N entier et $\Lambda = \mathbb{Z} / N$, c'est-à-dire les entiers mod N.

On suppose que

$$\beta(x, A) = \beta_k, \ \delta(x, A) = \delta_k \quad \text{lorsque} \quad |A \cap \partial x| = k,$$

l'interaction dépend donc de 6 paramètres.

Faisons l'hypothèse de réversibilité. Le théorème précédent implique qu'il existe une fonction réelle V sur Σ telle que :

$$V(S) = \begin{cases} \alpha & \text{si } |S| = 1 \\ \gamma & \text{si } |S| = 2 \\ 0 & \text{dans tous les autres cas} \end{cases}$$

et $\dfrac{\beta_k}{\delta_k} = \exp(\alpha + k\gamma)$, $k = 0, 1, 2$.

Le nombre de paramètres est donc réduit à 2.

Un théorème de Holley pour (I)

Nous revenons au cas d'une intéraction (I) sur un espace de phase Λ fini quelconque.

Théorème 5 [16]

Soit G_1 *et* G_2 *les générateurs de deux processus de naissance et mort,* (P_t^1), (P_t^2) *les semi-groupes associés. On suppose satisfaites les hypothèses*

(H_1) *si* $x \in A_2 \subset A_1$, $G_1 (A_1, A_2 \setminus x) \leq G_2 (A_2, A_2 \setminus x)$,

(H_2) *si* $A_2 \subset A_1$ *et si* $x \notin A_1$, $G_1 (A_1, A_2 \cup x) \geq G_2 (A_2, A_2 \cup x)$,

alors si $A_2 \subset A_1$ *et si* f *est une fonction réelle croissante sur* $\Gamma = 2^\Lambda$

$$P_t^2 f (A_2) \leq P_t^1 f (A_1),$$

en particulier pour $C \in \Gamma$,

$$\sum_{B, B \supset C} P_t^2 (A_2, B) \leq \sum_{B, B \supset C} P_t^1 (A_1, B).$$

Démonstration :

Le point important de cette preuve est la construction sur $\Gamma \times \Gamma$ d'un processus de Markov, associé au sous groupe (P_t) de générateur G, dont la projection sur le premier (resp. second) facteur est un processus de Markov associé au semi groupe (P_t^1) (resp. (P_t^2)) (couplage des deux processus).

Les éléments du premier facteur (resp. second) sont affectés des indices 1 (resp. 2).

On définit G par :

$x \in A_1 \cap A_2$ (1) $G(A_1,A_2 ; A_1 \setminus x, A_2 \setminus x) = \min_{i=1,2} G_i(A_i, A_i \setminus x)$

((2) $G(A_1,A_2 ; A_1 \setminus x, A_2) = G_1(A_1,A_1\setminus x)-G(A_1,A_2;A_1\setminus x,A_2\setminus x)$

(3) $G(A_1,A_2;A_1,A_2\setminus x) = G_2(A_2,A_2\setminus x)-G(A_1,A_2 ; A_1\setminus x,A_2\setminus x)$

$x \notin A_1$ et $x \notin A_2$ (4) $G(A_1,A_2 ; A_1 \cup x, A_2 \cup x) = \min_{i=1,2} G_i(A_i, A_i \cup x)$

(5) $G(A_1,A_2 ; A_1 \cup x,A_2) = G_1(A_1,A_1 \cup x)-G(A_1,A_2;A_1 \cup x,A_2 \cup x)$

(6) $G(A_1,A_2 ; A_1,A_2 \cup x) = G_2(A_2,A_2 \cup x)-G(A_1,A_2;A_1 \cup x,A_2 \cup x)$

$x \in A_1 \setminus A_2$ (7) $G(A_1, A_2 ; A_1 \setminus x, A_2) = G_1(A_1, A_1 \setminus x)$

(8) $G(A_1, A_2 ; A_1, A_2 \cup x) = G_2(A_2, A_2 \cup x)$

$x \in A_2 \setminus A_1$ (9) $G(A_1, A_2 ; A_1, A_2 \setminus x) = G_2(A_2, A_2 \setminus x)$

(10) $G(A_1, A_2 ; A_1 \cup x, A_2) = G_1(A_1, A_1 \cup x)$

$$\sum_{B_1,B_2} G(A_1, A_2 ; B_1, B_2) = 0$$

Nous établissons deux propriétés du semi-groupe (P_t).

(a) Pour $t \geq 0$ $\qquad \sum_{B_j} P_t (A_1, A_2 ; B_1, B_2) = P_t^i (A_i, B_i) \quad i,j = 1,2, i \neq j$

Considérons le cas $i = 1$, $j = 2$; il suffit de prouver que

$$(11) \qquad \sum_{B_2} G^n (A_1, A_2 ; B_1, B_2) = G_1^n (A_1, B_1)$$

Lorsque $n = 1$, cette relation résulte immédiatement de la définition de G : par exemple si $x \in A_1 \cap A_2$ et $B_1 = A_1 \setminus x$, (11) est la somme termes à termes de (1) et (2). On termine par récurrence sur n.

(b) Si $A_1 \supset A_2$ et $B_1 \not\supset B_2$ $\qquad P_t (A_1, A_2 ; B_1, B_2) = 0$

On procède comme pour (a). La condition

$A_1 \supset A_2$ et $B_1 \not\supset B_2$ implique $G (A_1, A_2 ; B_1, B_2) = 0$ et

équivaut à :

si $x \in A_2 \subset A_1$ $\qquad G (A_1, A_2 ; A_1 \setminus x, A_2) = 0$

si $x \notin A_1$, $A_2 \subset A_1$ $\qquad G (A_1, A_2 ; A_1, A_2 \cup x) = 0$

la première résulte de (2) et de (H1), la seconde de (6) et de (H2).

Soit maintenant f une fonction réelle croissante sur . Pour $A_1 \supset A_2$ utilisant successivement (a), (b), la croissance de f, (b) et (a) il vient :

$$\begin{aligned} P_t^1 f (A_1) &= \sum_{B_1} P_t^1 (A_1, B_1) f (B_1) = \sum_{B_1, B_2} P_t (A_1, A_2 ; B_1, B_2) f (B_1) \\ &= \sum_{\substack{B_1, B_2 \\ B_1 \supset B_2}} P_t (A_1, A_2 ; B_1, B_2) f (B_2) \geq \sum_{\substack{B_1, B_2 \\ B_1 \supset B_2}} P_t (A_1, A_2 ; B_1, B_2) f (B_2) \\ &= \sum_{B_1, B_2} P_t (A_1, A_2 ; B_1, B_2) f(B_2) = \sum_{B_2} P_t^2 (A_2, B_2) f(B_2) = P_t^2 f(A_2). \end{aligned}$$

Soit $C \in \Gamma$; appliquant le résultat ci-dessus à la fonction f définie par : $f (A) = 1$ si $A \supset C$, $f (A) = 0$ si $A \not\supset C$ on obtient la dernière assertion du théorème.

<u>Corollaire</u> (inégalité de Griffiths-Holley)

Soient μ_1 et μ_2 deux densités de probabilités strictement positives sur telles que :

$$A, B \in \Gamma , \mu_1 (A \cup B) \mu_2 (A \cap B) \geqslant \mu_1 (A) \mu_2 (B).$$

Si f est une fonction réelle croissante sur Γ

$$\sum_A f(A) \mu_1 (A) \geqslant \sum_A f(A) \mu_2 (A).$$

<u>Démonstration</u>

On associe aux fonctions μ_i , i = 1, 2 les générateurs G_i, i = 1, 2 définis par :

$$\text{Si } x \in A, G_i (A, A \setminus x) = \left(\frac{\mu_i (A \setminus x)}{\mu_i(A)}\right)^{1/2}$$

$$\text{et si } x \notin A, G_i (A, A \cup x) = \left(\frac{\mu_i (A \cup x)}{\mu_i(A)}\right)^{/12} .$$

Les hypothèses H1 et H2 s'écrivent respectivement

$$x \in A_2 \subset A_1 , \mu_1 (A_1) \mu_2 (A_2 \setminus x) \geqslant \mu_1 (A_1 \setminus x) \mu_2 (A_2)$$

$$A_2 \subset A_1, x \notin A_1 , \mu_1 (A_1 \cup x) \mu_2 (A_2) \geqslant \mu_1 (A_1) \mu_2 (A_2 \cup x)$$

et sont donc satisfaites. Il vient

$$A \in \Gamma , \sum_B P_t^1 (A, B) f(B) \geqslant \sum_B P_t^2 (A, B) f(B)$$

Puisqu'il est clair que μ_i est un équilibre (réversible) pour (P_t^i), le corollaire est établi par passage à la limite en t dans l'inégalité précédente (cf théorème 2).

3 - Cas d'un espace de phase infini

Λ est un ensemble infini dénombrable.

$\Omega = \{0, 1\}^{\Lambda}$, Ω est compact pour la topologie produit

$C(\Omega)$ est l'espace des fonctions continues sur Ω

$\mathfrak{F}$ est le sous-espace des fonctions sur Ω ne dépendant que d'un nombre fini de coordonnées, $\mathfrak{F}$ est dense dans $C(\Omega)$.

La généralisation des interactions (I) et (II) au cas $|\Lambda| = +\infty$ amène à considérer des opérateurs définis sur $\mathfrak{F}$.

Pour l'interaction (I)

$f \in \mathfrak{F}$, $A \in \Omega$

$$G f(A) = \sum_{x \notin A} \beta(x, A) \left[f(A \cup x) - f(A)\right] + \sum_{x \in A} \delta(x, A \setminus x) \left[f(A \setminus x) - f(A)\right]$$

où β et δ sont des fonctions positives.

Pour l'interaction (II)

$f \in \mathfrak{F}$, $A \in \Omega$

$$G f(A) = \sum_{x \in A,\ y \notin A} c(x, A)\, p(x, y) \left[f((A \setminus x) \cup y) - f(A)\right]$$

où p est une matrice stochastique sur Λ et c une fonction positive.

Liggett [18] et Holley [17] ont prouvé que, sous des hypothèses physiquement naturelles, ces opérateurs déterminent de façon unique des semi-groupes de Feller sur Ω.

Nous examinons maintenant lorsque $\Lambda = \mathbb{Z}_{\nu}$ les problèmes liés à l'existence d'états d'équilibres.

Interaction (I)

On suppose

$$\text{(L)} \quad \begin{aligned} &\beta(x, A) = \beta(x, A \cap \partial x) > 0, \ \delta(x, A) = \delta(x, A \cap \partial x) > 0, \\ &\beta(x, A) = \beta(x+z, A+z), \ \delta(x, A) = \delta(x+z, A+z) \text{ pour } z \in \mathbb{Z}_{\nu} \end{aligned}$$

généralisant le théorème 5 nous avons :

Théorème 6 (Logan) [19]

μ est un équilibre réversible pour (I) si, et seulement si, il existe un potentiel local U tel que $\mu \in \mathcal{G}_U$ et

$$(*) \quad \text{pour } x \notin A \quad \frac{\beta(x,A)}{\delta(x,A)} = \exp - \frac{1}{2} \left[U(A \cup x) - U(A)\right]$$

Si (*) n'est pas vérifiée on sait très peu de choses sur les mesures invariantes pour (I). Les meilleurs résultats sont dus à T. Harris [22].

Supposons * vérifiée, deux questions se posent :

Q_1 les éléments $\mu \in \mathcal{G}_U$ sont-ils les seules probabilités invariantes ?

Q_2 si $\mathcal{G}_U = \{\mu\}$ a-t-on

$$\lim_{t\to+\infty} \nu P_t = \mu$$

pour toute probabilité ν sur Ω ?

Des réponses affirmatives ont été données par Dobrushin [20] dans le cas d'interactions faibles, par Holley [16] dans le cas où le potentiel U est isotrope et attractif en utilisant une méthode de couplage analogue à celle utilisée dans la preuve du théorème 5, et dans le cas où les mesures d'équilibre sont invariantes par translation en généralisant la méthode du théorème 2 (b.ii) [21] . En revanche on ne sait pas la réponse à Q_1 et Q_2 dans le cas d'un potentiel répulsif même lorsque $\nu = 1$.

Pour l'interaction (II) sur $\mathbb{Z}_\nu$, les derniers résultats sont dans [23, 24 25, 33]. Ils sont assez complets dans le cas où P (x, y) = P (y, x), et la vitesse des sauts c (x, A) = constante.

Pour une introduction à d'autres évolutions temporelles markoviennes des systèmes finis ou infinis de particules, consultez [26].

CHAPITRE VII

CHAMPS DE MARKOV GAUSSIENS

Soit $(\xi(x)\ ;\ x \in \mathbb{Z}_\nu)$ une famille de variables aléatoires réelles gaussiennes centrées et $R(x, y) = E\ [\xi(x)\ \xi(y)]$ $x, y \in \mathbb{Z}_\nu$ la fonction de covariance.

Définition

$(\xi(x)\ ;\ x \in \mathbb{Z}_\nu)$ est un <u>champ de Markov gaussien</u> (notation C.M.G.) si :

(a) il est isotrope et invariant par translation, i.e. $R(x, y) = R(0, y-x)$ et $R(0, x)$ est invariant dans toute permutation des coordonnées de x.

(b) il a la propriété de Markov, i.e. la distribution de $\xi(x)$ conditionnelle à la connaissance de $\xi(.)$ sur un ensemble de $\mathbb{Z}_\nu \setminus \{x\}$, contenant ∂x, ne dépend que des valeurs de $\xi(.)$ sur ∂x.

(c) ξ est non singulier : si H désigne l'espace de Hilbert engendré par les $\{\xi(x)\ ;\ x \in \mathbb{Z}_\nu\}$ et H_Λ le sous-espace engendré par les $\{\xi(x)\ ;\ x \in \Lambda^c\}$, $H_\infty = \bigcap_\Lambda H_\Lambda \neq H$.

La caractérisation ci-dessous des covariances de C.M.G. suit Rozanov [27] avec les améliorations dues à Loren Pitt.

Théorème 1

$\{\xi(x)\ ;\ x \in \mathbb{Z}_\nu\}$ est un C.M.G. si et seulement si sa covariance est de la forme :

$$(1)\quad R(x, y) = A \sum_{n=0}^{+\infty} t^n P^n(x, y)$$

où : $A > 0$ est une constante,

$t \in]-1, +1[$ en dimension $\nu = 1$ ou 2,

$t \in [-1, +1]$ en dimension $\nu \geq 3$,

P^n est le $n^{ième}$ itéré du noyau

$$P(x, y) = \begin{cases} \dfrac{1}{2\nu} & \text{si } |x-y| = 1 \\ 0 & \text{sinon} \end{cases}$$

Démonstration

(a) et (b) montrent que l'espérance de ξ (o) conditionnelle à ξ (x) pour $x \neq 0$ doit être de la forme : $\frac{t}{2\nu} \sum_{|y|=1} \xi(y)$; on doit alors avoir

$\xi(o) - \frac{t}{2\nu} \sum_{y=1} \xi(y)$ orthogonal à $\xi(x)$, $\forall x \neq 0$ et en notant

$r(x) = R(o, x)$.

$$(2) \qquad r(x) - \frac{t}{2\nu} \sum_{|y|=1} r(x+y) = 0 \quad , \quad \forall x \neq 0 \ .$$

Pour résoudre l'équation (2) on remarque que, d'après le théorème de Bochner, $r(.)$ a une représentation de la forme $r(x) = \int_T e^{ix\theta} \mu(d\theta)$ et que l'on a :

$$(3) \qquad \int_T e^{ix\theta} \left[1 - \frac{t}{2\nu} \sum_{|y|=1} e^{iy\theta}\right] \mu(d\theta) = 0 \text{ pour } x \neq 0 \ .$$

En posant $p_t(\theta) = 1 - \frac{t}{2\nu} \sum_{|y|=1} e^{iy\theta}$ et $\nu(d\theta) = p_t(\theta)\, \mu(d\theta)$

(3) implique que la mesure ν a tous ses coefficients de Fourier nuls à l'exception du coefficient d'ordre 0 et par suite que ν est un multiple de la mesure de Lebesgue : $\nu(d\theta) = \lambda\, d\theta$. Deux cas se présentent :

(I) Si $\lambda \neq 0$, la relation $p_t(\theta)\, \mu(d\theta) = \lambda\, d\theta$ montre que μ est absolument continue, de densité $\frac{d\mu}{d\theta} = \frac{\lambda}{p_t(\theta)}$.

L'intégrabilité exige que t satisfasse aux hypothèses du théorème 1 et la relation (1) découle immédiatement du fait que :

$$\int_T e^{ix\theta} \sum_{n=0}^{+\infty} t^n P^n (0, x)\, dx = \frac{1}{1 - \frac{t}{2\nu} \sum_{|y|=1} e^{iy\theta}}$$

(II) Si $\lambda = 0$, nous devons montrer que la condition (c) exclut la possibilité pour μ d'être concentrée sur les zéros de $p_t\ (\theta)$. D'après le théorème de Bochner on a une isométrie entre H et $L^2\ (d\mu)$ telle que :

$$\xi\ (x) \longleftrightarrow e^{i\theta x} \quad \text{et } E\ (\eta_1\ \bar{\eta}_2) = \int_T f_1\ (\theta)\ \overline{f_2\ (\theta)}\ \mu\ (d\theta) \text{ si } \eta_i \longleftrightarrow f_i .$$

Mais la condition (c) implique qu'il existe dans H un élément $\eta \neq 0$ avec $\eta \notin H_\infty$, et donc un ensemble fini Λ tel que $\eta \notin H_\Lambda$. Alors il existe $\eta' \neq 0$ dans H orthogonal à H_Λ ; son image f par l'isométrie ci-dessus vérifie :

$$f \neq 0,\ f \in L^2\ (d\mu),\ f \perp e^{ix\theta},\ \forall x \in \Lambda^c, \text{ donc}$$

$$\int_T f\ (\theta)\ e^{-ix\theta}\ \mu\ (d\theta) = 0 \quad , \forall x \in \Lambda^c .$$

Posons $f\ (\theta)\ \mu\ (d\theta) = \nu\ (d\theta)$, alors $\hat{\nu}\ (x) = 0 \quad , \forall x \in \Lambda^c$, et par suite $\nu\ (d\theta) = \sum_{x \in \Lambda} \hat{\nu}\ (x)\ e^{ix\theta} = p\ (\theta)\ d\theta$ où $p\ (\theta)$ est un polynôme trigonométrique.

Finalement $f\ (\theta)\ \mu\ (d\theta) = p\ (\theta)\ d\theta$ et μ est absolument continue ce qui fournit une contradiction.

Ainsi tout C.M.G. satisfait à (1) et la réciproque est évidente.

Nous considérons maintenant un C.M.G. $(\xi\ (x)\ ;\ x \in \mathbb{Z}_\nu)$ fixé et sa fonction de covariance donnée par (1) ; nous supposons pour simplifier A = 1. On se propose de déterminer la distribution conditionnelle pour $x \in \Lambda$ connaissant $\xi\ (.)$ sur Λ^c ; nous montrerons, et d'une façon très intéressante, qu'elle ne dépend que des valeurs de ξ sur $\partial\Lambda$.

Il suffit de calculer :

$$M_\xi\ (x) = E\ \left[\xi\ (x) \middle|\ \xi\ (.) \text{ sur } \Lambda^c\right] \quad ;\ x \in \Lambda$$

et

$$\mathrm{Cov}_{\xi}\,(x,\,y) = E\left[(\xi\,(x)-M_{\xi}(x))\,(\xi(y)-M_{\xi}(y))\,\middle|\,\xi\,(.)\text{ sur }\Lambda^{c}\right]\ ;\ x,y\in\Lambda$$

Nous verrons que ces quantités sont liées à la mesure harmonique et à la fonction de Green d'un certain problème de Dirichlet.

Considérons la marche aléatoire (x_n) sur $\mathbb{Z}_\nu$ de noyau $P\,(x,\,y)$ et soit $T_{\partial\Lambda}$ le temps d'entrée dans $\partial\Lambda$, nous posons :

$$H_\Lambda\,(x,\,y) = \sum_{n=0}^{+\infty} t^n\, P^x\left[T_{\partial\Lambda} = n,\ x_n = y\right] \qquad x\in\Lambda\ ,\ y\in\partial\Lambda\ ,$$

$$g_\Lambda\,(x,\,y) = \sum_{n=0}^{+\infty} t^n\, P^x\left[T_{\partial\Lambda} > n,\ x_n = y\right] \qquad x\in\Lambda\ ,\ y\in\Lambda.$$

Théorème 2

Pour tout ensemble fini $\Lambda\subset\mathbb{Z}_\nu$:

(a) $M_\xi\,(x) = \sum_{y\in\partial\Lambda} H_\Lambda\,(x,\,y)\,\xi\,(y) \qquad ,\qquad x\in\Lambda\ ,$

(b) $\mathrm{Cov}_\xi\,(x,\,y) = g_\Lambda\,(x,\,y) \qquad ,\qquad x\in\Lambda,\ y\in\Lambda\ .$

Démonstration

(a) l'espérance conditionnelle $M_\xi\,(x)$ est caractérisée par :

$$\xi\,(x) - M_\xi\,(x)\perp\xi\,(z)\qquad\forall\, z\in\Lambda^c\ ;$$

il suffit de vérifier que

$$R\,(x,\,z) - \sum_{y\in\partial\Lambda} H_\Lambda\,(x,\,y)\,R\,(y,\,z) = 0\qquad\forall x\in\Lambda\ \text{ et }\ \forall z\in\Lambda^c\ ;$$

mais c'est une conséquence immédiate de la propriété de Markov de la marche aléatoire (x_n).

(b) Il est bien connu que la covariance conditionnelle ne dépend pas du conditionnement ; alors :

$$\mathrm{Cov}_\xi\,(x,\,y) = E\left\{\left[\xi\,(x) - M_\xi\,(x)\right]\,\left[\xi\,(y) - M_\xi\,(y)\right]\right\}$$
$$= R\,(x,\,y) - E\,(M_\xi\,(x)\,M_\xi\,(y))$$

$$= R(x, y) - E \sum_{u} \sum_{v} H_{\Lambda}(x, u) H_{\Lambda}(y, v) \xi(u) \xi(v)$$

$$= R(x, y) - \sum_{u \in \partial\Lambda} \sum_{v \in \partial\Lambda} H_{\Lambda}(x, u) H_{\Lambda}(y, v) R(u, v)$$

$$= R(x, y) - \sum_{v \in \partial\Lambda} R(x, v) H_{\Lambda}(y, v)$$

$$= R(y, x) - \sum_{v \in \partial\Lambda} H_{\Lambda}(y, v) R(v, x)$$

$$= g_{\Lambda}(y, x) = g_{\Lambda}(x, y)$$

Finalement on peut expliciter la densité conjointe conditionnelle sur Λ, connaissant $\xi(x) = \varphi(x)$ pour $x \in \partial\Lambda$.

Définissons : $\overline{\Lambda} = \Lambda \cup \partial\Lambda$

$$\text{et } \overline{\xi}(x) \begin{cases} \xi(x) & \text{si } x \in \Lambda \\ \varphi(x) & \text{si } x \in \partial\Lambda \end{cases}$$

Théorème 3

La densité conjointe conditionnelle $f : \mathbb{R}^{\Lambda} \to \mathbb{R}_{+}$ est donnée par :

$$f(\xi) = Z_{\Lambda}^{-1}(\varphi) \exp\left\{-\frac{1}{2} \sum_{x \in \overline{\Lambda}} \sum_{y \in \overline{\Lambda}} \left[\delta(x,y) - t\, P(x,y)\right] \overline{\xi}(x)\, \overline{\xi}(y)\right\}$$

Démonstration

Il faut montrer que la densité gaussienne ci-dessus a la moyenne et la covariance du théorème 2.

Pour la covariance on peut supposer que la condition de frontière φ est identiquement nulle. On peut écrire la fonction de Green $g_{\Lambda}(x, y)$ sous la forme :

$$g_{\Lambda}(x, y) = \sum_{n=0}^{+\infty} t^{n} P_{\Lambda}^{n}(x, y), \qquad x \in \Lambda \qquad y \in \Lambda$$

où P_{Λ} désigne la restriction de P à Λ ; ce qui montre que la forme quadratique figurant dans la densité f est bien l'inverse de la covariance du théorème 2.

Pour la moyenne on doit vérifier l'égalité de :

$$\sum_{x \in \bar{\Lambda}} \sum_{y \in \bar{\Lambda}} \left[\delta(x,y) - t\,P(x,y)\right] \bar{\xi}(x)\, \bar{\xi}(y)$$

et de

$$\sum_{x \in \Lambda} \sum_{y \in \Lambda} \left[\delta(x,y) - t\,P(x,y)\right] \left[\xi(x) - M_\varphi(x)\right] \left[\xi(y) - M_\varphi(y)\right] + R(\varphi)$$

où $R(\varphi)$ ne dépend que de la condition de frontière φ ; ce qui se ramène à vérifier :

$$\sum_{x \in \Lambda} \sum_{y \in \partial\Lambda} \left[\delta(x,y) - t\,P(x,y)\right] \xi(x)\, \varphi(y) = - \sum_{x \in \Lambda} \sum_{y \in \Lambda} \left[\delta(x,y) - tP(x,y)\right] \xi(x)\, M_\varphi(y)$$

il suffit donc de montrer que pour chaque $x \in \Lambda$ et chaque $\varphi : \partial\Lambda \to \mathbb{R}$,

$$t \sum_{y \in \partial\Lambda} P(x,y)\, \varphi(y) = \sum_{y \in \Lambda} \left[\delta(x,y) - tP(x,y)\right] M_\varphi(y)$$

Le second membre est :

$$M_\varphi(x) - t \sum_{y \in \Lambda} P(x,y) \sum_{z \in \partial\Lambda} H_\Lambda(y,z)\, \varphi(z) = \sum_{z \in \partial\Lambda} \left[H_\Lambda(x,z) - t \sum_{y \in \Lambda} P(x,y) H_\Lambda(y,z)\right] \varphi(z)$$

La démonstration est terminée en remarquant que la propriété de Markov implique :

$$H_\Lambda(x,z) = tP(x,z) + t \sum_{y \in \Lambda} P(x,y)\, H_\Lambda(y,z) \text{ pour } x \in \Lambda,\ z \in \partial\Lambda.$$

A l'aide des théorèmes 2 et 3 on peut expliquer le phénomène de transition de phase dans le modèle gaussien analysé pour la première fois par Kac et Berlin [28].

Si $\nu \geqslant 3$, pour $t = 1$, on peut ajouter une constante à un C.M.G sans changer les distributions conditionnelles. On obtient ainsi une infinité d'états avec des densités différentes.

Pour $t = -1$ on obtient une transition de phase de type anti feromagnétique.

$$\text{Si} \quad \psi(x) = \begin{cases} +1 & \text{si } \sum_i x_i \text{ est pair}, \\ -1 & \text{si } \sum_i x_i \text{ est impair}, \end{cases}$$

tous les champs du type $\xi(x) + c\, \psi(x)$ ont les mêmes distributions conditionnelles.

Pour conclure remarquons que les C.M.G. sur $\mathbb{Z}_\nu$ étudiés ci-dessus permettent d'approcher les C.M.G. généralisés sur $\mathbb{R}_\nu$ qui sont à l'heure actuelle d'un grand intérêt dans la théorie des champs quantiques [29], [30]. On sait que les champs sur $\mathbb{R}_\nu$ sont liés au problème de Dirichlet classique (pour le mouvement brownien) de la même façon que nos C.M.G. sont liés au problème de Dirichlet discret pour la marche aléatoire sur $\mathbb{Z}_\nu$.

BIBLIOGRAPHIE

[1] D. RUELLE, Statistical Mechanics, Benjamin, N.Y., 1969 l'édition Russe (Mir, Moscou, 1971) contient un chapitre sur les résultats plus récents par R.L. Dobrushin, R.A. Minlos, et Y.M. Sukhov.)

[2] G. GALLAVOTTI, Instabilities and Phase Transitions in the Ising Model. A Review, Rivista del Nuovo Cimento, 2, 1972.

[3] H.O. GEORGII, Phasenübergang 1. Art bei Gittergasmodellen, 16, Springer Lecture Notes in Physics, 1972 .

[4] R.L. DOBRUSHIN, Existence of Phase Transitions in the Two and Three dimensional Ising Models, Th. Prob. and Appl. 10, 1965.

[5] R.B. GRIFFITHS, Peierl's proof of spontaneous magnetization in a two-dimensional Ising Ferromagnet, Phys. Rev. (2) 136, 1964.

[6] L. ONSAGER, Crystal Statistics I, A two dimensional model with an Order - Disorder transition, Phys. Rev. 65, 1944.

[7] G.R. GRIMMETT, A theorem about Random Fields, Bull. London Math. Soc., 5, 1973.

[8] R.L. DOBRUSHIN, Description of a random field by means of conditional probabilities, Th. Prob. and Appl., 13, 1968.

[9] R.L. DOBRUSHIN, Gibbsian Random Fields for Lattice systems with pair interactions. Funct. An. and Appl. 2, 1968

[10] O.E. LANFORD and D. RUELLE, Observables at infinity and states with short range correlations in Statistical Mechanics, Comm. Math. Phys. 13, 1969.

[11] D. RUELLE, On the use of small external fields etc., Ann. of Phys., 69, 1972.

[12] R.L. DOBRUSHIN, The problem of uniqueness of a Gibbs random field. Funct. An. and Appl., 2, 1968.

[13] R.L. DOBRUSHIN, Coexistence of phases in the three dimensional Ising Model, Th. Prob. and Appl., 17, 1972.

[14] O.E. LANFORD, Entropy and equilibrum states in classical statistical mechanics, in Springer Lecture Notes in Physics, 20, 1973.

[15] R. THOMPSON, Cornell U. Ph. D. Thesis, 1973, to appear in Trans. A.M.S.

[16] R. HOLLEY, Recent results on the stochastic Ising Model, Rocky Mountain Math. J., to appear.

[17] R. HOLLEY, Markovian interaction processes with finite range interactions, Ann. Math. Stat., 43, 1972.

[18] T.M. LIGGETT, Existence theorems for infinite particle systems, Trans. A.M.S., 165, 1972.

[19] K. LOGAN, Cornell U. Ph. D. Thesis, 1974.

[20] R.L. DOBRUSHIN, Markov processes with a large number of locally independent components, Probl. Pered. Inf., 7, 1971.

[21] R. HOLLEY, Free energy in a Markovian model of a lattice spin system, Comm. Math. Phys., 23, 1971.

[22] T. HARRIS, Contact interactions on a lattice, preprint, 1973.

[23] T.M. LIGGETT, A characterization of the invariant measures for an infinite particle system with interactions, Trans. A.M.S., 179, 1973.

[24] T.M. LIGGETT, même sujet, va paraître aussi Trans. A.M.S., 1974.

[25] F. SPITZER, Recurrent random walk of an infinite particle system, à paraître, Trans. A.M.S., 1974.

[26] F. SPITZER, Interactions of Markov processes, Adv. in Math., 5, 1970.

[27] Yu. A. ROZANOV, On Gaussian fields with given conditional distributions, Th. Prob. and its appl. 12, 1967.

[28] T. BERLIN and M. KAC, The spherical model of a ferromagnet, Phys. Rev. 86, 1952.

[29] E. NELSON, Construction of Quantum Fields from Markoff Fields. J. Funct. An., 12, 1973.

[30] E. NELSON, The free Markoff Field, J. Funct. An., 12, 1973.

THE STOCHASTIC EVOLUTION OF INFINITE SYSTEMS OF INTERACTING PARTICLES

PAR T.M. LIGGETT

The preparation of these notes was supported in part by N.S.F. Grant No. MPS 72-04591, and by an Alfred P. Sloan Founc research fellowship.

Originally published in: *Ecole d'Eté de Probabilités de Saint-Flour VI – 1976*, Lecture Notes in Mathematics, Vol. **598**, 187–248, DOI: 10.1007/BFb0097493, © Springer-Verlag Berlin Heidelberg 1977, Reprint by Springer-Verlag Berlin Heidelberg 2012

INTRODUCTION

Classical statistical mechanics is concerned with the equilibrium theory of certain physical systems. During the past six or eight years, several classes of Markov processes have been proposed as models for the temporal evolution of such systems, and a significant amount of progress has been made in their study. Some of these models, in turn, have been given economic or sociological interpretations. These lectures are intended as an introduction to and survey of some of this recent work. We will say very little about statistical mechanics itself, and will treat our subject as a self-contained branch of probability theory. The reader who would like to know more about the statistical mechanics which lies in the background is referred to [30], [41], [42], and [47].

Even in terms of the temporal theory, our discussion will be far from exhaustive. There are many important and interesting results in the papers listed in the bibliography to which we will refer only briefly, if at all. As might be expected, the choice of material and the emphasis was influenced by my own personal interests. While some of the proofs have been modified, most of the results presented here have appeared elsewhere in the literature. The only material which is essentially new is that in chapter 4 of part II. One of the attractive features of this subject is that there are still a large number of important open problems. We will point some of these out as we proceed.

Two types of processes have received the most attention. The first is the spin-flip process, which is the object of study in part I. While a special case was considered by Glauber [7] several years earlier, the spin-flip process was first formally introduced and studied in some generality by Dobrushin [6]. The state space of the process is $X = \{0,1\}^S$, where S is a countable set of sites. The interpretation is that at each point of S there is a particle (e.g., an iron atom) which has a positive or negative spin (represented by 1 and 0 respectively) at each time. Thus X represents all possible configurations of spins, and the Markov process describes the evolution of the configuration of spins. The process is described in terms of a collection of nonnegative speed functions $c(x,\eta)$ defined for $x \in S$ and $\eta \in X$ which give the rate at which the spin at x flips from $\eta(x)$ to $1-\eta(x)$ when the configuration of the entire system is η. This process has the property that only one coordinate of η changes at a time.

In the second type of process, two coordinates change at once. The state space is X again, and the process is described in terms of nonnegative speed functions $c(x,y,\eta)$ which give the rate at which an interchange of the x and y coordinates of η will occur when the configuration of the system is η. There are two interpretations of this process which suggest two different forms for the functions $c(x,y,\eta)$. In the first, which was proposed by Spitzer [45], the system being modelled is a lattice gas. Here particles are distributed on S with at most one per site, and $\eta(x)$ is the number of particles at x. The function

$c(x,y,\eta)$ then gives the rate at which a particle at x will move to a vacant site y or vice versa. The second interpretation is that of a binary alloy, and appears, for example, in [1] and [2]. In this case there are two types of particles, and each site in S is occupied either by a particle of type 0 or a particle of type 1. Then $c(x,y,\eta)$ gives the rate at which two particles of opposite types at x and y will interchange their positions. These processes will be called exclusion processes, since multiple occupancy is excluded, and will be discussed in part II.

The functions $c(x,\eta)$ or $c(x,y,\eta)$ do not always determine a unique process on X. Further conditions on these functions must generally be imposed in order to guarantee the existence of a unique process with desirable properties. The conditions usually express the requirement that the rates not depend too heavily on distant parts of the configuration. The first problem which requires treatment, then, involves existence and uniqueness questions. This problem is by now rather well understood, although there are some important conjectures relating to it which remain to be settled. These questions will be discussed in detail in chapter 1 of part I, and more briefly in chapter 1 of part II.

Once existence and uniqueness questions are disposed of, the main questions of interest involve the ergodic theory of the process. These are motivated by the statistical mechanical origin of these processes, and can be stated briefly in the following way. If μ is a probability measure on X, let $\mu S(t)$ be the distribution of the process at time t when μ is the initial distribution. Let $\mathcal{J} = \{\mu : \mu S(t) = \mu \text{ for all } t > 0\}$ be the set of invariant measures for the process. Since X is compact and the process will have the Feller property in all the cases we will consider, $\mathcal{J}$ will always be a nonempty compact convex set. The first problem is then to determine the structure of $\mathcal{J}$. In the case of spin-flip processes, $\mathcal{J}$ will often be a singleton, in which case we will say that the process does not exhibit phase transition. Many of the results which we will discuss give sufficient conditions for the absence of phase transition. When phase transition does occur, the situation is not at all well understood. In the case of exclusion processes, there is typically at least a one parameter family of extremal invariant measures, which can often be exhibited explicitly. The problem is then to show that there are no other extremal invariant measures. The results here deal mainly with the simple exclusion process, in which $c(x,y,\eta)$ depends on η only through $\eta(x)$ and $\eta(y)$. Very little is known about the more general case.

After the structure of $\mathcal{J}$ is understood, the next step in the development of the ergodic theory of the process is to determine the domain of attraction of each element of $\mathcal{J}$. Here one wants to find for each $\nu \in \mathcal{J}$ the set of all probability measures μ on X so that $\mu S(t)$ converges weakly to ν as $t \to \infty$. In the case of a spin flip process without phase transition, of course, one wishes to prove that this occurs for all μ, in which case we will say that the process is ergodic. When $\mathcal{J}$ is not a singleton, it is unreasonable to expect to be able to determine the domains of attraction completely. There are a few cases where this is possible (see, for example, sections 3.3 of part I and 2.1 of part II), but normally one should be satisfied by finding the limit of $\mu S(t)$ for sufficiently nice μ.

The primary aim of the subject is to solve the problems described in the previous three paragraphs in the greatest possible generality. There are related problems which are also important, however. Recently, for example, Holley and Stroock [26] have used rates of convergence in $\mu S(t) \to \nu$ for the ergodic spin-flip process to obtain new information about the Gibbs measures of statistical mechanics.

The most direct relationship between the spin-flip and exclusion processes on the one hand and equilibrium statistical mechanics on the other, is that the Gibbs measures are contained in $\mathcal{J}$ when $c(x,\eta)$ or $c(x,y,\eta)$ is chosen to depend in an appropriate way on the potential involved in the definition of the Gibbs measures. This is of course one of the justifications for considering these processes as models for nonequilibrium statistical mechanics. This relationship will be seen in more detail in section 4.1 of part I and sections 1.3 and 4.2 of part II.

We will conclude this introduction by mentioning three techniques which have been used frequently in this subject, and which we will see more of in these lectures. One of their interesting features is that, while the two types of processes we will discuss in parts I and II are quite different in many respects, each of these techniques has proved very useful for both types of processes. Probably the most useful of these techniques is that of coupling two or more processes together. This will be a key tool in chapter 2 of part I and chapters 2, 3, and 4 of part II. A second tool involves obtaining relations between the infinite particle system of interest and an auxiliary finite system, thus reducing many questions to the study of the finite system. This will be used in chapter 3 of part I and chapter 2 of part II. Finally, there is the use of the monotonicity of the free energy of the system, which was exploited by Holley in [19] and [20], and will be referred to briefly in section 4.2 of part I and section 4.2 of part II.

PART I. SPIN-FLIP PROCESSES

Chapter 1. Existence results and first ergodic theorems. Several approaches to the existence problem have been proposed. In [12], Harris used a direct probabilistic construction of the process of interest in the exclusion context. In [6], Dobrushin obtained existence in the spin-flip context via finite approximations. More recently, Holley and Stroock [25] have changed the problem into one involving martingales, in much the same way that Stroock and Varadhan had done earlier in the context of diffusion processes. They then used compactness arguments to obtain a very general existence result. From their point of view, the principal problem then becomes one of proving the uniqueness of the process, since that does not follow directly from the martingale approach. We will concentrate primarily on the semigroup approach, which proceeds via the Hille-Yosida theorem. It requires the imposition of smoothness conditions on the speed functions, but has the advantage of yielding immediately a Feller process which is uniquely determined by them.

Section 1.1. The Hille-Yosida theorem. By a semigroup of operators on the Banach space W, we will mean a collection $\{S(t),\ t \geq 0\}$ of linear operators on W which satisfy the following conditions: $S(0) = I$, $S(t_1 + t_2) = S(t_1)S(t_2)$, $\|S(t)\| \leq 1$, and $S(t)f \to f$ as $t \to 0$ for all $f \in W$. If $W = C(X)$ where X is compact metric, a semigroup of operators will be said to be a Markov semigroup if $S(t)1 = 1$ and $S(t)f \geq 0$ whenever $f \geq 0$. As is well known, a Markov semigroup on $C(X)$ determines a unique strong Markov process η_t on X via $S(t)f(\eta) = E^{\eta}f(\eta_t)$.

A (possibly unbounded) linear operator Ω on W with domain $\mathcal{D}(\Omega)$ is said to be dissipative if $f - \lambda\Omega f = g$ implies $\|f\| \leq \|g\|$ whenever $f \in \mathcal{D}(\Omega)$ and $\lambda \geq 0$. Ω is closed if its graph is a closed subset of $W \times W$.

Theorem 1.1.1. (Hille-Yosida). There is a one-to-one correspondence between semigroups $S(t)$ of operators on W and closed dissipative operators Ω on W with dense domain which satisfy $R(I - \lambda\Omega) = W$ for all sufficiently small $\lambda > 0$. The correspondence is given by:

(a) $$S(t)f = \lim_{n\to\infty} \left(I - \frac{t}{n}\Omega\right)^{-n} f \quad \text{for } f \in W \text{ and } t \geq 0, \quad \text{and}$$

(b) $$\Omega f = \lim_{t \downarrow 0} \frac{S(t)f - f}{t} \qquad \text{for } f \in \mathcal{D}(\Omega).$$

Furthermore, $\frac{d}{dt} S(t)f = \Omega S(t)f = S(t)\Omega f$ for $f \in \mathcal{D}(\Omega)$. In case $W = C(X)$ for a compact X, $S(t)$ is Markov if and only if $\Omega 1 = 0$ and $(I - \lambda\Omega)^{-1}f \geq 0$ whenever $f \geq 0$ and $\lambda \geq 0$. Ω is called the generator of $S(t)$.

Remarks: If Ω is a semigroup generator, then $\mathcal{R}(I - \lambda\Omega) = W$ for all $\lambda \geq 0$. A bounded dissipative operator is of course automatically a semigroup generator.

Two easily verified facts which are useful in applying this theorem are: (a) if Ω is closed and dissipative, then $\mathcal{R}(I - \lambda\Omega)$ is closed in W for $\lambda > 0$, and (b) if Ω is dissipative and has dense domain, then the closure of Ω exists and is again a dissipative operator. As a consequence, if Ω is dissipative, has dense domain, and satisfies $\overline{\mathcal{R}(I - \lambda\Omega)} = W$ for all sufficiently small $\lambda > 0$, then the closure $\overline{\Omega}$ of Ω is a semigroup generator. A core of the generator Ω is a linear subspace D of $\mathcal{D}(\Omega)$ with the property that Ω is the closure of the restriction of Ω to D. Thus Ω is determined by its values on a core. In order for D to be a core, it is not in general sufficient that D be dense in W.

In case $W = C(X)$ for a compact X and $S(t)$ is a Markov semigroup, define $\mu S(t)$ for probability measures μ on X by

$$\int f \, d[\mu S(t)] = \int S(t) f \, d\mu .$$

Recalling that $\mathcal{I} = \{\mu : \mu S(t) = \mu \text{ for all } t \geq 0\}$, the Hille-Yosida theorem makes it possible to rewrite this definition in terms of the generator Ω of $S(t)$ in the following way:

$$\mathcal{I} = \left\{\mu : \int \Omega f \, d\mu = 0 \quad \text{for all} \quad f \in \mathcal{D}(\Omega)\right\} . \tag{1.1.2}$$

In order to verify that a given μ is in $\mathcal{I}$, it suffices to check that $\int \Omega f \, d\mu = 0$ for all f in a core for Ω. $\mathcal{I}$ is convex and is compact in the topology of weak convergence of measures, so that $\mathcal{I}$ is the closed convex hull of $\mathcal{I}_e$, its set of extreme points. Note that $\mathcal{I}$ is nonempty, since any weak limit of $\frac{1}{T}\int_0^T \mu S(t)dt$ as $T \to \infty$ is in $\mathcal{I}$ for any probability measure μ on X. Suppose Ω_n and Ω are Markov semigroup generators such that $\Omega_n f \to \Omega f$ for all f in a core for Ω. Then if $\mu_n \in \mathcal{I}_n$ and $\mu_n \to \mu$ weakly, it follows from (1.1.2) that $\mu \in \mathcal{I}$. It is not in general the case that $\mu \in \mathcal{I}$ implies that there exists $\mu_n \in \mathcal{I}_n$ so that $\mu_n \to \mu$ weakly, so that an explicit knowledge of $\mathcal{I}_n$ for each n leads to an identification of only a subset of $\mathcal{I}$.

We will have occasion to use the Trotter-Kurtz convergence theorem for semigroups. The following version [31] will suffice for our purposes.

<u>Theorem 1.1.3</u>. Suppose Ω_n and Ω are generators of semigroups $S_n(t)$ and $S(t)$ respectively. If there is a core V for Ω such that $V \subset \mathcal{D}(\Omega_n)$ for all n and $\Omega_n f \to \Omega f$ for all $f \in V$, then $S_n(t)f \to S(t)f$ for all $f \in W$ uniformly for t in compact sets.

<u>Section 1.2</u>. <u>Speed functions which are Lipschitz continuous</u>. Let S be a countable set and $X = \{0,1\}^S$ with the product topology. $C(X)$ is the space of continuous functions on X with the norm $\|f\| = \sup_\eta |f(\eta)|$. For $\eta \in X$ and $u \in S$, define $\eta_u \in X$ by

$$\eta_u(x) = \begin{cases} \eta(x) & \text{if } x \neq u \\ 1 - \eta(x) & \text{if } x = u . \end{cases}$$

For $u \in S$, define $\Delta_u : C(X) \to C(X)$ by $\Delta_u f(\eta) = f(\eta_u) - f(\eta)$, and put $C^1(X) = \{f \in C(X) : \quad |||f||| = \Sigma_u \|\Delta_u f\| < \infty\}$. Suppose that $c(x,\eta)$ is a uniformly bounded, nonnegative, continuous function on $S \times X$, and define Ω on $C^1(X)$ by

(1.2.1) $$\Omega f(\eta) = \sum_x c(x,\eta)\Delta_x f(\eta) .$$

$\Omega f \in C(X)$ for $f \in C^1(X)$ since the convergence is uniform. Ω is densely defined since $C^1(X) \supset \mathcal{F}$, the set of all functions which depend on finitely many coordinates. If $f \in C^1(X)$, $\lambda \geq 0$, and $f - \lambda\Omega f = g$, choose $\eta, \zeta \in X$ so that $f(\eta) = \inf\{f(\gamma) : \gamma \in X\}$ and $f(\zeta) = \sup\{f(\gamma): \gamma \in X\}$. Then $\Omega f(\eta) \geq 0$ and $\Omega f(\zeta) \leq 0$, so $f(\eta) \geq g(\eta)$ and $f(\zeta) \leq g(\zeta)$. Therefore Ω is dissipative and $g \geq 0$ implies $f \geq 0$. In order to conclude that the closure of Ω generates a unique Markov semigroup, it then suffices to verify that $\mathcal{R}(I - \lambda\Omega)$ is dense in $C(X)$ for all sufficiently small $\lambda \geq 0$.

We will carry out this verification under the assumption that $\{c(x,\eta), x \in S\}$ is a bounded subset of $C^1(X)$, and at the same time, we will obtain sufficient conditions for the ergodicity of the resulting process. The existence proof is based on [32], and the modification required to obtain the ergodicity result is based on [26]. For further generalizations, see [52] and [10].

The following lemma contains the essential a priori bound which is required for the proof. In order to state it, let $c(u) = \inf_\eta[c(u,\eta) + c(u,\eta_u)]$ and $\gamma(x,u) = \|\Delta_u c(x,\eta)\|$ for $x \neq u$.

<u>Lemma 1.2.2</u>. Suppose $f \in C^1(X)$, $\lambda \geq 0$, and $f - \lambda\Omega f = g$. Then

$$\|\Delta_u f\|[1 + \lambda c(u)] \leq \|\Delta_u g\| + \lambda \sum_{x \neq u} \gamma(x,u) \|\Delta_x f\| .$$

<u>Proof</u>. For $\eta \in X$ and $u \in S$,

$$\Delta_u f(\eta) = \Delta_u g(\eta) + \lambda \sum_{x\neq u} c(x,\eta_u)\Delta_u\Delta_x f(\eta)$$

$$+ \lambda \sum_{x\neq u} [\Delta_u c(x,\eta)]\Delta_x f(\eta) - \lambda[c(u,\eta_u) + c(u,\eta)]\Delta_u f(\eta) .$$

Choose $\zeta \in X$ so that $\Delta_u f(\zeta) = \|\Delta_u f\|$, which is possible since $\Delta_u f(\eta_u) = -\Delta_u f(\eta)$ for all η. Then $\Delta_u\Delta_x f(\zeta) \leq 0$, from which the result follows.

Now put $c = \inf_u c(u) \geq 0$, and assume that

(1.2.3) $$M = \sup_x \sum_{u\neq x} \gamma(x,u) \leq \sup_x |||c(x,\cdot)||| < \infty .$$

Then $(\Gamma\beta)(u) = \Sigma_{x\neq u}\, \beta(x)\gamma(x,u)$ defines a bounded operator on $\ell_1(S)$ with norm M.

<u>Corollary 1.2.4</u>. Suppose $f,g \in C^1(X)$, $\lambda \geq 0$, and $f - \lambda\Omega f = g$. Define $\beta \in \ell_1(S)$ by $\beta(u) = \|\Delta_u g\|$. Then

$$\|\Delta_u f\| \le [(1+\lambda c)I - \lambda\Gamma]^{-1}\beta(u)$$

provided that $\lambda M/(1+\lambda c) < 1$.

Theorem 1.2.5. Under assumption (1.2.3), the closure of Ω generates a unique Markov semigroup $S(t)$. For $g \in C^1(X)$, put $\beta(u) = \|\Delta_u g\|$. Then

$$\|\Delta_u S(t)g\| \le e^{t[\Gamma - cI]}\beta(u), \quad \text{so} \tag{1.2.6}$$

$$|||S(t)g||| \le e^{(M-c)t}|||g|||. \tag{1.2.7}$$

In particular, if $M < c$, the corresponding process is ergodic with unique invariant measure μ on X and

$$\left\|S(t)g - \int g\,d\mu\right\| \le \sup_x \|c(x,\cdot)\| \frac{e^{(M-c)t}}{c-M}|||g|||.$$

Proof. Let S_n be a sequence of finite sets such that $S_n \uparrow S$. Define $c_n(x,\eta) = c(x,\eta)$ for $x \in S_n$ and $c_n(x,\eta) = 0$ if $x \notin S_n$, and Ω_n and γ_n as before in terms of $c_n(x,\eta)$ instead of $c(x,\eta)$. Then Ω_n is a bounded dissipative operator, and therefore it is the generator of a Markov semigroup. Take $\lambda > 0$ so that $\lambda M < 1 + \lambda c$. We will show that $\mathcal{R}(I - \lambda\Omega)$ is dense in $C(X)$ for such a λ. Fix $g \in C^1(X)$ and define f_n by $f_n - \lambda\Omega_n f_n = g$. By the argument of lemma 1.2.2, $\|\Delta_u f_n\| \le \|\Delta_u g\| + \lambda\sum_{x\in S_n}\gamma(x,u)\|\Delta_x f\|$ for $u \notin S_n$, so $f_n \in C^1(X)$. Therefore by Corollary 1.2.4 and the fact that $\gamma_n(x,y) \le \gamma(x,y)$ for all $x,y \in S$, it follows that

$$\|\Delta_u f_n\| \le [(1+\lambda c)I - \lambda\Gamma]^{-1}\beta(u), \tag{1.2.8}$$

where $\beta(u) = \|\Delta_u g\|$. Put $g_n = f_n - \lambda\Omega f_n$. Then

$$\|g_n - g\| \le \lambda \sup_u \|c(u,\cdot)\| \sum_{x\notin S_n}\|\Delta_x f_n\|,$$

which tends to zero by (1.2.8). Since $g_n \in \mathcal{R}(I - \lambda\Omega)$, it follows that $g \in \overline{\mathcal{R}(I - \lambda\Omega)}$. Therefore $\overline{\mathcal{R}(I - \lambda\Omega)}$ contains $C^1(X)$, which is dense in $C(X)$, so $\bar\Omega$ generates a unique Markov semigroup $S(t)$ by the Hille-Yosida theorem. Furthermore, since $g_n \to g$ and $\|f_n - f_m\| \le \|g_n - g_m\|$, it follows that $f = \lim_n f_n$ exists, $\lim_n \Omega f_n$ exists, $f \in \mathcal{D}(\bar\Omega)$, $\bar\Omega f = \lim_n \Omega f_n$, and $f - \lambda\bar\Omega f = g$. Taking the limit in (1.2.8) yields

$$\|\Delta_u(I - \lambda\bar\Omega)^{-1}g\| \le [(1+\lambda c)I - \lambda\Gamma]^{-1}\beta(u).$$

This can then be iterated to obtain

$$\|\Delta_u(I - \lambda\bar{\Omega})^{-k} g\| \leq [(1 + \lambda c)I - \lambda\Gamma]^{-k} \beta(u) ,$$

from which (1.2.6) follows by using part (a) of theorem 1.1.1. Now suppose that $M < c$, and take $\tau < t$. Then for $g \in C^1(X)$,

$$\|S(t)g - S(\tau)g\| \leq \int_\tau^t \|\Omega S(s)g\| \, ds \tag{1.2.9}$$

$$\leq \sup_x \|c(x,\cdot)\| \int_\tau^t |||S(s)g||| \, ds$$

$$\leq \sup_x \|c(x,\cdot)\| \frac{e^{(M-c)\tau}}{c - M} |||g||| .$$

Therefore $\lim_{t\to\infty} S(t)g$ exists, and it is constant by (1.2.7). This constant must be $\int g \, d\mu$ for any $\mu \in \mathcal{J}$, since $\int S(t)g \, d\mu = \int g \, d\mu$ for all t. The final conclusion of the theorem is obtained by letting $t \to \infty$ in (1.2.9).

Remark: It is easy to see that $\mathfrak{F}$ is a core for $\bar{\Omega}$. Since we have proved that $C^1(X)$ is a core, it suffices to show that given $f \in C^1(X)$, there are $f_n \in \mathfrak{F}$ such that $f_n \to f$ and $\Omega f_n \to \Omega f$. To do this, it suffices to choose f_n so that $f_n \to f$ and $\|\Delta_u f_n\| \leq \|\Delta_u f\|$, which is always possible.

Section 1.3. Speed functions with absolutely convergent Fourier series.

This section is based on the work of Holley and Stroock [25], although the proofs are different. Let ν be the product measure on X with $\nu\{\eta : \eta(x) = 1\} = 1/2$ for all $x \in S$, and define $\chi_\emptyset(\eta) \equiv 1$ and $\chi_F(\eta) = \prod_{x\in F}[2\eta(x) - 1]$ for nonempty finite subsets F of S. Then $\{\chi_F\}$ is a complete orthonormal family in $L_2(\nu)$, and we may consider speed functions which have the representation

$$c(x,\eta) = \sum_F \hat{c}(x,F)\chi_F(\eta) ,$$

where $\sum_F |\hat{c}(x,F)| < \infty$ for each x. The convergence is then uniform, so $c(x,\eta)$ is continuous for each x. The basic assumption which will be made throughout this section is that there is an $0 < \alpha < 1$ so that

$$\sum_{F\neq\emptyset} |\hat{c}(x,F)| \leq \alpha\hat{c}(x,\emptyset). \tag{1.3.1}$$

Note that this automatically makes $c(x,\eta)$ nonnegative.

For finite subsets F and G of S, let $F \Delta G$ denote the symmetric difference of F and G, and $\beta(F) = \sum_{x\in F} \hat{c}(x,\emptyset)$. In the computations to follow, we will use the following easily verified facts: $\chi_F\chi_G = \chi_{F\Delta G}$,

$$\Delta_x \chi_F = \begin{cases} -2\chi_F & \text{if } x \in F \\ 0 & \text{if } x \notin F, \end{cases}$$

and

(1.3.2) $$\sum_{G \neq F} \sum_{x \in F} |\hat{c}(x, F \triangle G)| \leq \alpha\beta(F) .$$

Let D be the class of all functions f which have the representation

$$f(\eta) = \sum_F \hat{f}(F) \chi_F(\eta)$$

where $\Sigma_F[1 + \beta(F)]|\hat{f}(F)| < \infty$. For $f \in D$, the series

$$\Omega f(\eta) = \sum_x c(x,\eta) \Delta_x f(\eta)$$

(1.3.3) $$= -2 \sum_x \sum_G \hat{c}(x,G)\chi_G(\eta) \sum_{F \ni x} \hat{f}(F)\chi_F(\eta)$$

$$= -2 \sum_H \chi_H(\eta) \sum_F \hat{f}(F) \sum_{x \in F} \hat{c}(x, F \triangle H)$$

converges absolutely and uniformly by (1.3.2). Therefore Ω is a densely defined operator on $C(X)$, which is again dissipative. We will show that its closure is a semigroup generator under assumption (1.3.1). The required a priori bound is given in the following lemma. For finite subsets H and F of S, let $\gamma(H,F) = \Sigma_{x \in F}|\hat{c}(x, F \triangle H)|$ if $H \neq F$ and $\gamma(H,H) = 0$. Note that

(1.3.4) $$\sum_H \gamma(H,F) \leq \alpha\beta(F)$$

by (1.3.2).

Lemma 1.3.5. Suppose $f \in D$, $\lambda \geq 0$ and $f - \lambda\Omega f = g$. Then for any finite subset H of S,

$$|\hat{f}(H)|[1 + 2\lambda\beta(H)] \leq |\hat{g}(H)| + 2\lambda \sum_F \gamma(H,F)|\hat{f}(F)|$$

Proof. Separating out the terms on the right side of (1.3.3) for which $H = F$, we obtain

$$\sum_H \hat{f}(H)\chi_H(\eta)[1 + 2\lambda\beta(H)] = g(\eta) - 2\lambda \sum_{H \neq F} \left[\sum_{x \in F} \hat{c}(x, F \triangle H)\right] \hat{f}(F)\chi_H(\eta) .$$

Identifying the Fourier coefficients of both sides gives

$$\hat{f}(H)[1 + 2\lambda\beta(H)] = \hat{g}(H) - 2\lambda \sum_{F \neq H} \left[\sum_{x \in F} \hat{c}(x, F \triangle H)\right] \hat{f}(F)$$

from which the result follows by taking absolute values.

Theorem 1.3.6. Under assumption (1.3.1), the closure of Ω generates a unique Markov semigroup $S(t)$.

Proof. Let S_n be a sequence of finite sets so that $S_n \uparrow S$. Define speed functions $c_n(x,\eta)$ by $c_n(x,\eta) = 0$ if $x \notin S_n$ and $c_n(x,\eta) = \Sigma_{F \subset S_n} \hat{c}(x,F)\chi_F(\eta)$ if $x \in S_n$. Let Ω_n be defined as before with $c_n(x,\eta)$ replacing $c(x,\eta)$. Then Ω_n is a bounded dissipative operator on $C(X)$ which maps the class of functions which depend only on coordinates in S_n into itself. Therefore, if $\lambda \geq 0$ and g depends only on coordinates in S_n, there is a function f_n depending on the coordinates in S_n which solves $f_n - \lambda\Omega_n f_n = g$. So, take $g \in \mathfrak{F}$ and $\lambda > 0$ and define $f_n \in \mathfrak{F} \subset D$ for all sufficiently large n by $f_n - \lambda\Omega_n f_n = g$. By Lemma 1.3.5,

$$|\hat{f}_n(H)|[1 + 2\lambda\beta(H)] \leq |\hat{g}(H)| + 2\lambda \sum_F \gamma(H,F)|\hat{f}_n(F)| \tag{1.3.7}$$

for all finite $H \subset S$. Note that this needs to be verified only for $H \subset S_n$, since $\hat{f}_n(H) = 0$ otherwise. By (1.3.4),

$$\Gamma u(H) = \sum_{F:\beta(F)>0} \frac{\gamma(H,F)}{\beta(F)} u(F)$$

defines an operator of norm $\leq \alpha < 1$ on the ℓ_1 space on the finite subsets of S. Define $u(H) = (1/2\lambda)|\hat{g}(H)|$. Then $u \in \ell_1$, so $v = (I - \Gamma)^{-1}u \in \ell_1$ also, and $\beta(H)|\hat{f}_n(H)| \leq v(H)$ by (1.3.7). Put $g_n = f_n - \lambda\Omega f_n$, and observe from this last estimate and (1.3.3) that $\|g_n - g\| = \lambda\|\Omega_n f_n - \Omega f_n\| \to 0$ by the dominated convergence theorem. Therefore $g \in \overline{\mathcal{R}(I - \lambda\Omega)}$, so the result is a consequence of the Hille-Yosida theorem.

For the ergodic theorem, we will assume that

$$c = \inf_x \hat{c}(x,\emptyset) > 0 . \tag{1.3.8}$$

From lemma 1.3.5 and the above proof, it follows that if $\Sigma_F|\hat{g}(F)| < \infty$, then $f = (I - \lambda\Omega)^{-1}g \in D$ for $\lambda > 0$. Summing the inequality in the statement of lemma 1.3.5 over $H \neq \emptyset$, and using $\beta(H) \geq c$ for $H \neq \emptyset$, $\gamma(H,\emptyset) = 0$, and (1.3.4), we see that

$$[1 + 2\lambda c(1 - \alpha)] \sum_{H \neq \emptyset} |\hat{f}(H)| \leq \sum_{H \neq \emptyset} |\hat{g}(H)| .$$

Iterating this and using (a) of the Hille-Yosida theorem yields

$$\exp[2ct(1-\alpha)] \sum_{H\neq\emptyset} |\widehat{S(t)g}(H)| \leq \sum_{H\neq\emptyset} |\hat{g}(H)| ,$$

and hence

$$\|S(t)g - \widehat{S(t)g}(\emptyset)\| \leq \sum_{H} |\hat{g}(H)| \exp[-2ct(1-\alpha)] . \tag{1.3.9}$$

Therefore, if μ is any invariant measure for the process,

$$\left|\int g\, d\mu - \widehat{S(t)g}(\emptyset)\right| \leq \sum_{H} |\hat{g}(H)| \exp[-2ct(1-\alpha)] .$$

Using (1.3.9) again gives the following theorem.

<u>Theorem 1.3.10</u>. Under assumptions (1.3.1) and (1.3.8), the process is ergodic with unique invariant measure μ, and

$$\left\|S(t)g - \int g\, d\mu\right\| \leq 2 \sum_{H} |\hat{g}(H)| \exp[-2ct(1-\alpha)]$$

for all g for which $\Sigma_H |\hat{g}(H)| < \infty$.

In order to compare the results in this section with those of the previous one, note that

$$\sum_u \sup_\eta |c(x,\eta_u) - c(x,\eta)| \leq 2 \sum_F |F|\, |\hat{c}(x,F)| , \quad \text{and}$$

$$\inf_\eta [c(u,\eta) + c(u,\eta_u)] \geq 2\hat{c}(u,\emptyset) - 2 \sum_{\substack{F\not\ni u \\ F\neq\emptyset}} |\hat{c}(u,F)| .$$

Therefore the existence and uniqueness of the process follows from theorem 1.2.5 if $\sup_x \Sigma_F[|F| + 1]|\hat{c}(x,F)| < \infty$, while for the ergodicity,

$$\sup_x \sum_F |F|\, |\hat{c}(x,F)| + \sup_x \sum_{\substack{F\not\ni x \\ F\neq\emptyset}} |\hat{c}(x,F)| < \inf_x \hat{c}(x,\emptyset)$$

would be required.

In addition to the ergodicity result in theorem 1.3.10, Holley and Stroock also prove that the process is ergodic when (1.3.8) is replaced by the assumption that $\hat{c}(x,\emptyset) > 0$ for all $x \in S$, although then, of course, there is no exponential rate of convergence to the invariant measure. In [26], they have extended these results to the case in which ν is replaced by a product measure with density different from 1/2.

Section 1.4. The martingale approach; uniqueness problems. In [25], Holley and Stroock have replaced the problem of showing that the closure of Ω is a semigroup generator by the problem of proving existence and uniqueness for a related martingale problem. We will discuss their results briefly. Consider Λ, the canonical path space of right continuous functions with left limits with values in X, and the σ-algebra on Λ generated by the cylinder sets. Given a nonnegative function $c(x,\eta)$ on $S \times X$, define Ωf for $f \in \mathfrak{F}$ by (1.2.1). A probability measure P^η on Λ is said to solve the martingale problem for Ω with initial configuration η if $P^\eta[\eta(0) = \eta] = 1$ and $f(\eta(t)) - \int_0^t \Omega f(\eta(s))ds$ is a martingale for all $f \in \mathfrak{F}$ relative to P^η and the natural increasing class of σ-algebras.

The relationship between the semigroup and martingale problems in case $c(x,\eta)$ is continuous is the following: (a) If the closure of Ω generates a Markov semigroup, then the measure P^η defined by the resulting Feller process is the unique solution to the martingale problem for each $\eta \in X$. (b) If the martingale problem has a unique solution P^η for each $\eta \in X$, then P^η determines a Feller process, so some extension of Ω generates a Markov semigroup. It is not known whether uniqueness for the martingale problem implies that the closure of Ω itself generates a Markov semigroup. (See the recent paper [63], however.)

Holley and Stroock prove that the martingale problem always has a solution provided that $c(x,\eta)$ is continuous. From this point of view, the real problem is then to determine conditions on $c(x,\eta)$ which guarantee that the solution is unique. Essentially, uniqueness is known only under conditions of the type discussed in sections 1.2 and 1.3. It appears that the estimates required to prove uniqueness in the martingale context are quite similar to those which yield both existence and uniqueness in the semigroup context.

Uniqueness certainly does not always hold. In section 3 of [25], Holley and Stroock give an example of nonuniqueness in which $c(x,\eta)$ is uniformly bounded away from 0 and ∞, and $c(x,\eta) \in \mathfrak{F}$ for each x. In [10], Gray and Griffeath give another example which is related to the processes to be discussed in chapter 3.

Probably the most important open problem along these lines is to prove uniqueness in case $S = Z^d$, the d-dimensional integer lattice, and $c(x,\eta)$ is (a) translation invariant, (b) continuous, and (c) strictly positive for all x and η. The positivity of $c(x,\eta)$ is needed here, since, as was observed in [25], there are examples of nonuniqueness in this context if $c(x,\eta)$ is allowed to be zero for some η.

Chapter 2. The use of coupling techniques. One of the important tools which will be used in this and succeeding chapters is that of coupling two (or more) processes together. This simply means that two Markov processes are defined on the same probability space, and the transition mechanisms of the two processes are linked in a convenient way. In the case of spin-flip processes, the most natural way to do this is to insure that, insofar as possible, two coordinates which agree at a given time will flip together. This technique was introduced in [6] and [57], and has since been used in the spin-flip context in [11], [13], [21], and [23], for example.

Section 2.1. The basic coupling. Suppose $c_1(x,\eta)$ and $c_2(x,\eta)$ are uniformly bounded, nonnegative, continuous functions which satisfy (1.2.3). For functions $f(\eta,\zeta)$ on $X \times X$ which depend on finitely many coordinates, define

$$\tilde{\Omega}f(\eta,\zeta) = \sum_x c_1(x,\eta)[f(\eta_x,\zeta) - f(\eta,\zeta)] + \sum_x c_2(x,\zeta)[f(\eta,\zeta_x) - f(\eta,\zeta)]$$

$$+ \sum_{x:\eta(x)=\zeta(x)} \min[c_1(x,\eta),c_2(x,\zeta)][f(\eta_x,\zeta_x) - f(\eta_x,\zeta) - f(\eta,\zeta_x) + f(\eta,\zeta)] .$$

Using the techniques of section 1.2, one can prove that the closure of $\tilde{\Omega}$ is the generator of a Markov semigroup. The coupled process $\gamma_t = (\eta_t,\zeta_t)$ is the corresponding Markov process on $X \times X$.

For $i = 1$ and 2, let Ω_i be defined as in (1.2.1) in terms of the functions $c_i(x,\eta)$. Among the important properties of the coupled process are the following:

(a) If f depends only on η, then $\tilde{\Omega}f(\eta,\zeta) = \Omega_1 f(\eta,\zeta)$. Therefore η_t is the Feller process whose generator is the closure of Ω_1.

(b) If f depends only on ζ, then $\tilde{\Omega}f(\eta,\zeta) = \Omega_2 f(\eta,\zeta)$. Therefore ζ_t is the Feller process whose generator is the closure of Ω_2.

(c) If c_1 and c_2 satisfy

(2.1.1) $$[c_1(x,\eta) - c_2(x,\zeta)][\eta(x) + \zeta(x) - 1] \geq 0$$

for all $x \in S$ whenever $\eta \leq \zeta$ coordinatewise, then

$$P^{(\eta,\zeta)}[\eta_t \leq \zeta_t] = 1 \quad \text{for } \eta \leq \zeta .$$

The simplest way to prove facts such as (c) is often to prove them first in case S is finite, in which case γ_t is a finite Markov chain, and then to extend them to the case of infinite S via an application of theorem 1.1.3. For the proof in the finite case, note first that if f is constant on $M = \{(\eta,\zeta) : \eta \leq \zeta\}$, then $\tilde{\Omega}f = 0$ on M, and therefore $\tilde{\Omega}^n f = 0$ on M for $n \geq 1$. The result then follows from $\tilde{S}(t)1_M = \Sigma_{n=0}^{\infty} (t^n/n!)\tilde{\Omega}^n 1_M = 1$ on M.

Property (c) above suggests defining the following order relation on probability measures on X: We will say that $\mu_1 \leq \mu_2$ if there is a probability measure ν on $X \times X$ which has marginals μ_1 and μ_2 respectively and such that $\nu\{(\eta,\zeta) : \eta \leq \zeta\} = 1$. Note that $\mu_1 \leq \mu_2$ implies that $\mu_1\{\eta : \eta(x) = 1$ for $x \in T\} \leq \mu_2\{\eta : \eta(x) = 1$ for $x \in T\}$ for $T \subset S$. Property (c) then yields the following result.

Lemma 2.1.2. Suppose c_1 and c_2 satisfy (2.1.1), and let $S_1(t)$ and $S_2(t)$ be the corresponding semigroups. Then $\mu_1 \leq \mu_2$ implies $\mu_1 S_1(t) \leq \mu_2 S_2(t)$ for all $t \geq 0$

Section 2.2. Applications to attractive processes. This section is based on Holley's work in [21]. We will consider the Feller process η_t on X whose generator is the closure of the operator defined in (1.2.1). In addition to assuming the smoothness conditions on $c(x,\eta)$ from section 1.2 which guarantee the existence and uniqueness of the process, we will suppose throughout this section that the process is attractive in the sense that

$$[c(x,\eta) - c(x,\zeta)][\eta(x) + \zeta(x) - 1] \geq 0 \tag{2.2.1}$$

for all $x \in S$ and all $\eta,\zeta \in X$ such that $\eta \leq \zeta$ coordinatewise. The interpretation of this condition is that there is a tendency for coordinates to try to line up with their neighbors.

Theorem 2.2.2. If $\mathcal{I} = \{\nu\}$ is a singleton, then the process is ergodic.

Proof. Let ν_0 and ν_1 be the pointmasses on $\eta \equiv 0$ and $\eta \equiv 1$ respectively. Then $\nu_0 \leq \mu \leq \nu_1$ for all probability measures μ on X, so

$$\nu_0 S(t) \leq \mu S(t) \leq \nu_1 S(t) \tag{2.2.3}$$

for all such μ and all $t \geq 0$ by lemma 2.1.2. Since $\nu_0 \leq \nu_0 S(t)$ and $\nu_1 \geq \nu_1 S(t)$, it follows from the semigroup property that $\nu_0 S(t)$ increases and $\nu_1 S(t)$ decreases in t. Therefore the weak limits of $\nu_0 S(t)$ and $\nu_1 S(t)$ exist as $t \to \infty$, and thus both limits are ν by assumption, since they must be invariant. The result then follows by letting $t \to \infty$ in (2.2.3).

By the translation invariant case, we will mean the case in which $S = Z^d$ for some $d \geq 1$ and $c(x,\eta)$ is translation invariant in the sense that $c(x,\eta) = c(y,\zeta)$ if $\eta(x + u) = \zeta(y + u)$ for all $u \in S$. In the translation invariant case, let $\mathcal{S}$ be those probability measures on X which are invariant under translations in S. In some cases (see the comment at the end of section 4.2, for example) it is easier to show that $\mathcal{I} \cap \mathcal{S}$ is a singleton than to show that $\mathcal{I}$ is a singleton. The following strengthened form of theorem 2.2.2 is therefore sometimes useful.

Theorem 2.2.4. In the translation invariant case, if $\mathcal{S} \cap \mathcal{I}$ is a singleton, then so is $\mathcal{I}$, and the process is therefore ergodic.

Proof. It suffices to note that in the proof of theorem 2.2.2, $\nu_0 S(t)$ and $\nu_1 S(t)$ are in $\mathcal{S}$ for all $t \geq 0$, and therefore their limits are in $\mathcal{S} \cap \mathcal{I}$.

In order to obtain sufficient conditions for $\mathcal{I}$ to be a singleton, we will compare the process η_t with finite systems. Let S_n be finite sets which increase to S, and define for $i = 0$ and 1

$$c_i^n(x,\eta) = \begin{cases} c(x,\eta^i) & \text{if } x \in S_n \\ 0 & \text{if } x \notin S_n \text{ and } \eta(x) = i \\ M(x) & \text{if } x \notin S_n \text{ and } \eta(x) = 1 - i \end{cases}$$

where $M(x) = \sup_\zeta c(x,\zeta)$, $\eta^i(u) = \eta(u)$ for $u \in S_n$, and $\eta^i(u) = i$ for $u \notin S_n$. Then c_i^n satisfies (2.2.1), $[c_0^n(x,\eta) - c(x,\zeta)][\eta(x) + \zeta(x) - 1] \geq 0$, and $[c(x,\eta) - c_1^n(x,\zeta)][\eta(x) + \zeta(x) - 1] \geq 0$ whenever $\eta \leq \zeta$. Let $S_i^n(t)$ be the semi-group corresponding to c_i^n, and note that $\nu_i^n = \lim_{t\to\infty} \nu_i S_i^n(t)$ exists as in the proof of theorem 2.2.3. By lemma 2.1.2, ν_0^n increases in n and ν_1^n decreases in n, so that $\nu^i = \lim_{n\to\infty} \nu_i^n$ exists for $i = 0$ and 1. By the comments immediately preceeding theorem 1.1.3, ν^0 and ν^1 are in $\mathcal{I}$. In fact, by lemma 2.1.2, $\nu^0 \leq \nu \leq \nu^1$ for all $\nu \in \mathcal{I}$. Thus we have proved

Theorem 2.2.5. $\mathcal{I}$ is a singleton if and only if $\nu^0 = \nu^1$.

The usefulness of this theorem is especially evident in reversible cases (see section 4.1), where ν_i^n can be computed explicitly. One of the fundamental problems in classical statistical mechanics is to determine when $\nu^0 = \nu^1$ in certain reversible cases. Thus one can often use results from that field to determine whether particular attractive reversible processes are ergodic.

Section 2.3. Reduction to attractive processes. In this section, we will describe a technique which can be used to reduce the problem of proving ergodicity for a process with general $c(x,\eta)$ to a corresponding one for an associated attractive process which has a very simple invariant measure. There is in general a loss of information inherent in this technique, in the sense that the original process may in fact be ergodic, even though the associated attractive process is not. However, it is useful in many cases. We will assume throughout this section that $c(x,\eta)$ satisfies the smoothness assumptions of section 1.2.

The idea is to couple three processes η_t, ζ_t and ξ_t together, where η_t and ζ_t are two copies of the process with speed function $c(x,\eta)$ with different initial configurations, and ξ_t is an attractive process with $\xi \equiv 0$ as an absorbing point. The coupling between η_t and ζ_t will be as in section 2.1, and the process ξ_t will give a bound on the discrepancies between η_t and ζ_t. The speed function for the process ξ_t will be written in the form $\xi(x)\delta(x,\xi) + [1 - \xi(x)]\beta(x,\xi)$, where $\delta(x,\xi)$ and $\beta(x,\xi)$ are defined by

$$\beta(x,\xi) = \sup\{|c(x,\eta) - c(x,\zeta)| : |\eta(u) - \zeta(u)| \leq \xi(u) \text{ for all } u \in S\}$$

and

$$\delta(x,\xi) = \inf\{[c(x,\eta) + c(x,\zeta)] : \eta(x) \neq \zeta(x) \text{ and } |\eta(u) - \zeta(u)| \leq \xi(u) \text{ for all } u \in S\}.$$

It is easy to check that this speed function also satisfies the smoothness assumptions of section 1.2, that the process is attractive, and that $\beta(x,\xi) \equiv 0$ for $\xi \equiv 0$, so that the pointmass on $\xi \equiv 0$ is invariant for ξ_t. Therefore by theorem 2.2.2, ξ_t is ergodic if and only if this is the only invariant measure for ξ_t. The coupling will have the property that $|\eta_t(u) - \zeta_t(u)| \leq \xi_t(u)$ for all $t \geq 0$ and $u \in S$ if it has this property at $t = 0$. Therefore we obtain the following result.

Theorem 2.3.1. If ξ_t is ergodic, then η_t is ergodic.

Proof. If ξ_t is ergodic, then $P^{\xi}(\xi_t(u) = 1) \to 0$ as $t \to \infty$ for all $\xi \in X$ and $u \in S$. Therefore $P^{(\eta,\zeta)}(\eta_t(u) \neq \zeta_t(u)) \to 0$ for all $\eta, \zeta \in X$, when η_t and ζ_t are coupled via the basic coupling of section 2.1. If μ_1 and μ_2 are two probability measures on X and $S(t)$ is the semigroup corresponding to the speed function $c(x,\eta)$, it then follows that $\int f\,d\mu_1 S(t) - \int f\,d\mu_2 S(t) \to 0$ as $t \to \infty$ for any $f \in \mathfrak{F}$. To complete the proof, let μ_1 be invariant for η_t, so that we have $\mu_2 S(t) \to \mu_1$.

The coupling which has the properties required above is described in terms of the following list of transition rates at x as functions of the configuration (η,ζ,ξ). Here $a = 0$ or 1, and $\varepsilon = \varepsilon(x,\zeta,\eta,\xi)$ is defined by $\varepsilon = \delta(x,\xi)[c(x,\eta) + c(x,\zeta)]^{-1}$ if $c(x,\eta) + c(x,\zeta) > 0$, and $\varepsilon = 0$ if $c(x,\eta) + c(x,\zeta) = 0$.

$(a,a,0) \to$	$(1 - a, 1 - a, 0)$	$\min[c(x,\eta), c(x,\zeta)]$
	$(a, 1 - a, 1)$	$c(x,\zeta) - \min[c(x,\eta), c(x,\zeta)]$
	$(1 - a, a, 1)$	$c(x,\eta) - \min[c(x,\eta), c(x,\zeta)]$
	$(a, a, 1)$	$\beta(x,\xi) - |c(x,\eta) - c(x,\zeta)|$

$(a,a,1) \to$	$(1-a,\ 1-a,\ 1)$	$\min[c(x,\eta),\ c(x,\zeta)]$
	$(a,\ 1-a,\ 1)$	$c(x,\zeta) - \min[c(x,\eta),\ c(x,\zeta)]$
	$(1-a,\ a,\ 1)$	$c(x,\eta) - \min[c(x,\eta),\ c(x,\zeta)]$
	$(a,\ a,\ 0)$	$\delta(x,\xi)$
$(a,1-a,1) \to$	$(a,\ a,\ 1)$	$c(x,\zeta)(1-\varepsilon)$
	$(a,\ a,\ 0)$	$c(x,\zeta)\varepsilon$
	$(1-a,\ 1-a,\ 1)$	$c(x,\eta)(1-\varepsilon)$
	$(1-a,\ 1-a,\ 0)$	$c(x,\eta)\varepsilon$.

These rates apply only to configurations (η,ζ,ξ) for which $|\eta(u)-\zeta(u)| \leq \xi(u)$ for all $u \in S$. Since this set of configurations is closed for the coupled process, there is no need to define the process on its complement. It is clear that all of the above rates are nonnegative, and it is in fact this nonnegativity requirement which led to our definition of $\beta(x,\xi)$ and $\delta(x,\xi)$. It is not hard to verify that the rates are such as to yield the right marginal processes.

It should be noted that if $\beta_i(x,\xi)$ and $\delta_i(x,\xi)$ for $i = 1,2$ satisfy $\beta_2(x,\xi) \geq \beta_1(x,\xi)$ and $\delta_2(x,\xi) \leq \delta_1(x,\xi)$, $\beta_2(x,\xi) = 0$ for $\xi = 0$, and both processes are attractive, then a simple coupling argument shows that the ergodicity of the process with speed function $\xi(x)\delta_2(x,\xi) + [1-\xi(x)]\beta_2(x,\xi)$ implies the ergodicity of the process with speed function $\xi(x)\delta_1(x,\xi) + [1-\xi(x)]\beta_1(x,\xi)$. Thus, in applying theorem 2.3.1., it is often sufficient to consider functions β and δ which have a certain structure (such as that of the proximity processes of section 3.1) and which majorize the given functions in the above sense.

Chapter 3. The use of duality techniques. Suppose η_t and ζ_t are Markov processes with state spaces U and V respectively, and let $H(\eta,\zeta)$ be a bounded measurable function on $U \times V$. Then η_t and ζ_t are said to be dual to one another with respect to H if

$$E^{\eta}H(\eta_t,\zeta) = E^{\zeta}H(\eta,\zeta_t)$$

for all $\eta \in U$ and $\zeta \in V$. Given a process η_t of interest, it is often useful to find a suitable function H and a process ζ_t which is dual to η_t with respect to H. Many problems involving η_t can then be recast in terms of ζ_t, and often solved more easily in this new context. Various forms of duality (i.e., using various functions H) and various resulting dual processes have been applied in [14],[24],[27],[40], and [58] to obtain ergodic theorems for certain spin-flip processes. In this chapter, we will concentrate primarily on the material in [24]. The processes which we will consider are among those which can arise when the coupling in section 2.3 is carried out.

Section 3.1. The duality theorem for proximity processes. By a proximity process, we will mean a spin-flip process η_t with speed function of the form

$$c(x,\eta) = c(x)\left\{[1 - \eta(x)] + [2\eta(x) - 1] \sum_F p(x,F)\chi_F(\eta)\right\} ,$$

where the sum is over finite subsets F of S, and $\chi_F(\eta) = \Pi_{x\in F}[1 - \eta(x)]$ for $F \neq \emptyset$ and $\chi_\emptyset(\eta) \equiv 1$. We will assume that $c(x)$ and $p(x,F)$ satisfy $c(x) > 0$, $p(x,F) \geq 0$, and $\Sigma_F\, p(x,F) = 1$ for all $x \in S$, and

$$\sup_x c(x) \sum_F p(x,F)[|F| + 1] < \infty , \tag{3.1.1}$$

where $|F|$ denotes the cardinality of F. Note that for $u \neq x$,

$$\sup_\eta |c(x,\eta_u) - c(x,\eta)| = c(x) \sum_{F\ni u} p(x,F) , \tag{3.1.2}$$

so (3.1.1) guarantees that $c(x,\eta)$ is uniformly bounded and satisfies (1.2.3).

For each finite $G \subset S$, $\chi_G(\eta)$ is in the domain of the generator Ω of η_t, and

$$\Omega\chi_G(\eta) = \sum_x c(x,\eta)\, \Delta_x\chi_G(\eta)$$

$$= \sum_{x\in G} c(x,\eta)[2\eta(x) - 1]\chi_{G\backslash x}(\eta) \tag{3.1.3}$$

$$= \sum_{x\in G} c(x) \sum_F p(x,F)[\chi_{(G\backslash x)\cup F}(\eta) - \chi_G(\eta)] .$$

This suggests the possibility that η_t is dual with respect to $H(\eta,F) = \chi_F(\eta)$ to the Markov chain whose state space is the set of all finite subsets of S and which has the following description: if the chain is in state G, then (a) each $x \in G$ has an exponential lifetime with rate $c(x)$, (b) when the particle at x dies, it is replaced with probability $p(x,F)$ by all the points in F, and (c) only one particle, rather than two, remains at points in $F \cap (G\backslash x)$. This Markov chain will be called a branching process with interference (BPI). The interference comes from property (c) above.

Consider the Q matrix on the finite subsets of S given by

$$q(A,B) = \sum_{x\in A} c(x) \sum_{F:(A\backslash x)\cup F=B} p(x,F)$$

for $A \neq B$. We want to show that there is a unique nonexplosive Markov chain A_t corresponding to this Q matrix, and that it satisfies $E^A|A_t| \leq e^{\omega t}|A|$, where

$$\omega = \sup_x c(x) \sum_F p(x,F)[|F| - 1] .$$

In order to do this, observe first that

$$(3.1.4) \qquad \sum_B q(A,B) = \sum_{x\in A} c(x)$$

and

$$(3.1.5) \qquad \sum_B q(A,B)[|B| - |A|] \leq \sum_{x\in A} c(x) \sum_F p(x,F)[|F| - 1] \leq \omega|A| .$$

Define

$$q_n(A,B) = \begin{cases} q(A,B) & \text{if } |A| \leq n \\ 0 & \text{if } |A| > n . \end{cases}$$

By (3.1.4), $\sup_A \Sigma_B q_n(A,B) < \infty$, so this Q matrix determines a unique Markov chain A_t^n. By (3.1.5), this chain satisfies

$$(3.1.6) \qquad E^A|A_t^n|e^{-t\omega^+} \leq E^A\left[|A_t^n|e^{-\omega(t\wedge\tau_n)}\right] \leq |A|$$

where $\tau_n = \inf\{t > 0 : |A_t^n| > n\}$ and $\omega^+ = \omega \vee 0$. Therefore

$$P^A(\tau_n \leq t) \leq \frac{1}{n} E^A|A_t^n| \leq \frac{|A|}{n} e^{t\omega^+} .$$

Since $\lim_{n\to\infty} P^A(\tau_n \leq t) = 0$, the minimal solution to the backward equation corresponding to $q(A,B)$ is stochastic, and therefore the chain A_t is nonexplosive. Taking a limit in (3.1.6) yields

$$(3.1.7) \qquad E^A|A_t| \leq |A|e^{\omega t} .$$

<u>Theorem 3.1.8</u>. Let η_t be a proximity process whose parameters satisfy (3.1.1), and let A_t be the corresponding BPI. Then for any $\eta \in X$ and finite $A \subset S$,

$$E^\eta \chi_A(\eta_t) = E^A \chi_{A_t}(\eta) .$$

<u>Proof</u>. Let $u_\eta(t,A) = E^\eta \chi_A(\eta_t) = S(t)\chi_A(\eta)$. Since χ_A is in the domain of Ω for each finite $A \subset S$, the Hille-Yosida theorem and (3.1.3) yield

$$\frac{d}{dt} u_\eta(t,A) = S(t)\Omega\chi_A(\eta) = \sum_B q(A,B)[u_\eta(t,B) - u_\eta(t,A)] .$$

The result then follows from the fact that $E^A \chi_{A_t}(\eta)$ is the unique bounded solution to this differential equation with initial condition $u_\eta(0,A) = \chi_A(\eta)$. (The proof of this uniqueness is similar to the proof that whenever the minimal solution to the backward equation is stochastic, it is the only stochastic solution.)

Corollary 3.1.9. If μ is a probability measure on X, then

$$\mu S(t)\{\eta : \eta(x) = 1 \text{ on } A\} = \sum_B P^A[A_t = B]\, \mu\{\eta : \eta(x) = 0 \text{ on } B\}$$

for all finite $A \subset S$.

Section 3.2. Some applications. Since a proximity process is attractive and $\eta \equiv 0$ is absorbing for it, the process is ergodic if and only if $P^\eta[\eta_t(x)=1] \to 0$ as $t \to \infty$ for $\eta \equiv 1$ and all $x \in S$. Thus the following result is an immediate consequence of theorem 3.1.8.

Theorem 3.2.1. Let η_t be a proximity process and A_t the corresponding BPI. Then η_t is ergodic if and only if $P^A[A_t \neq \emptyset] \to 0$ as $t \to \infty$ for all finite $A \subset S$.

Corollary 3.2.2. Suppose $\omega = \sup_x[c(x) \Sigma_F\, p(x,F)[|F| - 1]] < 0$. Then η_t is ergodic.

Proof. $P^A[A_t \neq \emptyset] \leq E^A|A_t|$, so the result follows from the previous theorem and (3.1.7).

Even if $\omega = 0$ above, one can often conclude that η_t is ergodic. If $\omega = 0$, (3.1.7) says that $|A_t|$ is a (nonnegative) supermartingale, so $\lim_{t\to\infty}|A_t|$ exists with probability one. Therefore if $\inf_x c(x)p(x,\emptyset) > 0$, for example, it follows that $|A_t| = \emptyset$ eventually.

It is of interest to compare the result in corollary 3.2.2 with the ergodicity part of theorem 1.2.5. By 3.1.2, the assumption $M < c$ in that theorem becomes in the present context

$$\sup_x c(x)\left\{\sum_F p(x,F)[|F| - 1] + \sum_{F \not\ni x} p(x,F)\right\} < \inf_x c(x) \sum_{F \not\ni x} p(x,F) ,$$

which is in general a stronger assumption than $\omega < 0$, although the two conditions coincide in the translation invariant case.

As an example of the application of corollary 3.2.2, consider Harris' contact processes ([13]; see also [11]) in which $S = Z^d$, $c(x,\eta) = \lambda$ if $\eta(x) = 1$ and $c(x,\eta) = \Sigma_{|y-x|=1}\, \eta(y)$ if $\eta(x) = 0$. This speed function can be put in the form of a proximity process by letting $c(x) = 2d + \lambda$, $p(x,\emptyset) = \lambda/(2d+\lambda)$, and

$p(x,F) = 1/(2d+\lambda)$ for $F = \{x,y\}$ with $|x - y| = 1$. Then corollary 3.2.2 and the remark which follows it imply that the process is ergodic for $\lambda \geq 2d$. Harris has proved that the process is not ergodic for sufficiently small values of λ. A simple coupling argument shows that there is a $\lambda_0 > 0$ so that the process is ergodic for $\lambda > \lambda_0$ and is not ergodic for $\lambda < \lambda_0$. The above result of course implies that $\lambda_0 \leq 2d$.

In corollary 3.2.2, no advantage was taken of the interference in the BPI. In other words, the estimate (3.1.7) which was used in it ignored property (c) in the description of the BPI. While it is usually difficult to take advantage of the interference effectively, one can sometimes do better by applying theorem 3.2.1 directly. In the case of Harris' process above, for example, one can prove ergodicity for $\lambda > 2d - 1$ in the following way: for finite $A, B \subset S$,

$$\begin{aligned} P^{A\cup B}[A_t \neq \emptyset] &= P^1[\eta_t \not\equiv 0 \text{ on } A \cup B] \\ &\leq P^1[\eta_t \not\equiv 0 \text{ on } A] + P^1[\eta_t \not\equiv 0 \text{ on } B] - P^1[\eta_t \not\equiv 0 \text{ on } A \cap B] \\ &= P^A[A_t \neq \emptyset] + P^B[A_t \neq \emptyset] - P^{A\cap B}[A_t \neq \emptyset] . \end{aligned}$$

Here 1 represents the configuration $\eta \equiv 1$. Let $\sigma(A) = \lim_{t\to\infty} P^A[A_t \neq \emptyset]$, which is a harmonic function for A_t. Then

$$\sigma(A \cup B) + \sigma(A \cap B) \leq \sigma(A) + \sigma(B) , \quad \text{and} \tag{3.2.3}$$

$$\sigma(A) = \frac{1}{|A|} \sum_{x\in A} \left[\frac{\lambda}{\lambda + 2d}\sigma(A\backslash x) + \frac{1}{\lambda + 2d} \sum_{|y-x|=1} \sigma(A \cup y)\right] . \tag{3.2.4}$$

In order to apply theorem 3.2.1, we need to show that $\sigma(A) = 0$ for all finite A. By (3.2.3), it suffices to prove this when A is a singleton. Since A_t is invariant under translations and rotations, $\sigma_1 = \sigma(\{x\})$ and $\sigma_2 = \sigma(\{x,y\})$ for $|x - y| = 1$ are independent of x and y. Applying (3.2.4) to a singleton A, we obtain $\sigma_1 = [2d/(\lambda+2d)]\sigma_2$. Applying it to a doubleton A containing two nearest neighbors and using (3.2.3) yields

$$\sigma_2 \leq \frac{\lambda}{\lambda + 2d}\sigma_1 + \frac{1}{\lambda + 2d}\sigma_2 + \frac{(2d - 1)}{\lambda + 2d}(2\sigma_2 - \sigma_1) .$$

Simplification gives $(\lambda - 2d + 1)\lambda\sigma_1 \leq 0$, so $\sigma_1 = 0$ if $\lambda > 2d - 1$.

For further improvements of these results, and other applications of this duality, see [11], [13], and [24]. For other general forms of duality and their applications, see [14] and [27]. Of course, any time one proves an ergodic theorem for a proximity process, one can automatically conclude that more general spin-flip processes are ergodic via the coupling of section 2.3.

Section 3.3. The voter model. One of the unpleasant properties of the applications of duality we have seen so far is that in no case do the results appear to be even close to best possible. There is one class of processes, however, where the exploitation of the duality theorem gives complete results. We will refer to this class as the voter model, and will discuss the results briefly, since very similar techniques will be used to analyze the symmetric simple exclusion process in chapter 2 of part II. The details in the case of the voter model can be found in [24].

The voter model is the proximity process in which $p(x,F) = 0$ whenever $|F| \neq 1$. We will write $p(x,y) = p(x,\{y\})$, and will assume that $\sup_x c(x) < \infty$ in order to guarantee (3.1.1). The interpretation of the process is that individuals (voters) are located at the points of S, and that at each time, each individual holds one of two possible positions (denoted by 0 and 1 respectively) on some issue. At exponential times with rate $c(x)$, the individual at x reassesses his position. He does this by choosing a y according to the probabilities $p(x,y)$, and then changes his position to that of the individual at y. Let ν_0 and ν_1 be the pointmasses on $\eta \equiv 0$ and $\eta \equiv 1$ respectively. Then it is clear that $\nu_0, \nu_1 \in \mathcal{I}$, and the main problem is to determine when $\mathcal{I}_e = \{\nu_0, \nu_1\}$. In other words, we want to determine when all the equilibria represent total consensus.

Suppose $p_t(x,y)$ is the transition function for the Markov chain on S which has Q matrix $c(x)p(x,y)$ for $x \neq y$. The BPI A_t which corresponds to the voter model can then be thought of in the following way: The points in A_t move independently according to the Markov chain on S which has transition probabilities $p_t(x,y)$ until the first time that a point moves to an occupied site. At that time, the two particles coalesce, and the system proceeds as before except that its size has been reduced by one. The important property of A_t which leads to a complete analysis of the voter model is that $|A_t|$ never increases. We will assume throughout that $p_t(x,y) > 0$ for all $t > 0$, and $x,y \in S$.

Let $Y_1(t)$ and $Y_2(t)$ be independent copies of the Markov chain with transition probabilities $p_t(x,y)$. The structure of $\mathcal{I}_e$ depends on whether or not $Y_1(t)$ and $Y_2(t)$ hit in finite time with probability one, since that determines whether or not A_t is eventually of cardinality one.

Theorem 3.3.1. Suppose

$$(3.3.2)\qquad P(Y_1(t) = Y_2(t) \text{ for some } t > 0) = 1$$

for all initial states $Y_1(0)$ and $Y_2(0)$. Then $\mathcal{I}_e = \{\nu_0, \nu_1\}$. Furthermore, if μ is a probability measure on X such that $\mu\{\eta : \eta(x) = 1\} = \lambda$ for all $x \in S$, then $\mu S(t) \to \lambda\nu_1 + (1 - \lambda)\nu_0$ as $t \to \infty$.

Proof. Suppose $\mu \in \mathcal{I}$. Then $\mu\{\eta : \eta(x) = 1\}$ is a (bounded) harmonic function for $Y_1(t)$ by corollary 3.1.9. A simple coupling argument using (3.3.2) shows that $Y_1(t)$ has no nonconstant bounded harmonic functions, so $\mu\{\eta : \eta(x) = 1\} = \lambda$ is independent of x. From (3.3.2), it also follows that $P^A[|A_t| = 1] \to 1$ as $t \to \infty$ for all $A \neq \emptyset$. Corollary 3.1.9 gives

$$\mu\{\eta : \eta(x) = 1 \text{ on } A\} = \lambda P^A[|A_t| = 1] + \sum_{|B| \neq 1} P^A[A_t = B]\,\mu\{\eta : \eta(x) = 1 \text{ on } B\}$$

for $A \neq \emptyset$, so it follows that $\mu\{\eta : \eta(x) = 1 \text{ on } A\} = \lambda$. Therefore $\mu = \lambda\nu_1 + (1 - \lambda)\nu_0$. This proves the first part of the theorem. The proof of the second part is similar.

A similar, but more involved, analysis of corollary 3.1.9 yields the following result.

Theorem 3.3.3. Suppose

$$(3.3.4)\qquad P(Y_1(t) = Y_2(t) \text{ for some } t > 0) < 1$$

for some (and hence all) initial states $Y_1(0) \neq Y_2(0)$, and that $p_t(x,y)$ has no nonconstant bounded harmonic functions. Then

(a) $\mu_\rho = \lim_{t\to\infty} \nu_\rho S(t)$ exists for each $\rho \in [0,1]$, where ν_ρ is the product measure on X with $\nu_\rho\{\eta : \eta(x) = 1\} = \rho$ for all $x \in S$.

(b) $\mathcal{I}_e = \{\mu_\rho,\ 0 \leq \rho \leq 1\}$.

(c) In the translation invariant case, μ_ρ is translation invariant and ergodic for each $\rho \in [0,1]$.

(d) In the translation invariant case, if μ is any translation invariant and ergodic probability measure on X, then $\mu S(t) \to \mu_\rho$, where $\rho = \mu\{\eta : \eta(x) = 1\}$.

In the translation invariant case, if $p(x,y)$ satisfies appropriate moment conditions, theorem 3.3.1 applies when $d = 1$ or 2 and theorem 3.3.3 applies when $d \geq 3$. Thus in one and two dimensions, the system approaches a consensus, while in higher dimensions, there are many dynamic equilibria which do not represent consensus.

In the above theorems, we have given only sufficient conditions for convergence to an element of $\mathcal{I}_e$. Necessary and sufficient conditions are given in [24] for both cases. In [40], Matloff has used similar techniques applied to a different dual process to analyze the antivoter model, in which individuals adopt the opposite position from their neighbor's position, instead of imitating the neighbor. Another approach to the antivoter model is in [27].

Chapter 4. Other results. When it is impossible to determine $\mathcal{I}$ completely, it is sometimes possible to identify certain natural subsets of $\mathcal{I}$. In this chapter, we study briefly two results of this type.

Section 4.1. Reversible measures. Suppose Ω is the generator of a Markov semi-group of operators on $C(X)$. A probability measure μ on X is said to be reversible for the corresponding process η_t if

$$\int f\Omega g \, d\mu = \int g\Omega f \, d\mu \tag{4.1.1}$$

for all $f,g \in \mathcal{D}(\Omega)$. In verifying that μ is reversible, it of course suffices to verify (4.1.1) for f and g in a core for Ω. We will denote by $\mathcal{R}$ the set of all reversible measures. By taking $g \equiv 1$ in (4.1.1), it is seen that $\mathcal{R} \subset \mathcal{I}$. While $\mathcal{I}$ is always nonempty for a Feller process on a compact set, $\mathcal{R}$ is frequently empty. In this section, we will see that for speed functions of a certain form, $\mathcal{R} \neq \emptyset$ for the corresponding spin-flip process, and in fact, $\mathcal{R}$ can be determined completely in that case. These results have been proved under varying assumptions in [6], [17], [39], [41] and [46].

Consider first the case of finite S, and suppose the speed function is of the form

$$c(x,\eta) = \exp\left[\sum_{A \ni x} \Phi(A)\chi_A(\eta)\right] \tag{4.1.2}$$

where $\chi_A(\eta) = \prod_{y \in A} [2\eta(y) - 1]$ and $\Phi(A)$ is a real valued function on the power set of S. Then η_t is an irreducible finite state Markov chain, and a simple computation shows that the probability measure

$$\nu\{\eta\} = K \cdot \exp\left[-\sum_A \Phi(A)\chi_A(\eta)\right],$$

where K is a normalizing constant, is reversible for η_t. It is of course the unique invariant measure for η_t in this case.

Now take S to be countable, and let Φ be a real valued function on the finite subsets of S which satisfies

$$\sup_x \sum_{A \ni x} |A|\,|\Phi(A)| < \infty .$$

Then the function $c(x,\eta)$ defined as in (4.1.2) satisfies assumption (1.2.3), so we may consider the process η_t whose generator is the closure of the operator defined in (1.2.1).

We define the class $\mathcal{G}$ of Gibbs measures corresponding to Φ in the following way. If T is a finite subset of S and $\zeta \in \{0,1\}^{S\setminus T}$, then $\nu_{T,\zeta}$ is the probability measure on $\{0,1\}^T$ given by

$$\nu_{T,\zeta}\{\eta\} = K(T,\zeta) \exp\left[-\sum_{A\cap T\neq\emptyset} \Phi(A)\chi_A(\eta\zeta)\right],$$

where $K(T,\zeta)$ is again a normalizing constant and $\eta\zeta$ is the configuration in $\{0,1\}^S$ which agrees with η on T and with ζ on $S\setminus T$. This measure can also be thought of as a probability measure on $X = \{0,1\}^S$ by setting

$$\nu_{T,\zeta}\{\eta \in X : \eta(x) = \zeta(x) \text{ for } x \in S\setminus T\} = 1 .$$

Note that $\nu_{T,\zeta}$ is reversible for the process on $\{0,1\}^T$ with speed function $c_{T,\zeta}(x,\eta) = c(x,\eta\zeta)$. Let $\mathcal{G}_T$ be the closed convex hull of $\{\nu_{T,\zeta} : \zeta \in \{0,1\}^{S\setminus T}\}$. Then $\mathcal{G}$ is defined as the set of all possible weak limits of μ_n as $n \to \infty$ where $\mu_n \in \mathcal{G}_{T_n}$ and $T_n \uparrow S$. $\mathcal{G}$ is nonempty since X is compact. Note that if $\Phi(A) = 0$ for $|A| > 1$, then $\mathcal{G}$ is a singleton product measure, as would be expected.

Theorem 4.1.3. $\mathcal{R} = \mathcal{G}$.

Proof. To show that $\mathcal{G} \subset \mathcal{R}$, fix $\mu \in \mathcal{G}$ and $f,g \in \mathcal{F}$. Let $\Omega_{T,\zeta}$ be the generator for the process on $\{0,1\}^T$ with speed function $c_{T,\zeta}(x,\eta)$ for T a finite subset of S and $\zeta \in \{0,1\}^{S\setminus T}$. Then if T contains the coordinates on which f and g depend, we have

$$\int_{\{0,1\}^T} f\Omega_{T,\zeta} g \, d\nu_{T,\zeta} = \int_{\{0,1\}^T} g\Omega_{T,\zeta} f \, d\nu_{T,\zeta} .$$

Since $\mu \in \mathcal{G}$, there are $T_n \uparrow S$ and $\mu_n \in \mathcal{G}_{T_n}$ so that $\mu_n \to \mu$. Using

$$\lim_{n\to\infty} \sup_{\zeta\in\{0,1\}^{S\setminus T_n}} \|\Omega_{T_n,\zeta} h - \Omega h\| = 0$$

for $h = f$ and g, we obtain $\int_X f\Omega g \, d\mu = \int_X g\Omega f \, d\mu$. For the reverse containment, suppose $\mu \in \mathcal{R}$. For a finite subset T of S, and $x \in T$, put $f(\eta) = \prod_{y\in T} \eta(y)$

and $g(\eta) = f(\eta_x)$. Then

$$g(\eta)\Omega f(\eta) = f(\eta_x) \sum_{y\in T} c(y,\eta)[f(\eta_y) - f(\eta)] = c(x,\eta)f(\eta_x) , \text{ and}$$

$$f(\eta)\Omega g(\eta) = f(\eta) \sum_{y\in T} c(y,\eta)[g(\eta_y) - g(\eta)] = c(x,\eta)f(\eta) .$$

Since $\mu \in \mathcal{R}$,

$$(4.1.4) \qquad \int_X c(x,\eta)[f(\eta_x) - f(\eta)]d\mu = 0 .$$

Using linearity and the fact that $\mathcal{F}$ is dense in $C(X)$, we conclude that (4.1.4) holds for all $f \in C(X)$. Suppose now that f depends only on the coordinates in T and that g is a continuous function on X which depends only on the coordinates in $S\backslash T$. Then we may rewrite (4.1.4) as

$$(4.1.5) \qquad \int_X c(x,\eta)[f(\eta_x) - f(\eta)]g(\eta)d\mu = 0 .$$

The conditional measure $\mu(d\eta|\zeta)$ on $\{0,1\}^T$ exists for a.e. $\zeta \in \{0,1\}^{S\backslash T}$ with respect to μ, and we may conclude from (4.1.5) that

$$\int_{\{0,1\}^T} c(x,\eta\zeta)[f(\eta_x) - f(\eta)]\mu(d\eta|\zeta) = 0$$

for all f which depend on the coordinates in T and for a.e. ζ. Therefore $\mu(d\eta|\zeta)$ is invariant for the process with generator $\Omega_{T,\zeta}$, and hence $\mu(d\eta|\zeta) = \nu_{T,\zeta}(d\eta)$ for a.e. ζ. It follows that μ is in $\mathcal{G}_T$ when regarded as a measure on $\{0,1\}^T$, and therefore that $\mu \in \mathcal{G}$.

We conclude this section with several remarks. First, that $\mu \in \mathcal{R}$ implies (4.1.4), while $\mu \in \mathcal{I}$ implies only $\Sigma_x \int_X c(x,\eta)[f(\eta_x) - f(\eta)]d\mu = 0$, is a good indication of why $\mathcal{R}$ is much easier to treat than is $\mathcal{I}$. Secondly, even though we considered only speed functions of the form (4.1.2), there are others which are covered by this result. In order to see this, it suffices to note that if $c_1(x,\eta)$ and $c_2(x,\eta)$ are positive and are such that $c_1(x,\eta)/c_2(x,\eta)$ is independent of $\eta(x)$ for each $x \in S$, then the generators with these speed functions have the same class of reversible measures. Finally, if $c(x,\eta)$ is of the form (4.1.2) and is attractive, then the measures ν^0 and ν^1 of theorem 2.2.5 are both in $\mathcal{G}$, so in this case, $\mathcal{I}$ is a singleton if and only if $\mathcal{G}$ is a singleton.

One interesting open problem is to determine conditions under which $\mathcal{I} = \mathcal{G}$ in the context of this section. As mentioned above, this is known in the attractive case when $\mathcal{G}$ is a singleton. Holley has proved that $\mathcal{I} = \mathcal{G}$ in the one-dimensional, translation invariant, finite range case, using the free energy technique which will be mentioned in the next section.

Section 4.2. Translation invariant measures. In this section, we mention briefly some cases in which the set of invariant measures which are also translation invariant can be identified. Let $S = Z^d$, and let $\mathcal{S}$ be the class of translation invariant measures on X.

In section 9 of [14], Harris has used duality arguments to give sufficient conditions for a class of attractive processes to satisfy $|(\mathcal{I} \cap \mathcal{S})_e| \leq 2$. Under similar assumptions, Holley and Stroock [27] have shown that it is possible to have $|\mathcal{I}_e| > 2$.

Now suppose $c(x,\eta)$ is given by (4.1.2) where Φ is translation invariant and satisfies

$$(4.2.1) \qquad \sum_{A \ni 0} |A| \; |\Phi(A)| < \infty .$$

The following theorem was proved by Holley [19] in case Φ has finite range (i.e., for some finite $T \subset S$, $\Phi(A) > 0$ and $A \ni 0$ implies $A \subset T$) and was extended to Φ's satisfying (4.2.1) by Higuchi and Shiga [17]. The technique of proof exploits a functional $\varphi(\mu)$, called the free energy, which is defined on the set of probability measures μ on X. It is proved that $\varphi[\mu S(t)]$ is monotone in t for each μ, and is strictly monotone if $\mu \in \mathcal{S} \backslash \mathcal{G}$.

Theorem 4.2.2. (a) $\mathcal{I} \cap \mathcal{S} = \mathcal{G} \cap \mathcal{S}$,

(b) If $\mu \in \mathcal{S}$, $t_n \to \infty$, and $\mu S(t_n) \to \nu$, then $\nu \in \mathcal{G}$.

In the attractive case, this result can be combined with theorem 2.2.4 to conclude that if $\mathcal{G} \cap \mathcal{S}$ is a singleton, then $\mathcal{I}$ is a singleton (and hence, of course, $\mathcal{G}$ is a singleton).

Section 4.3. The one dimensional case. One of the most important open problems in the subject is to prove the ergodicity of all spin-flip processes in the following class: $S = Z^1$, $c(x,\eta)$ is translation invariant, $c(x,\eta) \in \mathcal{F}$, and $c(x,\eta) > 0$. The techniques we have discussed give partial results in this direction, of course, but they seem to be far from sufficient to prove the complete conjecture. This conjecture is suggested by the fact that the statistical mechanical analogue is correct [43].

There are several special cases of this conjecture which are of interest, and which may be more tractable than the general problem. One class of processes for which the result is not known, for example, is that in which $c(x,\eta)$ is attractive. In considering this case, one should presumably take advantage of the results and techniques of section 2.2. Another class of interest is that in which $c(x,\eta)$ is independent of $\eta(x)$. Here, of course, the product measure with density 1/2 is invariant.

A third special case is that in which $c(x,\eta)$ depends on η only through $\{\eta(y),\ y \geq x\}$. Holley and Stroock [27] initiated the study of this one sided process using duality techniques. A particular case of the one sided process which illustrates some of the difficulties involved is that in which $c(x,\eta) = \alpha \geq 1$ if $\eta(x) = \eta(x+1) = 1$ and $c(x,\eta) = 1$ otherwise. When theorem 1.2.5 is applied here, one obtains ergodicity for $\alpha < 3$. The results of Holley and Stroock yield ergodicity for $\alpha < 4$. Furthermore, it can be shown that ergodicity holds for $\alpha = \infty$ in the following sense: there is a uniquely determined probability measure μ such that whenever $\alpha_n \to \infty$ and μ_n is invariant for the process with $\alpha = \alpha_n$, then μ_n converges to μ as $n \to \infty$.

In the intersection of the second and third cases above, it is not hard to prove ergodicity directly, even when $c(x,\eta)$ is not translation invariant and does not have finite range.

<u>Theorem 4.3.1</u>. Suppose $S = Z^1$, $c(x,\eta)$ depends on η only through $\{\eta(y),\ y > x\}$, $c(x,\eta) > 0$, and $c(x,\eta)$ satisfies the conditions for existence and uniqueness of the process from chapter 1. Then the process is ergodic with invariant measure ν, the product measure with $\nu\{\eta : \eta(x) = 1\} = 1/2$.

<u>Proof</u>. Since $c(x,\eta)$ depends on the coordinates of η which are strictly to the right of x,

$$P^{\eta}\Big[\eta_t(y) = \zeta(y) \quad \text{for} \quad x \leq y \leq x+n\Big] = P^{\eta_x}\Big[\eta_t(y) = \zeta_x(y) \quad \text{for} \quad x \leq y \leq x+n\Big]$$

for all choices of $x \in S$, $\eta, \zeta \in X$, and $n \geq 0$. A simple coupling argument, together with the positivity of $c(x,\eta)$ yields

$$P^{\eta}\Big[\eta_t(y) = \zeta(y) \quad \text{for} \quad x \leq y \leq x+n\Big] - P^{\eta_x}\Big[\eta_t(y) = \zeta(y) \quad \text{for} \quad x \leq y \leq x+n\Big] \to 0$$

as $t \to \infty$ for all choices of x, η, ζ, and $n \geq 0$. Combining these two observations leads to

$$P^{\eta}\Big[\eta_t(y) = \zeta(y) \quad \text{for} \quad x \leq y \leq x+n\Big] - P^{\eta}\Big[\eta_t(y) = \zeta_x(y) \quad \text{for} \quad x \leq y \leq x+n\Big] \to 0$$

as $t \to \infty$. For $n = 0$, this gives $P^{\eta}[\eta_t(x) = 1] \to 1/2$. An induction argument then yields the convergence of $P^{\eta}[\eta_t(y) = \zeta(y)$ for $x \leq y \leq x+n]$ to $(1/2)^{n+1}$, which is the desired result. It is easy to see from the coupling argument that the convergence actually occurs at an exponential rate.

PART II. EXCLUSION PROCESSES

Chapter 1. Existence results and identification of simple invariant measures.

Section 1.1. Existence results. Again we take S to be a countable set and put $X = \{0,1\}^S$. For $\eta \in X$ and $u \neq v \in S$, define $\eta_{uv} \in X$ by

$$\eta_{uv}(x) = \begin{cases} \eta(x) & \text{if } x \neq u,v \\ \eta(u) & \text{if } x = v \\ \eta(v) & \text{if } x = u \end{cases}$$

so that η_{uv} is obtained from η by interchanging the u and v coordinates. Consider a nonnegative continuous function $c(x,y,\eta)$ on $S \times S \times X$ which satisfies $c(x,x,\eta) \equiv 0$, $c(x,y,\eta) = c(y,x,\eta)$, and $c(x,y,\eta) \leq c(x,y)$ for some symmetric function $c(x,y)$ on $S \times S$ which satisfies

(1.1.1) $$\sup_x \sum_y c(x,y) < \infty .$$

For $f \in \mathfrak{F}$, define Ωf by

(1.1.2) $$\Omega f(\eta) = \frac{1}{2} \sum_{x,y} c(x,y,\eta)[f(\eta_{xy}) - f(\eta)] .$$

By assumption (1.1.1), this series converges uniformly and absolutely, and so defines a continuous function. Note that only the values of $c(x,y,\eta)$ for η such that $\eta(x) \neq \eta(y)$ play any role in this expression. The following existence and uniqueness theorem is proved by the same techniques as those used in section 1.2 of part I. The details may be found in [32].

Theorem 1.1.3. In addition to the above conditions, assume that there exists a constant L so that

$$\sum_u \sup_\eta |c(x,y,\eta_u) - c(x,y,\eta)| \leq L\, c(x,y) .$$

Then the closure of Ω generates a unique Markov semigroup $S(t)$ on $C(X)$.

As in the case of spin-flip processes, some conditions are required in order to get a unique process with reasonable properties, although the above conditions are presumably not the best possible. One of the difficulties which occurs here, but does not occur in the spin-flip context, can be seen in the following example. Let $c(x,y,\eta) = \eta(x)q(x,y) + \eta(y)q(y,x)$, where $q(x,y)$ are the transition probabilities for a discrete time Markov chain. If $\Sigma_x q(x,y) = \infty$ for some $y \in S$, and the process begins in configuration η where $\eta(u) = 1$ for $u \neq y$ and

$\eta(y) = 0$, there is no reasonable way to define the process for small positive times. This example helps explain why a condition like (1.1.1) is needed.

As mentioned in the introduction, two particular forms of the function $c(x,y,\eta)$ have been proposed to model different physical phenomena. In both cases, $c(x,y,\eta)$ is given in terms of nonnegative functions $q(x,y)$ on $S \times S$ with $q(x,x) = 0$ and $c(x,\eta)$ on $S \times X$. In Spitzer's model of a lattice gas [45], $c(x,y,\eta)$ has the form

$$c(x,y,\eta) = \eta(x)c(x,\eta)q(x,y) + \eta(y)c(y,\eta)q(y,x) . \tag{1.1.4}$$

In this case, sufficient conditions for theorem 1.1.3 to apply are

$$\sup_x \sum_y [q(x,y) + q(y,x)] < \infty \tag{1.1.5}$$

$$\sup_{x,\eta} c(x,\eta) < \infty , \quad \text{and} \tag{1.1.6}$$

$$\sup_x \sum_u \sup_\eta |c(x,\eta_u) - c(x,\eta)| < \infty . \tag{1.1.7}$$

In [1] and [2], on the other hand, a special case of the exclusion process with $c(x,y,\eta)$ of the form

$$c(x,y,\eta) = \eta(x) \frac{c(x,\eta)}{c(x,\eta) + c(y,\eta_{xy})} q(x,y) + \eta(y) \frac{c(y,\eta)}{c(y,\eta) + c(x,\eta_{xy})} q(y,x) \tag{1.1.8}$$

has been considered to model the evolution of a binary alloy. In this case, sufficient conditions for theorem 1.1.3 to apply are (1.1.5), (1.1.7), and

$$\inf_{x,\eta} c(x,\eta) > 0 . \tag{1.1.9}$$

It is interesting to note that in both cases (1.1.4) and (1.1.8),

$$\frac{c(x,y,\eta_{xy})}{c(x,y,\eta)} = \eta(y) \frac{c(x,\eta_{xy})q(x,y)}{c(y,\eta)q(y,x)} + \eta(x) \frac{c(y,\eta_{xy})q(y,x)}{c(x,\eta)q(x,y)} \tag{1.1.10}$$

for $\eta(x) \neq \eta(y)$. It is this property, in fact, that leads both processes to have the same reversible measures.

The simple exclusion process is obtained by taking $c(x,y,\eta)$ to depend on only through $\eta(x)$ and $\eta(y)$, so that

$$c(x,y,\eta) = \eta(x)q(x,y) + \eta(y)q(y,x) . \tag{1.1.11}$$

This is of course the simplest special case of both (1.1.4) and (1.1.8). It is not itself of great physical interest, since it corresponds to the situation in which the temperature is infinite in either the lattice gas or binary alloy context. However it is of considerable mathematical interest, and it provides some insight into the more difficult physically relevant processes. A sufficient condition for theorem 1.1.3 to apply is given by (1.1.5).

Throughout the remainder of part II, we will always assume that the appropriate conditions which guarantee the existence and uniqueness of the process via theorem 1.1.3 are satisfied.

An interesting open problem is to prove existence and uniqueness of the process in either the lattice gas or binary alloy context in the translation invariant case under the assumptions that $c(x,\eta)$ is positive and continuous and q satisfies $\Sigma_y q(0,y) < \infty$. This problem is of course closely related to that mentioned at the end of section 1.4 of part I.

Section 1.2. Invariant measures for the simple exclusion process.

One of the pleasant characteristics of the simple exclusion process is that there are often product measures on X which are invariant for the process. Given a function $\alpha(\cdot)$ on S such that $0 \leq \alpha(x) \leq 1$ for all x, let ν_α be the product measure on X with probabilities given by

$$\nu_\alpha\{\eta : \eta(x) = 1 \text{ for all } x \in T\} = \prod_{x\in T} \alpha(x)$$

for all $T \subset S$. In order to determine those simple exclusion processes for which a given ν_α is invariant, we need the following computation, which will be used again in the proof of the duality theorem in chapter 2. Put $\chi_A(\eta) = \Pi_{u\in A} \eta(u)$ for finite subsets A of S.

Lemma 1.2.1. If A is a finite subset of S, then

$$\Omega\chi_A(\eta) = \sum_{\substack{x\in A\\ y\notin A}} \{q(y,x)[1 - \eta(x)]\chi_{(A\cup y)\backslash x}(\eta) - q(x,y)[1 - \eta(y)]\chi_A(\eta)\} .$$

Proof. By (1.1.2) and (1.1.11),

$$\Omega\chi_A(\eta) = \frac{1}{2} \sum_{x,y} [\eta(x)q(x,y) + \eta(y)q(y,x)][\chi_A(\eta_{xy}) - \chi_A(\eta)]$$

$$= \sum_{\substack{x\in A\\ y\notin A}} [\eta(x)q(x,y) + \eta(y)q(y,x)][\eta(y) - \eta(x)]\chi_{A\backslash x}(\eta)$$

$$= \sum_{\substack{x\in A\\ y\notin A}} \{\eta(y)[1 - \eta(x)]q(y,x) - \eta(x)[1 - \eta(y)]q(x,y)\}\chi_{A\backslash x}(\eta)$$

$$= \sum_{\substack{x\in A\\ y\notin A}} \{q(y,x)[1-\eta(x)]\chi_{(A\cup y)\setminus x}(\eta) - q(x,y)[1-\eta(y)]\chi_A(\eta)\} .$$

Theorem 1.2.2. Suppose $0 < \alpha(x) < 1$ for all $x \in S$. Then $\nu_\alpha \in \mathcal{I}$ if and only if

$$\frac{\alpha(y)}{1-\alpha(y)} q(y,x) = \frac{\alpha(x)}{1-\alpha(x)} q(x,y) \quad \text{whenever } \alpha(x) \neq \alpha(y) \tag{1.2.3}$$

and

$$\sum_{y:\alpha(y)=\alpha(x)} q(y,x) = \sum_{y:\alpha(y)=\alpha(x)} q(x,y) \quad \text{for all } x \in S .$$

Proof. By lemma 1.2.1, $\nu_\alpha \in \mathcal{I}$ if and only if for all finite $A \subset S$,

$$\sum_{\substack{x\in A\\ y\notin A}} q(y,x)\alpha(y)\frac{1-\alpha(x)}{\alpha(x)} = \sum_{\substack{x\in A\\ y\notin A}} q(x,y)[1-\alpha(y)] . \tag{1.2.4}$$

For $A = \{u\}$, (1.2.4) becomes

$$[1-\alpha(u)]\sum_y q(y,u)\alpha(y) = \alpha(u)\sum_y q(u,y)[1-\alpha(y)] . \tag{1.2.5}$$

Rewriting (1.2.4) for general A yields

$$\sum_{x\in A}\frac{1-\alpha(x)}{\alpha(x)}\left\{\sum_y q(y,x)\alpha(y) - \sum_{y\in A} q(y,x)\alpha(y)\right\}$$

$$= \sum_{x\in A}\left\{\sum_y q(x,y)[1-\alpha(y)] - \sum_{y\in A} q(x,y)[1-\alpha(y)]\right\} ,$$

so using (1.2.5), we have

$$\sum_{x,y\in A}\left\{\frac{\alpha(y)}{1-\alpha(y)} q(y,x) - \frac{\alpha(x)}{1-\alpha(x)} q(x,y)\right\}\frac{[1-\alpha(x)][1-\alpha(y)]}{\alpha(x)} = 0 .$$

Interchanging the roles of x and y, and then adding yields

$$\sum_{x,y\in A}\left\{\frac{\alpha(y)}{1-\alpha(y)} q(y,x) - \frac{\alpha(x)}{1-\alpha(x)} q(x,y)\right\}[\alpha(y)-\alpha(x)]\frac{[1-\alpha(x)][1-\alpha(y)]}{\alpha(x)\alpha(y)} = 0.$$

The above sum is zero for all finite A if and only if the summands are zero for all $x,y \in S$. Therefore $\nu_\alpha \in \mathcal{I}$ if and only if (1.2.3) and (1.2.5) hold. To complete the proof let $\{\alpha_n\}$ be the distinct values of $\{\alpha(x)\}$, and let $C_n = \{x \in S : \alpha(x) = \alpha_n\}$. If $u \in C_m$, (1.2.5) can be written as

$$(1-\alpha_m)\sum_n \alpha_n \sum_{y\in C_n} q(y,u) = \alpha_m \sum_n (1-\alpha_n)\sum_{y\in C_n} q(u,y) .$$

Using (1.2.3), the right hand side becomes

$$\alpha_m(1-\alpha_m)\sum_{y\in C_m} q(u,y) + (1-\alpha_m)\sum_{n\neq m}\alpha_n \sum_{y\in C_n} q(y,u) .$$

Thus, in the presence of (1.2.3), (1.2.5) becomes

$$\sum_{y\in C_m} q(y,u) = \sum_{y\in C_m} q(u,y)$$

for $u \in C_m$.

This result gives two cases in which there is a simple one parameter family of product measures in $\mathcal{I}$.

Corollary 1.2.6. (a) If $\Sigma_y q(x,y) = \Sigma_y q(y,x)$ for all $x \in S$, then $\{\nu_\rho, 0 \leq \rho \leq 1\} \subset \mathcal{I}$, where $\nu_\rho = \nu_\alpha$ for $\alpha(x) \equiv \rho$.

(b) If there is a positive function $\pi(x)$ on S such that $\pi(x)q(x,y) = \pi(y)q(y,x)$, then $\{\tilde{\nu}_\rho, 0 \leq \rho \leq \infty\} \subset \mathcal{I}$, where $\tilde{\nu}_\rho = \nu_\alpha$ for $\alpha(x) = \rho\pi(x)/(1+\rho\pi(x))$.

It is easy to check that in case (b), the measures $\tilde{\nu}_\rho$ are actually reversible for the process, while in case (a), the measures ν_ρ are reversible if and only if q is symmetric. When q is symmetric, which is the case to be studied in chapter 2, both (a) and (b) of the corollary hold, though the two classes of invariant measures it produces are identical. Another important case to which (a) applies is the one in which $S = Z^d$ and $q(x,y)$ is translation invariant. On the other hand, (b) applies for example when $S = Z^1$ and $q(x,y) = 0$ for $|x - y| \geq 2$. One case in which both apply and produce different classes of invariant measures is the asymmetric one dimensional simple random walk, which will be discussed in section 3.4. This gives an example in which $\mathcal{R} \neq \emptyset$ but $\mathcal{R} \neq \mathcal{I}$.

Section 1.3. Invariant measures for the exclusion process with speed change. In this section, we will take $c(x,y,\eta)$ to be of the form (1.1.4) or (1.1.8) where $q(x,y) = q(y,x)$ and $c(x,\eta)$ is given by

$$c(x,\eta) = \exp\left[2\sum_{A\ni x}\Phi(A)\chi_A(\eta)\right] , \tag{1.3.1}$$

where $\chi_A(\eta) = \Pi_{u\in A}[2\eta(u) - 1]$ and $\Phi(A)$ is a real valued function on the finite subsets of S which satisfies $\sup_x \Sigma_{A\ni x} |A|\,|\Phi(A)| < \infty$. This last condition guarantees that theorem 1.1.3 applies to $c(x,y,\eta)$, provided that

$$\sup_x \sum_y q(x,y) < \infty , \tag{1.3.2}$$

which we also assume.

In order to construct a class of measures which are invariant (and in fact reversible) for the process, we proceed in a manner similar to that of section 4.1 of part I. The major difference comes from the fact that there is no creation or destruction of particles in the exclusion process. The effect of this can be seen most clearly in case that S is finite. The typical spin-flip system is then an irreducible Markov chain on X, and therefore has a unique invariant measure. The typical exclusion process, on the other hand, has $|S|+1$ closed irreducible classes, corresponding to the fact that the cardinality of η_t does not change in time.

Returning to general S, define probability measures $\nu_{T,\zeta,k}$ on $\{0,1\}^T$ for a finite subset T of S, $\zeta \in \{0,1\}^{S\setminus T}$, and $0 \le k \le |T|$ by

$$\nu_{T,\zeta,k}\{\eta\} = K(T,\zeta,k)\exp[-\Sigma_{A\cap T\neq\emptyset}\ \Phi(A)X_A(\eta\zeta)]$$

if $\Sigma_{x\in T}\ \eta(x)=k$ and $\nu_{T,\zeta,k}\{\eta\}=0$ otherwise, where $K(T,\zeta,k)$ is a normalizing constant and $\eta\zeta$ is the configuration on $\{0,1\}^S$ which agrees with η on T and with ζ on $S\setminus T$. This measure can also be thought of as a probability measure on X by setting $\nu_{T,\zeta,k}\{\eta\in X: \eta=\zeta \text{ on } S\setminus T\} = 1$. Note that $\nu_{T,\zeta,k}$ is reversible for the process on $\{\eta\in\{0,1\}^T: \Sigma_{x\in T}\ \eta(x) = k\}$ with speed function $c_{T,\zeta}(x,y,\eta) = c(x,y,\eta\zeta)$, since

$$c_{T,\zeta}(x,y,\eta)\nu_{T,\zeta,k}\{\eta\} = c_{T,\zeta}(x,y,\eta_{xy})\nu_{T,\zeta,k}\{\eta_{xy}\}$$

for $x,y \in T$ by (1.1.10). Let $\tilde{\mathcal{G}}_T$ be the closed convex hull of $\{\nu_{T,\zeta,k}$: $\zeta \in \{0,1\}^{S\setminus T}$, $0 \le k \le |T|\}$, and let $\tilde{\mathcal{G}}$ be the set of all possible weak limits of μ_n as $n \to \infty$ where $\mu_n \in \tilde{\mathcal{G}}_{T_n}$ and $T_n \uparrow S$. $\tilde{\mathcal{G}}$ is the set of canonical Gibbs measures corresponding to Φ. For further information on $\tilde{\mathcal{G}}$, as well as a discussion of its relationship to the Gibbs measures of section 4.1 of part I, see [8], [9] and [16].

Using the same techniques as in theorem 4.1.3 of part I, one easily proves that $\tilde{\mathcal{G}} \subset \mathcal{R} \subset \mathcal{I}$. In [16] and [39], it is proved that $\mathcal{R} = \tilde{\mathcal{G}}$ under appropriate conditions. If $\Phi(A) \equiv 0$, so $c(x,\eta) \equiv 1$, it is easy to see that $\tilde{\mathcal{G}}$ is the set of all exchangeable probability measures on X. This case should be compared with part (a) of corollary 1.2.6 and the results of chapter 2 in order to get some feeling for what should be expected in the general case in which $c(x,\eta)$ is not constant.

It should be noted that while $\tilde{\mathcal{G}}$ depends on Φ, and therefore on $c(x,\eta)$, it does not depend on $q(x,y)$. In general, it should be expected that $\mathcal{I}$ will depend on $q(x,y)$ as well as on $c(x,\eta)$. As will be seen in chapter 2, this can be the case even when $c(x,\eta) \equiv 1$. It is easily seen, though, that if $\mathcal{I}_q$ denotes the set of invariant measures for the process with speed function given by (1.1.4) or (1.1.8) with $c(x,\eta)$ given by (1.3.1) then $\tilde{\mathcal{G}} = \cap\, \mathcal{I}_q$, where the intersection is over all symmetric functions q which satisfy (1.3.2).

Chapter 2. The symmetric simple exclusion process. Throughout this chapter, we consider the process η_t whose generator is the closure of the operator defined in (1.1.2) with $c(x,y,\eta) = [\eta(x) + \eta(y)]p(x,y)$ for $x \neq y$. We assume that $p(x,y)$ is symmetric, that the Markov chain on S with Q matrix $p(x,y)$ is irreducible, and that $\sup_x \Sigma_y p(x,y) < \infty$ so that theorem 1.1.3 applies. With this condition, we may assume without loss of generality that $\Sigma_y p(x,y) = 1$ for all x. In order to see this, note that the process obtained by replacing $p(x,y)$ by

$$\bar{p}(x,y) = \begin{cases} \frac{1}{c}\, p(x,y) & \text{for } x \neq y \\ 1 - \frac{1}{c} \sum_{z \neq x} p(x,z) & \text{for } x = y \end{cases}$$

where $c = \sup_x \Sigma_z p(x,z)$ is the same as the original process except for a constant time change. By the translation invariant case, we will mean the case in which $S = Z^d$ for some $d \geq 1$ and $p(x,y) = p(0,y-x)$. The results of this chapter are based on [34], [35] and [48].

<u>Section 2.1.</u> <u>Statement of results</u>. Let $\mathcal{H}$ be defined by

$$\mathcal{H} = \left\{\alpha(\cdot) \text{ on } S : \sum_y p(x,y)\alpha(y) = \alpha(x) \text{ and } 0 \leq \alpha(x) \leq 1\right\} .$$

Then the invariant measures for η_t can be described in the following way. The measures ν_α are those defined in section 1.2.

<u>Theorem 2.1.1.</u> (a) $\mu_\alpha = \lim_{t\to\infty} \nu_\alpha S(t)$ exists for each $\alpha \in \mathcal{H}$.

(b) $\mu_\alpha\{\eta : \eta(x) = 1\} = \alpha(x)$ for all $x \in S$.

(c) The map $\alpha \to \mu_\alpha$ gives a one-to-one and onto correspondence between $\mathcal{H}$ and $\mathcal{I}_e$.

(d) $\mu_\alpha = \nu_\alpha$ if and only if α is constant.

<u>Corollary 2.1.2.</u> If $\mathcal{H}$ consists only of constants, then $\mathcal{I}_e = \{\nu_\rho, 0 \leq \rho \leq 1\}$.

In the translation invariant case, $\mathcal{H}$ consists only of constants by the Choquet-Deny theorem, so it follows that all invariant measures for η_t are exchangeable in that case. One interesting case in which $\mathcal{H}$ has many nonconstant elements and therefore $\mathcal{I}$ contains many nonexchangeable measures, is that of the simple random walk on the connected graph in which there are no loops and each vertex is connected by an edge to exactly $k \geq 3$ other vertices.

In order to state the convergence result, let

$$p_t(x,y) = e^{-t} \sum_{n=0}^{\infty} \frac{t^n}{n!} p^{(n)}(x,y) ,$$

where $p^{(n)}(x,y)$ are the n-step transition probabilities corresponding to $p(x,y)$.

<u>Theorem 2.1.3.</u> Suppose $\alpha \in \mathcal{H}$ and μ is a probability measure on X which satisfies

(2.1.4) $$\sum_y p_t(x,y)[\eta(y) - \alpha(y)] \to 0$$

in probability with respect to μ as $t \to \infty$ for each $x \in S$. Then $\mu S(t) \to \nu_\alpha$ as $t \to \infty$.

Condition (2.1.4) is in fact also necessary for $\mu S(t) \to \nu_\alpha$ in case $p_t(x,y)$ is transient [34]. It is probably necessary in general, though this has not yet been proved. In [48], Spitzer gives another condition which is necessary and sufficient in general, but which is less useful than (2.1.4).

Corollary 2.1.5. Consider the translation invariant case. If μ is translation invariant and ergodic, then $\mu S(t) \to \nu_\rho$ as $t \to \infty$, where $\rho = \mu\{\eta : \eta(x) = 1\}$.

In order to simplify matters, we will present proofs of these results under the assumption that $\mathcal{H}$ consists only of constants. The proofs in case p has nonconstant bounded harmonic functions (in which case, of course, p must be transient) can be found in [34]. The key to the proofs is Spitzer's observation in [45] that the symmetric simple exclusion process is self-dual, in a sense that will be described in section 2.2. This permits the translation of problems concerning the invariant measures for the infinite system into problems involving the bounded harmonic functions for the corresponding finite system. Once this is done the proofs follow different lines in the two cases which must be considered separately. We will say that case I occurs if two independent copies of the chain with transition probabilities $p_t(x,y)$ will hit in finite time with probability one, and that case II occurs otherwise. The proof in case I is based on a coupling argument, which enables one to prove that the finite particle system has no nonconstant bounded harmonic functions. For case II, the finite particle system is compared with the corresponding system without the exclusion interaction to obtain the same conclusion under the assumption that $\mathcal{H}$ consists only of constants. The proof in case I is due to Spitzer [48] and in case II to Liggett [34, 35]. One of the interesting features of the proofs is that the two cases require two essentially different techniques; neither technique works in the other case.

In [44], Schwartz considered the case in which the simple exclusion process is modified by letting particles be created at vacant sites and annihilated at occupied sites. A similar analysis leads to a complete identification of $\mathcal{I}$ in case the creation and annihilation rates are independent of η.

Section 2.2. The duality theorem. This section is devoted to the proof of the following result which is due to Spitzer [45], and to some of its immediate consequences. The symmetry of $p(x,y)$ is essential here.

Theorem 2.2.1. Suppose $\eta, \zeta \in X$ and $\sum_x \zeta(x) < \infty$. Then

$$(2.2.2) \qquad P^{\eta}[\eta_t \geq \zeta] = P^{\zeta}[\zeta_t \leq \eta] \,,$$

where the inequalities are interpreted coordinatewise.

Proof. Let $f(\eta,\zeta) = 1_{\{\eta \geq \zeta\}}$. As a function of η, $f(\eta,\zeta)$ is in the domain of Ω for each finite configuration ζ. Therefore if we define $u_\eta(t,\zeta) = E^{\eta} f(\eta_t,\zeta) = S(t) f(\eta,\zeta)$, it follows from lemma 1.2.1 that

$$\frac{d}{dt} u_\eta(t,\zeta) = S(t)\Omega f(\eta,\zeta)$$

$$= S(t) \sum_{\substack{\zeta(x)=1 \\ \zeta(y)=0}} p(x,y)[f(\eta,\zeta_{xy}) - f(\eta,\zeta)]$$

$$= \sum_{\substack{\zeta(x)=1 \\ \zeta(y)=0}} p(x,y)[u_\eta(t,\zeta_{xy}) - u_\eta(t,\zeta)] \,.$$

Since ζ_t is a nonexplosive Markov chain, the unique bounded solution to this equation with initial condition $u_\eta(0,\zeta) = f(\eta,\zeta)$ is given by $E^{\zeta} f(\eta,\zeta_t)$, which gives the desired result.

By integrating both sides of (2.2.2) with respect to μ, we obtain:

Corollary 2.2.3. If μ is any probability measure on X and ζ is any finite configuration, then

$$\mu S(t)\{\eta : \eta \geq \zeta\} = \sum_{\gamma} P^{\zeta}[\zeta_t = \gamma]\, \mu\{\eta : \eta \geq \gamma\} \,.$$

Corollary 2.2.4. Suppose that for each $n \geq 1$, the Markov chain ζ_t restricted to $\{\gamma \in X : |\gamma| = \Sigma_x \gamma(x) = n\}$ has only constant bounded harmonic functions. Then $\mathcal{J}$ is the set of exchangeable measures on X, and therefore $\mathcal{J}_e = \{\nu_\rho, 0 \leq \rho \leq 1\}$. Furthermore, if μ is any probability measure on X, $t_n \to \infty$, and $\lim_{n\to\infty} \mu S(t_n)$ exists, then this limit is exchangeable.

The only part of this corollary which is not an immediate consequence of corollary 2.2.3 is the final statement. It is proved by applying the following result, which appears in [40], to the Markov chains corresponding to the finite particle systems.

Lemma 2.2.5. Suppose $q(x,y)$ is the Q matrix for a Markov chain $X(t)$ on S which satisfies $\sup_x \Sigma_{y\neq x} q(x,y) < \infty$. If f is a bounded function on S and $g(x) = \lim_{n\to\infty} E^x f(X(t_n))$ exists for all $x \in S$ and some sequence $t_n \to \infty$, then g is harmonic for $X(t)$.

Proof. Let $c = \sup_x \Sigma_{y\neq x} q(x,y)$, $\tilde{q}(x,y) = (1/c)q(x,y)$ for $x \neq y$, and $\tilde{q}(x,x) = 1 - (1/c)\Sigma_{y\neq x} q(x,y)$. Then $\tilde{q}(x,y)$ is a stochastic transition function, and the transition probabilities $q_t(x,y)$ for the chain $X(t)$ are given by

$$q_t(x,y) = e^{-ct} \sum_{n=0}^{\infty} \frac{(ct)^n}{n!} \tilde{q}^{(n)}(x,y) .$$

It suffices to prove that

$$\lim_{t\to\infty} \sum_y \left| \sum_x q(z,x)q_t(x,y) - q_t(z,y) \right| = 0$$

for all $z \in S$. But this expression is dominated by

$$\lim_{t\to\infty} e^{-ct} \sum_{n=1}^{\infty} \left|\frac{(ct)^{n-1}}{(n-1)!} - \frac{(ct)^n}{n!}\right| ,$$

which is zero.

While the rest of the proofs will be carried out under the assumption that $\mathcal{H}$ consists only of constants, we should note that the existence of the limit in part (a) of theorem 2.1.1 is an immediate consequence of corollary 2.2.3 for all $\alpha \in \mathcal{H}$. The reason is that, in fact, $\nu_\alpha S(t)\{\eta : \eta(x_i) = 1 \text{ for } 1 \leq i \leq n\}$ is nonincreasing in t for any choice of $n \geq 1$ and distinct $x_1,\dots,x_n \in S$. In order to see this, it suffices to show that $Qf \leq 0$, where $f(x_1,\dots,x_n) = \Pi_{i=1}^n \alpha(x_i)$ and Q is the generator for the n-particle simple exclusion process. This is the result of the following computation, which of course uses the harmonicity of α for $p(x,y)$, as well as the symmetry of $p(x,y)$:

$$\begin{aligned}
Qf(\vec{x}) &= \sum_{i=1}^n \sum_{y\notin\{x_1,\dots,x_n\}} p(x_i,y)[f(x_1,\dots,x_{i-1},y,x_{i+1},\dots,x_n) - f(\vec{x})] \\
&= \sum_{i=1}^n \left[\prod_{k\neq 1} \alpha(x_k)\right]\left[\alpha(x_i) - \sum_{j=1}^n p(x_i,x_j)\alpha(x_j) - \alpha(x_i)\left\{1 - \sum_{j=1}^n p(x_i,x_j)\right\}\right] \\
&= -\sum_{i=1}^n \left[\prod_{k\neq i} \alpha(x_k)\right] \sum_{j=1}^n p(x_i,x_j)[\alpha(x_j) - \alpha(x_i)] \\
&= -\frac{1}{2} \sum_{i,j=1}^n \left[\prod_{k\neq i,j} \alpha(x_k)\right] p(x_i,x_j)[\alpha(x_j) - \alpha(x_i)]^2 \leq 0 .
\end{aligned}$$

Section 2.3. The finite process. Throughout this section, we will assume that $\mathcal{H}$ consists only of constants. Let $X_n(t)$ be the Markov chain on S^n in which the individual coordinates undergo independent Markov chains on S with transition probabilities $p_t(x,y)$, and let $U_n(t)$ be the corresponding transition semigroup:

$$U_n(t)g(\vec{x}) = E^{\vec{x}}g(X_n(t)) = \sum_{\vec{y}} \prod_{i=1}^{n} p_t(x_i,y_i)g(\vec{y}) .$$

Lemma 2.3.1. Suppose g is a bounded function on S^n such that $U_n(t)g = g$ for all $t \geq 0$. Then g is constant.

Proof. Define $W_i g$ for $1 \leq i \leq n$ by

$$W_i g(\vec{x}) = \sum_u p(x_i,u)g(x_1,\ldots,x_{i-1},u,x_{i+1},\ldots,x_n) .$$

Then the operators W_i commute, and any bounded harmonic function g of $X_n(t)$ satisfies

$$g = \frac{1}{n}\sum_{i=1}^{n} W_i g . \tag{2.3.2}$$

Let K be the set of all g which satisty (2.3.2) and $|g(\vec{x})| \leq 1$ for all $\vec{x} \in S^n$. Then K is a weak* compact and convex subset of $\ell_\infty(S^n)$, so K is the closed convex hull of its extreme points by the Krein-Milman theorem. Since W_i maps K into itself for each i, it follows that $W_i g = g$ for all i if g is extreme in K. Since $\mathcal{H}$ consists only of constants, it then follows that every extreme g in K is constant, and therefore that K consists only of constants.

Put $D = \{(x,x),\ x \in S\} \subset S^2$, and define

$$f_2(\vec{x}) = P^{\vec{x}}\,[X_2(t) \in D \text{ for some } t > 0] .$$

It is an immediate consequence of the irreducibility of $p(x,y)$ that if $f_2(\vec{x}) < 1$ for some $\vec{x} \notin D$, then $f_2(\vec{x}) < 1$ for all $\vec{x} \notin D$. Of course $f_2(\vec{x}) = 1$ for $\vec{x} \in D$. We will consider two cases:

Case I: $f_2(\vec{x}) = 1$ for all $\vec{x} \in S^2$.

Case II: $f_2(\vec{x}) < 1$ for all $\vec{x} \in S^2\setminus D$.

The results of this section will be used primarily for the proofs of the main results in case II. Lemma 2.3.4 below, however, will also be used in case I. For $n \geq 2$, let

$$f_n(\vec{x}) = \sum_{1\leq i<j\leq n} f_2(x_i,x_j) .$$

Lemma 2.3.3. In case II, $\lim_{t\to\infty} U_n(t)f_n(\vec{x}) = 0$ for all $n \geq 2$ and $\vec{x} \in S^n$.

Proof. It suffices to prove the result for $n = 2$ since

$$U_n(t)f_n(\vec{x}) = \sum_{\vec{y}} \prod_{k=1}^{n} p_t(x_k,y_k) \sum_{1\leq i<j\leq n} f_2(y_i,y_j)$$

$$= \sum_{1\leq i<j\leq n} p_t(x_i,y_i)p_t(x_j,y_j)f_2(y_i,y_j)$$

$$= \sum_{1\leq i<j\leq n} U_2(t)f_2(x_i,x_j) .$$

But $U_2(t)f_2(x,y) = P^{(x,y)}[X_2(s) \in D$ for some $s > t]$, so

$$U_2(t)f_2(x,y) \to P^{(x,y)}[X_2(s) \in D \text{ for arbitrarily large } s] .$$

By lemma 2.3.1 and proposition 5.19 of [29], this last probability is either zero or one. Since case II holds, it must be zero.

We will say that a bounded symmetric function $g(x,y)$ on S^2 is positive definite if $\Sigma_{x,y}\, \beta(x)\beta(y)g(x,y) \geq 0$ whenever $\Sigma_x\, |\beta(x)| < \infty$ and $\Sigma_x\, \beta(x) = 0$. For $n > 2$, a bounded symmetric function on S^n will be called positive definite if it is a positive definite function of each pair of variables.

Let $T_n = \{\vec{x} \in S^n : x_i \neq x_j$ for all $i \neq j\}$, and let $Y_n(t)$ be the Markov chain on T_n which corresponds to the n particle interacting system. This chain is obtained from $X_n(t)$ by suppressing transitions to points in $S^n\setminus T^n$. Let $V_n(t)$ be the transition semigroup corresponding to $Y_n(t)$.

Lemma 2.3.4. Suppose g is a bounded, symmetric, positive definite function on S^n for $n \geq 2$. Then

$$V_n(t)g(\vec{x}) \leq U_n(t)g(\vec{x}) \quad \text{for } \vec{x} \in T_n .$$

Proof. For $t \geq 0$, $U_n(t)g$ is positive definite also since

$$\sum_{x_1,x_2} \beta(x_1)\beta(x_2)U_n(t)g(\vec{x}) = \sum_{j\neq 1,2} \sum_{y_j} \prod_{i\neq 1,2} p_t(x_i,y_i) \sum_{y_1,y_2} \gamma(y_1)\gamma(y_2)g(\vec{y}) ,$$

where $\gamma(y) = \Sigma_x \beta(x)p_t(x,y)$. Therefore

$$U_n(t)g(x_1,x_1,x_3,\dots,x_n) + U_n(t)g(x_2,x_2,x_3,\dots,x_n) - 2U_n(t)g(x_1,x_2,\dots,x_n) \geq 0 .$$

for $\vec{x} \in T_n$, and hence $(Q_U - Q_V)U_n(t)g(\vec{x}) \geq 0$ for $\vec{x} \in T_n$, where Q_U and Q_V are the generators of $U_n(t)$ and $V_n(t)$ respectively. The desired result then follows from

$$U_n(t) - V_n(t) = \int_0^t V_n(t - s)(Q_U - Q_V)U_n(s)\, ds$$

and the fact that $V_n(t)$ maps functions which are nonnegative on T_n to functions which are nonnegative on T_n.

Lemma 2.3.5. f_n is positive definite for all $n \geq 2$.

Proof. Suppose $\Sigma_x |\beta(x)| < \infty$ and $\Sigma_x \beta(x) = 0$. Then

$$\sum_{x_1,x_2} \beta(x_1)\beta(x_2)f_n(\vec{x}) = \sum_{x_1,x_2} \beta(x_1)\beta(x_2)f_2(x_1,x_2) ,$$

so it suffices to prove the result for $n = 2$. Let $H = \{t_1,\dots,t_n\}$ be a finite set of times with $0 < t_1 < \cdots < t_n$, and define

$$h(x,y) = P^{(x,y)}[X_2(t) \in D \text{ for some } t \in H] .$$

The decomposition according to the time and place of the last visit to D then yields

$$h(x,y) = \sum_{i=1}^{n} P^{(x,y)}[X_2(t_i) \in D,\ X_2(t_j) \notin D \text{ for all } j > i]$$

$$= \sum_{i=1}^{n} \sum_{u\in S} p_{t_i}(x,u)p_{t_i}(y,u)P^{(u,u)}[X_2(t_j - t_i) \notin D \text{ for all } j > i] .$$

Therefore

$$\sum_{x,y} \beta(x)\beta(y)h(x,y) = \sum_{i=1}^{n} \sum_{u\in S} \left|\sum_x \beta(x)p_{t_i}(x,u)\right|^2 P^{(u,u)}[X_2(t_j - t_i) \notin D \text{ for } j > i]$$

which is nonnegative. Therefore $h(x,y)$ is positive definite. Now let H_k be an increasing sequence of finite sets of times for which $\cup_k H_k$ is dense in $[0,\infty)$, and let $h_k(x,y)$ be defined as before in terms of H_k. Then $h_k \uparrow f_2$, so f_2 is positive definite also.

The previous three lemmas combine to give the following corollary.

Corollary 2.3.6. In case II, $\lim_{t\to\infty} V_n(t)f_n(\vec{x}) = 0$ for all $n \geq 2$ and $\vec{x} \in T_n$.

Section 2.4. Proofs of the main results. We continue to assume that $\mathcal{H}$ consists only of constants. In order to prove corollary 2.1.2 (which is equivalent to theorem 2.1.1 under this assumption), it suffices by corollary 2.2.4 to prove the following result.

Theorem 2.4.1. Suppose g is a bounded symmetric function on T_n for $n \geq 2$ such that $V_n(t)g = g$. Then g is constant.

Proof in case I [48]. In this case, the proof is based on a simple coupling argument. In order to prove that $g(\vec{x}) = g(\vec{y})$ for all $\vec{x},\vec{y} \in T_n$, it suffices to show this when $\vec{x}$ and $\vec{y}$ differ at only one coordinate. To prove it in that case, consider the continuous time Markov chain $(Z_1(t),\ldots,Z_{n+1}(t))$ on $A = \{\vec{z} \in S^{n+1} : z_i \neq z_j \text{ for } i \neq j \text{ and } \{i,j\} \neq \{n,n+1\}\}$ which can be described in the following way: (a) the set $\{\vec{z} \in A : z_n = z_{n+1}\}$ is closed for the process, and on this set, the process $(Z_1(t),\ldots,Z_n(t))$ has the same transition law as the chain $Y_n(t)$, and (b) on the set $\{\vec{z} \in A : z_n \neq z_{n+1}\}$, the process has the following transition intensities:

$\vec{z} \to (z_1,\ldots,z_{i-1},u,z_{i+1},\ldots,z_n,z_{n+1})$	$p(z_i,u)$ for $u \neq z_j\ \forall j$
$\vec{z} \to (z_1,\ldots,z_{n-1},z_n,z_{n+1})$	$p(z_{n+1},z_n)$
$\vec{z} \to (z_1,\ldots,z_{n-1},z_{n+1},z_{n+1})$	$p(z_n,z_{n+1})$
$\vec{z} \to (z_1,\ldots,z_{i-1},z_n,z_{i+1},\ldots,z_{n-1},z_i,z_{n+1})$	$p(z_i,z_n)$ for $i < n$
$\vec{z} \to (z_1,\ldots,z_{i-1},z_{n+1},z_{i+1},\ldots,z_{n-1},z_n,z_i)$	$p(z_i,z_{n+1})$ for $i < n$.

Note that $\{Z_1(t),\ldots,Z_{n-1}(t),Z_n(t)\}$ and $\{Z_1(t),\ldots,Z_{n-1}(t),Z_{n+1}(t)\}$ are both Markovian and have the same transition law as the chain $\{Y_n^{(1)}(t),\ldots,Y_n^{(n)}(t)\}$, where these chains are regarded as having as state space the collection of all subsets of S of size n. Furthermore, using the symmetry of $p(x,y)$, it can be seen that the process $(Z_n(t),Z_{n+1}(t))$ is also Markovian and has the same transition law as $X_2(t)$ until the first time that $Z_n(t) = Z_{n+1}(t)$. After that time, of course, $Z_n \equiv Z_{n+1}$. Now suppose $\vec{x},\vec{y} \in T_n$ are such that $x_i = y_i$ for $i < n$ and $x_n \neq y_n$. Then

$$g(\vec{x}) = V_n(t)g(\vec{x}) = Eg(Z_1(t),\ldots,Z_{n-1}(t),Z_n(t)), \text{ and}$$

$$g(\vec{y}) = V_n(t)g(\vec{y}) = Eg(Z_1(t),\ldots,Z_{n-1}(t),Z_{n+1}(t)),$$

where the process $(Z_1(t),\ldots,Z_{n+1}(t))$ has initial point $(x_1,\ldots,x_{n-1},x_n,y_n)$. It follows from $f_2 \equiv 1$ that $P[Z_n(t) = Z_{n+1}(t)$ for all large $t] = 1$. Therefore since g is bounded

$$Eg(Z_1(t),\ldots,Z_{n-1}(t),Z_n(t)) - Eg(Z_1(t),\ldots,Z_{n-1}(t),Z_{n+1}(t))$$

tends to zero as $t \to \infty$, and hence $g(\vec{x}) = g(\vec{y})$.

Proof in case II [34, 35]. We will assume $|g| \leq 1$ for simplicity. Then

$$|V_n(t)g(\vec{x}) - U_n(t)g(\vec{x})| \leq f_n(\vec{x})$$

for $\vec{x} \in T_n$, so

$$|g(\vec{x}) - U_n(t)g(\vec{x})| \leq f_n(\vec{x})$$

for $\vec{x} \in T_n$. By taking a pointwise limit of $U_n(t)g$ along a sequence $t_n \to \infty$, it follows from lemmas 2.2.5 and 2.3.1 that there is a constant c so that $|g(\vec{x}) - c| \leq f_n(\vec{x})$ for $\vec{x} \in T_n$. Applying $V_n(t)$ to this, we see that

$$|g(\vec{x}) - c| = |V_n(t)g(\vec{x}) - c| \leq V_n(t)f_n(\vec{x})$$

for $\vec{x} \in T_n$. Therefore $g(\vec{x}) = c$ for all $\vec{x} \in T_n$ by corollary 2.3.6.

Proof of theorem 2.1.3. Suppose μ is a probability measure on X which satisfies (2.1.4) with $\alpha(x) \equiv \rho$, a constant. The function $h(x,y) = \mu\{\eta: \eta(x)=1, \eta(y)=1\}$ is positive definite, so

$$V_2(t)h(x,y) \leq U_2(t)h(x,y) \tag{2.4.2}$$

for $x \neq y$ by lemma 2.3.4. By (2.1.4),

$$\lim_{t\to\infty} \sum_y p_t(x,y)\,\mu\{\eta: \eta(y) = 1\} = \rho \tag{2.4.3}$$

for each $x \in S$ and $\lim_{t\to\infty} U_2(t)h(x,y) = \rho^2$ for $x,y \in S$. Therefore by (2.4.2),

$$\overline{\lim_{t\to\infty}}\; V_2(t)h(x,y) \leq \rho^2 \tag{2.4.4}$$

for $x \neq y$. But by corollary 2.2.3, $\Sigma_y\, p_t(x,y)\mu\{\eta: \eta(y) = 1\} = \mu S(t)\{\eta: \eta(x) = 1\}$ for $x \in S$ and $V_2(t)h(x,y) = \mu S(t)\{\eta : \eta(x) = 1, \eta(y) = 1\}$ for $x \neq y$. Therefore all weak limits ν of $\mu S(t)$ as $t \to \infty$ have the following properties:

(a) $\nu\{\eta : \eta(x) = 1\} = \rho$ for $x \in S$, by (2.4.3).

(b) $\nu\{\eta : \eta(x) = 1, \eta(y) = 1\} \leq \rho^2$ for all $x \neq y$, by (2.4.4).

(c) ν is exchangeable by corollary 2.2.4 and theorem 2.4.1.

By deFinetti's theorem, every exchangeable probability measure on X can be written in the form $\int_0^1 \nu_\rho \lambda(d\rho)$ for some probability measure λ on $[0,1]$. Therefore, the only probability measure which satisfies (a), (b) and (c) above is ν_ρ.

<u>Proof of corollary 2.1.5</u>. Consider the translation invariant case, and let μ be a translation invariant and ergodic probability measure on X such that $\mu\{\eta : \eta(x) = 1\} = \rho$ for all $x \in S = Z^d$. By Bochner's theorem, there is a finite measure $\lambda(d\sigma)$ on $[0,1)^d$ so that

$$\mu\{\eta : \eta(u) = 1, \eta(v) = 1\} = \int \exp[2\pi i\langle u - v, \sigma\rangle]\lambda(d\sigma) ,$$

where $\langle\ ,\ \rangle$ is the usual inner product on R^d. Let $Z_t(x,\eta) = \Sigma_y\, p_t(x,y)\eta(y)$ and $h(\sigma) = \Sigma_x\, p(0,x)\exp[2\pi i\langle x,\sigma\rangle]$, and compute

$$\begin{aligned}\int_X Z_t(x,\eta)Z_s(x,\eta)\mu(d\eta) &= \sum_{u,v} p_t(x,u)p_s(x,v)\,\mu\{\eta : \eta(u) = \eta(v) = 1\} \\ &= \int_{[0,1)^d} \left[\sum_u p_t(x,u)e^{2\pi i\langle u,\sigma\rangle}\right]\left[\sum_v p_s(x,v)e^{-2\pi i\langle v,\sigma\rangle}\right] d\lambda \\ &= \int_{[0,1)^d} \exp[-(t+s)(1 - h(\sigma))]d\lambda .\end{aligned}$$

Now $|h(\sigma)| \leq 1$ and $h(\sigma) = 1$ if and only if $\sigma = 0$ since $p(x,y)$ is irreducible, so

$$\lim_{s,t\to\infty} \int Z_t(x,\eta)Z_s(x,\eta)\mu(d\eta) = \lambda(\{0\}) .$$

Therefore $Z_t(x,\eta)$ is Cauchy in $L_2(\mu)$, so $Z(x,\eta) = \lim_{t\to\infty} Z_t(x,\eta)$ exists in $L_2(\mu)$. By definition, $Z_{t+s}(x,\eta) = \Sigma_y p_s(x,y)Z_t(y,\eta)$, so $Z(x,\eta) \in \mathcal{H}$ for almost every η with respect to μ. Therefore $Z(x,\eta) = Z(0,\eta)$ for a.e. η, and hence $Z(0,\eta)$ is an invariant random variable with respect to the shifts in Z^d. Since μ is ergodic, it then follows that $Z(0,\eta)$ is constant, and therefore $Z(x,\eta) = \rho$ a.e. since $\int Z(x,\eta)d\mu = \rho$. Therefore $Z_t(x,\eta)$ converges to ρ in probability for each $x \in S$, and thus the result follows from theorem 2.1.3.

<u>Chapter 3</u>. <u>The asymmetric simple exclusion process</u>. Throughout this chapter, we consider the process η_t whose generator is the closure of the operator defined in (1.1.2) with $c(x,y,\eta) = \eta(x)p(x,y) + \eta(y)p(y,x)$ for $x \neq y$, where $p(x,y)$ are the transition probabilities for an irreducible Markov chain on S. We assume of course that $\sup_y \Sigma_x p(x,y) < \infty$, so that theorem 1.1.3 applies. When $p(x,y)$ is not symmetric, theorem 2.2.1 fails, and there is virtually no hope of obtaining results in the general case which are as complete as those obtained in chapter 2 for the symmetric case. Since duality techniques appear to be inapplicable, we will discuss here some of those results which can be obtained via coupling. While the symmetric case is essentially completely understood, there remain many important open problems in the asymmetric case. The results in this chapter are based on [36], [37], and [38].

<u>Section 3.1</u>. <u>The basic coupling</u>. By the coupled process, we will mean the Markov process $\gamma_t = (\eta_t,\zeta_t)$ with state space $X \times X$ which has the following properties: (a) the marginal processes η_t and ζ_t are Markovian and have the transition law of the simple exclusion process corresponding to $p(x,y)$, and (b) whenever $\eta_t(x) = \zeta_t(x) = 1$ for some $x \in S$, the two marginal processes will use the same random mechanisms to decide when the particle at x will attempt a transition, and where it will attempt to go. Thus the generator $\bar{\Omega}$ of γ_t takes the following form when restricted to functions $f(\eta,\zeta)$ which depend on only finitely many coordinates:

$$\bar{\Omega}f(\eta,\zeta) = \sum_{\eta(x)=1,\eta(y)=0} p(x,y)[f(\eta_{xy},\zeta)-f(\eta,\zeta)] + \sum_{\zeta(x)=1,\zeta(y)=0} p(x,y)[f(\eta,\zeta_{xy})-f(\eta,\zeta)]$$

$$+ \sum_{\substack{\eta(x)=\zeta(x)=1\\ \eta(y)=\zeta(y)=0}} p(x,y)[f(\eta_{xy},\zeta_{xy}) - f(\eta_{xy},\zeta) - f(\eta,\zeta_{xy}) + f(\eta,\zeta)] .$$

One of the important properties of the coupled process is that $\{\eta = \zeta\}$, $\{\eta \le \zeta\}$ and $\{\eta \ge \zeta\}$ are closed for the motion. This observation leads to the following result, in which we use an upper bar to denote symbols pertaining to the coupled process (so that $\bar{\mathcal{I}}$ is the set of invariant measures for γ_t, and $\bar{\mathcal{S}}$ is the set of translation invariant probability measures on $X \times X$ in the translation invariant case).

Lemma 3.1.1. If $\nu \in \bar{\mathcal{I}}_e$, then $\nu\{\eta = \zeta\}$, $\nu\{\eta \le \zeta\}$ and $\nu\{\eta \ge \zeta\}$ are each either zero or one. The same is true for $\nu \in (\bar{\mathcal{I}} \cap \bar{\mathcal{S}})_e$ in the translation invariant case.

The approach we will follow is to use information concerning $\bar{\mathcal{I}}$ to draw conclusions about $\mathcal{I}$. In order to do this, we need to obtain relations between $\bar{\mathcal{I}}$ and $\mathcal{I}$. These are given in the next lemma.

Lemma 3.1.2. (a) If $\nu \in \bar{\mathcal{I}}$, then the marginals μ_1 and μ_2 of ν are in $\mathcal{I}$. (b) If $\mu_1, \mu_2 \in \mathcal{I}$, then there is a $\nu \in \bar{\mathcal{I}}$ with marginals μ_1 and μ_2. (c) If $\mu_1, \mu_2 \in \mathcal{I}_e$, then ν can be taken in $\bar{\mathcal{I}}_e$.

Proof. (a) is immediate from the fact that the marginals of γ_t are Markovian with the right transition law. For (b), let $\tilde{\nu}$ be the product measure $\mu_1 \times \mu_2$ on $X \times X$. Then $\tilde{\nu}\bar{S}(t)$ has marginals μ_1 and μ_2 for all $t > 0$, so we can take ν to be any weak limit as $t \to \infty$ of $(1/t)\int_0^t \tilde{\nu}\bar{S}(s)ds$. In order to prove (c), let

$$\mathcal{C} = \{\nu \in \bar{\mathcal{I}} : \nu \text{ has marginals } \mu_1 \text{ and } \mu_2\} .$$

Then $\mathcal{C} \neq \emptyset$ by part (b), and $\mathcal{C}$ is compact and convex, so $\mathcal{C}_e \neq \emptyset$ by the Krein-Milman theorem. The result then follows from $\mathcal{C}_e \subset \bar{\mathcal{I}}_e$, which is a consequence of the assumption that μ_1 and μ_2 are extremal in $\mathcal{I}$.

Lemma 3.1.3. Consider the translation invariant case. (a) If $\nu \in \bar{\mathcal{I}} \cap \bar{\mathcal{S}}$, then the marginals μ_1 and μ_2 of ν are in $\mathcal{I} \cap \mathcal{S}$. (b) If $\mu_1, \mu_2 \in \mathcal{I} \cap \mathcal{S}$, then there is a $\nu \in \bar{\mathcal{I}} \cap \bar{\mathcal{S}}$ with marginals μ_1 and μ_2. (c) If $\mu_1, \mu_2 \in (\mathcal{I} \cap \mathcal{S})_e$, then ν can be taken in $(\bar{\mathcal{I}} \cap \bar{\mathcal{S}})_e$.

The proof of the above lemma is omitted, since it is similar to the proof of the previous one. The fact that $\{\eta \le \zeta\}$ is closed for the motion of the coupled process leads immediately to the following result.

Lemma 3.1.4. If $\mu_1 \le \mu_2$, then $\mu_1 S(t) \le \mu_2 S(t)$ for all $t \ge 0$.

Section 3.2. The reversible positive recurrent case. In addition to the basic assumptions of this chapter, we will assume throughout this section that there is a positive function $\pi(x)$ on S such that $\Sigma_x \pi(x) < \infty$ and

$$\pi(x)p(x,y) = \pi(y)p(y,x) . \tag{3.2.1}$$

By part (b) of corollary 1.2.6, the product measures $\tilde{\nu}_\rho$ on X with $\tilde{\nu}_\rho\{\eta: \eta(x) = 1\} = \rho\pi(x)/(1+\rho\pi(x))$ are invariant for the process for all $\rho \in [0,\infty]$. Since $\Sigma_x \pi(x) < \infty$, $\tilde{\nu}_\rho$ concentrates on $\{\eta: \Sigma_x \eta(x) < \infty\}$ and $\tilde{\nu}_\rho\{\eta: \Sigma_x\eta(x)=n\} > 0$ for $\rho \in (0,\infty)$ and $0 \leq n < \infty$. Therefore the process η_t restricted to $\{\eta: \Sigma_x \eta(x) = n\}$ is an irreducible positive recurrent Markov chain, whose stationary distribution is given by the conditional measure $\nu_n(\cdot) = \tilde{\nu}_\rho(\cdot \mid \Sigma_x\eta(x)=n)$, which is independent of ρ. Let ν_∞ be the point mass on $\eta \equiv 1$. The following result is from [36].

Theorem 3.2.2. (a) $\mathcal{I}_e = \{\nu_n, 0 \leq n \leq \infty\}$. (b) If $\mu\{\eta: \Sigma_x \eta(x) = n\} = 1$, then $\mu S(t) \to \nu_n$ for $0 \leq n \leq \infty$.

Proof. Part (a) follows from part (b), and part (b) for $n < \infty$ is just the ordinary convergence theorem for positive recurrent Markov chains. To prove (b) for $n = \infty$, take μ such that $\mu\{\eta: \Sigma_x \eta(x) = \infty\} = 1$, and choose μ_n so that $\mu_n \leq \mu$ and $\mu_n\{\eta: \Sigma_x \eta(x) = n\} = 1$. By lemma 3.1.4, $\mu_n S(t) \leq \mu S(t)$ for all $t \geq 0$. Since $\mu_n S(t) \to \nu_n$ as $t \to \infty$, it follows that $\nu_n \leq \nu$ for all n if ν is any weak limit of $\mu S(t)$ as $t \to \infty$. But

$$\frac{\rho\pi(x)}{1 + \rho\pi(x)} = \tilde{\nu}_\rho\{\eta : \eta(x) = 1\} = \sum_{n=0}^{\infty} \nu_n\{\eta : \eta(x) = 1\} \tilde{\nu}_\rho\{\eta : \Sigma_x \eta(x) = n\} ,$$

so $\nu\{\eta: \eta(x) = 1\} \geq \rho\pi(x)/(1+\rho\pi(x))$ for all $\rho \in (0,\infty)$. Letting $\rho \to \infty$, it follows that $\nu\{\eta: \eta(x) = 1\} = 1$ for all $x \in S$, and therefore $\nu = \nu_\infty$.

Similar techniques apply if $\Sigma_x \pi(x) < \infty$ is replaced by $\Sigma_x 1/\pi(x) < \infty$. Then the extremal invariant measures concentrate on $\{\eta: \Sigma_x [1 - \eta(x)] < \infty\} \cup \{\eta \equiv 0\}$. This amounts, of course, merely to an interchange in the roles of the zeros and ones, since one can think of the zeros moving according to the transpose of p instead of the ones moving according to p. If $\pi(x)$ satisfies (3.2.1), $\Sigma_x \pi(x) = \infty$ and $\Sigma_x 1/\pi(x) = \infty$, the situation is more complex, and is not fully understood. One such case, which illustrates some of the difficulties involved, will be discussed in section 3.4. Another class of cases (that in which $S = \{0,1,2,\cdots\}$ and $p(x,y) = 0$ for $|x - y| > 1$) is analyzed via coupling techniques in [38].

Section 3.3. The translation invariant case. Throughout this section, we will assume that $S = Z^d$ and $p(x,y) = p(0,y-x)$. In this case, part (a) of corollary 1.2.6 gives $\nu_\rho \in \mathcal{I}$ for all $\rho \in [0,1]$. As will be seen in the next section, it is not necessarily the case that $\mathcal{J}_e = \{\nu_\rho, \ 0 \le \rho \le 1\}$. It is true, though, that $(\mathcal{J} \cap \mathcal{S})_e = \{\nu_\rho, \ 0 \le \rho \le 1\}$, and the proof is illustrative of the way coupling techniques can be used to determine $\mathcal{J}_e$ completely in some cases. The first step in the proof is given by the following lemma.

Lemma 3.3.1. If $\nu \in \bar{\mathcal{J}} \cap \bar{\mathcal{S}}$, then

$$\nu\{(\eta,\zeta) : \eta \le \zeta \ \text{ or } \ \eta \ge \zeta\} = 1. \tag{3.3.2}$$

Proof. Let $f_u(\eta,\zeta) = |\eta(u) - \zeta(u)|$. Then $f_u \in \mathcal{D}(\bar{\Omega})$, so evaluating $\bar{\Omega} f_u$ by using the expression for $\bar{\Omega}$ in section 3.1 yields

$$\begin{aligned}
\bar{\Omega} f_u(\eta,\zeta) &= [2\zeta(u) - 1]\left\{\sum_y \eta(u)[1 - \eta(y)]p(u,y) - \sum_x \eta(x)[1 - \eta(u)]p(x,u)\right\} \\
&\quad + [2\eta(u) - 1]\left\{\sum_y \zeta(u)[1 - \zeta(y)]p(u,y) - \sum_x \zeta(x)[1 - \zeta(u)]p(x,u)\right\} \\
&\quad - 2\eta(u)\zeta(u)\sum_y p(u,y)[1-\eta(y)][1-\zeta(y)] - 2[1-\eta(u)][1-\zeta(u)]\sum_x p(x,u)\eta(x)\zeta(x) \\
&= \sum_y p(u,y)\left[1_{\{\eta(u)=\zeta(u)=1,\eta(y)\ne\zeta(y)\}} - 1_{\{\eta(y)=\zeta(y)=0,\eta(u)\ne\zeta(u)\}}\right] \\
&\quad + \sum_y p(y,u)\left[1_{\{\eta(u)=\zeta(u)=0,\eta(y)\ne\zeta(y)\}} - 1_{\{\eta(y)=\zeta(y)=1,\eta(u)\ne\zeta(u)\}}\right] \\
&\quad - \sum_y [p(u,y) + p(y,u)]\, 1_{\{\eta(u)=\zeta(y)\ne\eta(y)=\zeta(u)\}} .
\end{aligned}$$

Since $\nu \in \bar{\mathcal{J}}$, $\int \bar{\Omega} f_u(\eta,\zeta)d\nu = 0$, so

$$\begin{aligned}
0 &= \sum_y p(u,y)[\nu\{\eta(u) = \zeta(u) = 1, \eta(y) \ne \zeta(y)\} - \nu\{\eta(y) = \zeta(y) = 0, \eta(u) \ne \zeta(u)\}] \\
&\quad + \sum_y p(y,u)[\nu\{\eta(u) = \zeta(u) = 0, \eta(y) \ne \zeta(y)\} - \nu\{\eta(y) = \zeta(y) = 1, \eta(u) \ne \zeta(u)\}] \\
&\quad - \sum_y [p(u,y) + p(y,u)]\, \nu\{\eta(u) = \zeta(y) \ne \eta(y) = \zeta(u)\} .
\end{aligned} \tag{3.3.3}$$

Since $\nu \in \bar{\mathcal{S}}$, the first two terms above are zero, and hence

$$\sum_y [p(u,y) + p(y,u)]\, \nu\{\eta(u) = \zeta(y) \ne \eta(y) = \zeta(u)\} = 0 ,$$

so

$$(3.3.4) \qquad \nu\{\eta(u) = \zeta(y) \neq \eta(y) = \zeta(u)\} = 0$$

whenever $p(u,y) + p(y,u) > 0$. Using the irreducibility of $p(x,y)$ and the invariance of ν, it is not hard to deduce from this that (3.3.4) holds for all $u \neq y$, and therefore that (3.3.2) holds.

Corollary 3.3.5. (a) If $\nu \in (\bar{\mathcal{I}} \cap \bar{\mathcal{S}})_e$, then either $\nu\{\eta \leq \zeta\} = 1$ or $\nu\{\eta \geq \zeta\} = 1$. (b) If $\mu_1, \mu_2 \in (\mathcal{I} \cap \mathcal{S})_e$, then either $\mu_1 \leq \mu_2$ or $\mu_2 \leq \mu_1$.

Proof. The first statement follows from lemmas 3.1.1 and 3.3.1, while the second is a consequence of the first and (c) of lemma 3.1.3.

Theorem 3.3.6. $(\mathcal{I} \cap \mathcal{S})_e = \{\nu_\rho, \ 0 \leq \rho \leq 1\}$.

Proof. Since ν_ρ is translation invariant and ergodic, $\nu_\rho \in \mathcal{S}_e$. Also $\nu_\rho \in \mathcal{I}$, so that $\nu_\rho \in (\mathcal{I} \cap \mathcal{S})_e$. By corollary 3.3.5, if $\mu \in (\mathcal{I} \cap \mathcal{S})_e$ and $\rho \in [0,1]$, then either $\mu \leq \nu_\rho$ or $\nu_\rho \leq \mu$. Therefore if $\mu \in (\mathcal{I} \cap \mathcal{S})_e$, there is a $\rho_0 \in [0,1]$ so that $\mu \geq \nu_\rho$ for $\rho < \rho_0$ and $\mu \leq \nu_\rho$ for $\rho > \rho_0$. It then follows that $\int f\, d\mu \geq \int f\, d\nu_\rho$ for $\rho < \rho_0$ and $\int f\, d\mu \leq \int f\, d\nu_\rho$ for $\rho > \rho_0$ for any f of the form $f(\eta) = \prod_{x \in T} \eta(x)$ for finite $T \subset S$. Therefore $\int f\, d\mu = \int f\, d\nu_{\rho_0}$ for all such f, and hence $\mu = \nu_{\rho_0}$.

Using similar techniques, one can prove the following somewhat stronger result.

Theorem 3.3.7. If $\mu \in \mathcal{S}$, then all weak limits of $\mu S(t)$ as $t \to \infty$ are exchangeable.

This suggests the conjecture that if $\mu \in \mathcal{S}_e$, then $\mu S(t) \to \nu_\rho$ as $t \to \infty$, where $\rho = \mu\{\eta: \eta(x) = 1\}$. Of course in the symmetric case, this is just corollary 2.1.5. The conjecture would follow from theorem 3.3.7 if one could prove that all weak limits of $\mu S(t)$ are ergodic when $\mu \in \mathcal{S}_e$, but this appears to be difficult. Clearly a proof of the conjecture would lead to an identification of $\lim_{t\to\infty} \mu S(t)$ for all $\mu \in \mathcal{S}$.

A more careful use of (3.3.3) leads to a proof of the following result [38] in one dimension, which we conjecture to be true for $d > 1$.

Theorem 3.3.8. Suppose $d = 1$, $\sum_x |x| p(0,x) < \infty$, and $\sum_x x p(0,x) = 0$. Then $\mathcal{I}_e = \{\nu_\rho, \ 0 \leq \rho \leq 1\}$.

As will be seen in the next section, the mean zero assumption is needed for the result to hold. In order to give an indication of the relevance of the mean zero assumption, we will prove a lemma which is required in the proof of theorem 3.3.8. The conclusion of the lemma says that there is no net "flow of discrepancies" (i.e., of x's such that $\eta(x) \neq \zeta(x)$) across zero in equilibrium.

Lemma 3.3.9. Suppose ν is a probability measure on $X \times X$ with exchangeable marginals such that $\nu\{\eta \leq \zeta\} = 1$, and assume $p(x,y)$ satisfies the assumptions of theorem 3.3.8. Then

$$\sum_{x<0\leq y} p(x,y)\,\nu\{\eta(x) \neq \zeta(x), \eta(y) = \zeta(y) = 0\} + \sum_{x<0\leq y} p(y,x)\,\nu\{\eta(x) \neq \zeta(x), \eta(y) = \zeta(y) = 1\}$$

$$= \sum_{x<0\leq y} p(x,y)\nu\{\eta(y) \neq \zeta(y), \eta(x) = \zeta(x) = 1\} + \sum_{x<0\leq y} p(y,x)\nu\{\eta(y) \neq \zeta(y), \eta(x) = \zeta(x) = 0\}$$

Proof. By adding

$$\sum_{x<0\leq y} p(x,y)\nu\{\eta(x) = \zeta(x) = 1, \eta(y) = \zeta(y) = 0\} + \sum_{x<0\leq y} p(y,x)\nu\{\eta(x) = \zeta(x) = 0, \eta(y) = \zeta(y) = 1\}$$

to both sides of the equality, and using $\nu\{\eta \leq \zeta\} = 1$, we can rewrite the conclusion of the lemma in the following way:

$$\sum_{x<0\leq y} p(x,y)\,\nu\{\zeta(x) = 1, \zeta(y) = 0\} + \sum_{x<0\leq y} p(y,x)\,\nu\{\eta(x) = 0, \eta(y) = 1\}$$

$$= \sum_{x<0\leq y} p(x,y)\,\nu\{\eta(x) = 1, \eta(y) = 0\} + \sum_{x<0\leq y} p(y,x)\,\nu\{\zeta(x) = 0, \zeta(y) = 1\}.$$

But this is true since ν has exchangeable marginals and p has mean zero.

Section 3.4. The asymmetric simple random walk on Z^1. In this section, we will describe briefly and without proofs the known results in case $S = Z^1$, $p(x,x+1) = p$ and $p(x,x-1) = q$ for all $x \in S$, where $p+q = 1$. These illustrate some of the results which may be expected in the general translation invariant case, and therefore lead to various natural conjectures. We will assume $p \neq 1/2$, since otherwise $p(x,y)$ is symmetric and the results of chapter 2 apply. By reflection, we may assume without loss of generality that $p > 1/2$. The case $p = 1$ is included, even though the irreducibility assumption is not satisfied.

In order to describe $\mathcal{I}_e$, suppose first that $1/2 < p < 1$. Then the transition probabilities satisfy $\pi(x)p(x,y) = \pi(y)p(y,x)$ for $\pi(x) = (p/q)^x$, so that $\tilde{\nu}_\rho \in \mathcal{I}$ for each $\rho \in [0,\infty]$ by part (b) of corollary 1.2.6. Let $A_n = \{\eta \in X : \Sigma_{x<n}\, \eta(x) = \Sigma_{x\geq n}\, [1 - \eta(x)] < \infty\}$ and $A = \{\eta \in X : \Sigma_{x<0}\, \eta(x) < \infty$ and $\Sigma_{x>0}\, [1 - \eta(x)] < \infty\}$. Then A is countable, $A = \bigcup_{n=-\infty}^{\infty} A_n$, $\tilde{\nu}_\rho(A) = 1$ for all $\rho \in (0,\infty)$, and $\tilde{\nu}_\rho(A_n) > 0$ for all $\rho \in (0,\infty)$ and $-\infty < n < \infty$. Since $\{A_n\}$ are the closed irreducible classes for the Markov chain η_t on A, it then follows that η_t is positive recurrent on A and that its stationary distribution ν_n on

A_n is given by the conditional measure

(3.4.1) $$\nu_n(\cdot) = \tilde{\nu}_\rho(\cdot \mid A_n),$$

which is of course independent of ρ. For $p = 1$, let ν_n be the pointmass on the configuration η_n where $\eta_n(x) = 1$ for $x \geq n$ and $\eta_n(x) = 0$ for $x < n$. Then ν_n is clearly invariant for the process, since no motion takes place if the initial distribution is ν_n. By part (a) of corollary 1.2.6, $\nu_\rho \in \mathcal{I}$ for $0 \leq \rho \leq 1$.

The following theorem is proved in [38] by using coupling techniques similar to those used in section 3.3. The main added ingredient can be described in the following way. The arguments of section 3.3 are based on the idea that the function $f(\eta_t, \zeta_t) = \sum_{u=-n}^{n} |\eta_t(u) - \zeta_t(u)|$ is essentially monotone, in the sense that transitions within $[-n,n]$ can only decrease $f(\eta_t, \zeta_t)$. The only transitions which can increase $f(\eta_t, \zeta_t)$ are those which cross the boundaries $-n$ and n. In the case considered in this section, the function $g(\eta_t, \zeta_t)$ = number of sign changes of $\eta_t(u) - \zeta_t(u)$ in $[-n,n]$ is essentially monotone in the same sense. This monotonicity property of both functions is exploited in the proof.

<u>Theorem 3.4.2.</u> $\mathcal{I}_e = \{\nu_\rho, \ 0 \leq \rho \leq 1\} \cup \{\nu_n, \ -\infty < n < \infty\}$.

We conjecture that a similar result holds whenever $S = Z^1$, $p(x,y) = p(0,y-x)$, $\sum_x |x| p(0,x) < \infty$, and $\sum_x x p(0,x) > 0$. It would be necessary to prove first that the process with initial configuration η_n is a positive recurrent Markov chain, so that ν_n would be identified as the resulting limiting distribution. This is harder than it is in the case considered in this section, because ν_n will not have a representation in the form (3.4.1) relative to a product measure. Then a replacement for the monotonicity of the function $g(\eta_t, \zeta_t)$ would have to be found.

The problem of proving convergence has been resolved only for very simple initial distributions. Even these simple cases, however, illustrate the increased complexity of the process in the asymmetric case. The following theorem is proved in [37] by comparing the process with a related finite process.

<u>Theorem 3.4.3.</u> Suppose that μ is a product measure on X for which the following limits exist:

$$\lambda = \lim_{x \to -\infty} \mu\{\eta: \eta(x) = 1\}, \qquad \rho = \lim_{x \to +\infty} \mu\{\eta: \eta(x) = 1\}.$$

(a) If $\lambda \geq 1/2$ and $\rho \leq 1/2$, then $\lim_{t \to +\infty} \mu S(t) = \nu_{1/2}$.

(b) If $\rho \geq 1/2$ and $\lambda + \rho > 1$, then $\lim_{t \to \infty} \mu S(t) = \nu_\rho$.

(c) If $\lambda \leq 1/2$ and $\lambda + \rho < 1$, then $\lim_{t\to\infty} \mu S(t) = \nu_\lambda$.

(d) If $\lambda < 1/2$ and $\lambda + \rho = 1$, then $\lim_{t\to\infty} \mu S(t)$ may fail to exist.

For purposes of comparison, note that if $p = 1/2$, then theorem 2.1.3 gives $\lim_{t\to\infty} \mu S(t) = \nu_{((\lambda+\rho)/2)}$ under these assumptions for any choice of λ and ρ.

<u>Chapter 4.</u> <u>The exclusion process with speed change.</u> One of the key properties of the coupling used in the previous chapter was that $\{\eta \leq \zeta\}$ was closed for the motion of the coupled process. For the exclusion process with nontrivial speed change, the natural coupling (i.e., the one in which particles move together whenever they can) fails to have this property, and is therefore apparently not very useful. For Spitzer's exclusion process with speed change, however, there is another coupling which does have the desired property, provided that $c(x,\eta)$ and and $q(x,y)$ in (1.1.4) are related in an appropriate manner In this chapter, we will describe this coupling, and will illustrate its use by determining $\mathcal{I}_e$ completely for a class of one-dimensional processes.

<u>Section 4.1.</u> <u>The basic coupling.</u> Throughout this chapter, η_t will be the exclusion process with speed change with $c(x,y,\eta)$ given by (1.1.4), and $S(t)$ will be the corresponding semigroup. We will assume that $q(x,y)$ satisfies (1.1.5), and for technical reasons, that there is an $\varepsilon > 0$ so that for each $x,y \in S$, either $q(x,y) = 0$ or $q(x,y) \geq \varepsilon$. In order that a coupling with the desired properties exist, we must assume that $c(x,\eta)$ satisfies

$$(4.1.1) \qquad c(x,\eta) \leq c(x,\zeta) \text{ and } c(x,\eta) \sum_{\eta(z)=0} q(x,z) \geq c(x,\zeta) \sum_{\zeta(z)=0} q(x,z)$$

for $\eta \leq \zeta$. Note that this implies that $c(x,\eta)$ can only depend on coordinates $\eta(z)$ for which $q(x,z) > 0$, so condition (1.1.7) is a consequence of (1.1.6), which we assume. Conditions (4.1.1) are quite restrictive, of course, and the results obtained here should be regarded as only a first step toward understanding the general process with nontrivial speed change.

For $\eta,\zeta \in X$, let

$$A = A(\eta,\zeta) = \{x \in S : \eta(x) = \zeta(x) = 1\}$$

$$B = B(\eta,\zeta) = \{x \in S : \eta(x) = 1, \zeta(x) = 0\}$$

$$C = C(\eta,\zeta) = \{x \in S : \eta(x) = 0, \zeta(x) = 1\}$$

$$D = D(\eta,\zeta) = \{x \in S : \eta(x) = \zeta(x) = 0\}$$

$$G(x,\eta,\zeta) = [c(x,\eta) - c(x,\zeta)]^+ \left\{ \sum_{u\in B} q(x,u) \right\}^{-1}$$

$$H(x,\eta,\zeta) = [c(x,\zeta) - c(x,\eta)]^+ \left\{ \sum_{u\in C} q(x,u) \right\}^{-1} .$$

The generator $\bar{\Omega}$ of the coupled process then takes the following form on functions f which depend on finitely many coordinates: $\bar{\Omega}f(\eta,\zeta) =$

$$\sum_{x\in A, y\in D} [c(x,\eta) \wedge c(x,\zeta)] q(x,y)[f(\eta_{xy},\zeta_{xy}) - f(\eta,\zeta)]$$

$$+ \sum_{x\in A, y\in D, z\in B} G(x,\eta,\zeta) q(x,y) q(x,z)[f(\eta_{xy},\zeta_{xz}) - f(\eta,\zeta)]$$

$$+ \sum_{x\in A, y\in D, z\in C} H(x,\eta,\zeta) q(x,y) q(x,z)[f(\eta_{xz},\zeta_{xy}) - f(\eta,\zeta)]$$

$$+ \sum_{x\in A, y\in B} \left[c(x,\zeta) - G(x,\eta,\zeta) \sum_{u\in D} q(x,u)\right] q(x,y)[f(\eta,\zeta_{xy}) - f(\eta,\zeta)]$$

$$+ \sum_{x\in A, y\in C} \left[c(x,\eta) - H(x,\eta,\zeta) \sum_{u\in D} q(x,u)\right] q(x,y)[f(\eta_{xy},\zeta) - f(\eta,\zeta)]$$

$$+ \sum_{x\in B, y\in C\cup D} q(x,y) c(x,\eta)[f(\eta_{xy},\zeta) - f(\eta,\zeta)]$$

$$+ \sum_{x\in C, y\in B\cup D} q(x,y) c(x,\zeta)[f(\eta,\zeta_{xy}) - f(\eta,\zeta)]$$

The existence of the coupled process is guaranteed by theorem 2.8 of [32], and the corresponding semigroup will be denoted by $\bar{S}(t)$. The nonnegativity of the coefficients in the fourth and fifth terms follows from (4.1.1), as can be seen, for example, from the following computation for the fourth term:

$$c(x,\zeta) \sum_{u\in B\cup D} q(x,u) \geq c(x,\zeta \vee \eta) \sum_{u\in D} q(x,u) \geq c(x,\eta) \sum_{u\in D} q(x,u).$$

It is not hard to check that if $f(\eta,\zeta) = g(\eta)$, then $\bar{\Omega}f(\eta,\zeta) = \Omega g(\eta)$, while if $f(\eta,\zeta) = g(\zeta)$, then $\bar{\Omega}f(\eta,\zeta) = \Omega g(\zeta)$. Therefore the marginals of the coupled process are Markovian and have the transition law of η_t. Finally, note that the set $\{\eta \leq \zeta\}$ is closed for the motion of the coupled process. Therefore the statements of lemmas 3.1.1, 3.1.2, 3.1.3, and 3.1.4 carry over to the present context.

Section 4.2. The invariant measures. Now we specialize further to the case in which $S = Z^1$, $q(x,y)$ is symmetric, translation invariant and irreducible, and

$c(x,\eta)$ takes the form (1.3.1) where Φ is translation invariant. In this case, the canonical Gibbs states corresponding to Φ which were defined in section 1.3 are invariant for the process η_t, and it is known ([9], [43]) that $\tilde{\mathcal{G}}_e = \{\gamma_\rho, \ 0 \le \rho \le 1\}$ where γ_ρ is translation invariant and ergodic for each $\rho \in [0,1]$, γ_ρ is weakly continuous in ρ, and $\gamma_\rho\{\eta: \eta(x) = 1\} = \rho$. Our aim is to use the coupled process to prove that $\mathcal{J} = \tilde{\mathcal{G}}$. The first step is to prove that $\mathcal{J} \cap \mathcal{S} = \tilde{\mathcal{G}}$. This was proved in [20] under somewhat different assumptions via free energy techniques. Our proof is similar to that of theorem 3.3.6.

<u>Theorem 4.2.1</u>. Suppose $\nu \in \bar{\mathcal{S}}$, and put $\nu_t = \nu\bar{S}(t)$. Then

(4.2.2) $$\lim_{t\to\infty} \nu_t\{(\eta,\zeta) : \eta(x) = \zeta(y) \ne \eta(y) = \zeta(x)\} = 0$$

for all $x,y \in S$. Therefore if ν_∞ is any weak limit of ν_t as $t \to \infty$, $\nu_\infty\{(\eta,\zeta) : \eta \le \zeta \text{ or } \eta \ge \zeta\} = 1$.

<u>Proof</u>. Define $f_y(\eta,\zeta) = 1 - |\eta(y) - \zeta(y)|$. Then

(4.2.3) $$\bar{\Omega} f_y(\eta,\zeta) = \sum_x q(y,x)\{[c(x,\zeta)+c(y,\eta)]1_C(x)1_B(y)+[c(x,\eta)+c(y,\zeta)1_C(y)1_B(x)\}$$

$$+ \sum_x q(y,x)[\eta(x)+\eta(y) - 1]\{c(x,\zeta)\zeta(x)[1-\zeta(y)] - c(y,\zeta)\zeta(y)[1-\zeta(x)]\}$$

$$+ \sum_x q(y,x)[\zeta(x)+\zeta(y) - 1]\{c(x,\eta)\eta(x)[1-\eta(y)] - c(y,\zeta)\eta(y)[1-\eta(x)]\}$$

$$- \sum_x q(y,x)\{|c(x,\eta)-c(x,\zeta)|1_A(x)1_D(y) - |c(y,\eta)-c(y,\zeta)|1_A(y)1_D(x)\}.$$

Since f is in the domain of $\bar{\Omega}^n$ for all n,

(4.2.4) $$\frac{d^n}{dt^n} \int f \, d\nu_t = \int \bar{\Omega}^n f \, d\nu_t ,$$

and therefore each derivative of $\int f\,d\nu_t$ is uniformly bounded in t. Since $\nu_t \in \bar{\mathcal{S}}$ for all t, and the integral of each of the last three terms in (4.2.3) with respect to any measure in $\bar{\mathcal{S}}$ is zero, it follows from (4.2.4) with $n = 1$ that $\int f\,d\nu_t$ is nondecreasing in t. Therefore

(4.2.5) $$\lim_{t\to\infty} \int \bar{\Omega}^n f \, d\nu_t = 0$$

for all $n \ge 1$. Using (4.2.5) for $n = 1$ and (4.2.3), we see that (4.2.2) holds whenever $q(y,x) > 0$. Using (4.2.5) for successively larger values of n, and recalling that q is irreducible, it follows that (4.2.2) holds for all $x,y \in S$.

Corollary 4.2.6. Suppose $\mu \in \mathcal{S}$. Then any weak limit of $\mu S(t)$ as $t \to \infty$ is in $\tilde{\mathcal{G}}$. In particular, $\mathcal{I} \cap \mathcal{S} = \tilde{\mathcal{G}}$.

Proof. For any $\rho \in [0,1]$, let ν be the product measure $\mu \times \gamma_\rho$. By theorem 4.2.1, any weak limit of $\nu\bar{S}(t)$ as $t \to \infty$ concentrates on $\{(\eta,\zeta) : \eta \leq \zeta$ or $\eta \geq \zeta\}$. Therefore, if μ_∞ is any weak limit of $\mu S(t)$ as $t \to \infty$, there exists $\alpha(\rho),\beta(\rho) \geq 0$ and $\mu_1^\rho,\mu_2^\rho \in \mathcal{S}$ so that $\alpha(\rho) + \beta(\rho) = 1$, $\mu_\infty = \alpha(\rho)\mu_1^\rho + \beta(\rho)\mu_2^\rho$, and $\mu_1^\rho \leq \gamma_\rho \leq \mu_2^\rho$. Since $\mu_\infty \in \mathcal{S}$, $L = \lim_{n\to\infty} (1/n)\sum_{x=1}^n \eta(x)$ exists a.s. with respect to μ_∞ by the ergodic theorem. Since $L = \rho$ a.s. with respect to γ_ρ, it follows that $\mu_1^\rho(\cdot) = \mu_\infty(\cdot\,|L \leq \rho)$, $\mu_2(\cdot) = \mu_\infty(\cdot\,|L \geq \rho)$, and $\alpha(\rho) = \mu_\infty\{L \leq \rho\}$ for all ρ for which $\mu_\infty\{L = \rho\} = 0$. Therefore, since γ_ρ is ergodic,

$$\gamma_{\rho_1} \leq \mu_\infty(\cdot \mid \rho_1 < L < \rho_2) \leq \gamma_{\rho_2}$$

whenever $\rho_1 < \rho_2$ and $\mu_\infty\{\rho_1 < L < \rho_2\} > 0$. Since γ_ρ is continuous in ρ, it then follows that $\mu_\infty(\cdot\,|L) = \gamma_L$ a.s. with respect to μ_∞. Therefore, $\mu_\infty = \int \gamma_\rho \mu_\infty(L \in d\rho)$, so that $\mu_\infty \in \tilde{\mathcal{G}}$.

Lemma 4.2.7. Suppose ν is a probability measure on $X \times X$ such that both marginals of ν are in $\tilde{\mathcal{G}}$ and ν concentrates on $\{\eta \leq \zeta\}$ or on $\{\eta \geq \zeta\}$. Then for each $x,y \in S$,

$$\begin{aligned}
&\int [\eta(x)+\eta(y)-1]\{c(x,\zeta)\zeta(x)[1-\zeta(y)] - c(y,\zeta)\zeta(y)[1-\zeta(x)]\}d\nu \\
&+\int [\zeta(x)+\zeta(y)-1]\{c(x,\eta)\eta(x)[1-\eta(y)] - c(y,\eta)\eta(y)[1-\eta(x)]\}d\nu \\
&=\int \eta(x)\zeta(x)[1-\eta(y)][1-\zeta(y)]\,|c(x,\eta)-c(x,\zeta)|\,d\nu \\
&-\int \eta(y)\zeta(y)[1-\eta(x)][1-\zeta(x)]\,|c(y,\eta)-c(y,\zeta)|\,d\nu
\end{aligned}$$

Proof. We may assume that ν concentrates on $\{\eta \leq \zeta\}$. Let

$$F(\zeta) = c(x,\zeta)\zeta(x)[1-\zeta(y)] - c(y,\zeta)\zeta(y)[1-\zeta(x)] .$$

Since the marginals of ν are in $\tilde{\mathcal{G}}$, $\int F(\eta)\,d\nu = \int F(\zeta)\,d\nu = 0$. Therefore, since $\eta(x) = \eta(y) = 1$ implies $\zeta(x) = \zeta(y) = 1$ and hence $F(\zeta) = 0$, we have

$$\int [\eta(x) + \eta(y) - 1]F(\zeta)d\nu = -\int [1 - \eta(x)][1 - \eta(y)]F(\zeta)d\nu$$

$$= \int [1 - \eta(x)]\eta(y)F(\zeta)d\nu + \int [1 - \eta(y)]\eta(x)F(\zeta)d\nu$$

$$= -\int c(y,\zeta)\eta(y)\zeta(y)[1 - \eta(x)][1 - \zeta(x)]d\nu$$

$$+ \int c(x,\zeta)\eta(x)\zeta(x)[1 - \eta(y)][1 - \zeta(y)]d\nu .$$

Similarly,

$$\int [\zeta(x) + \zeta(y) - 1]F(\eta)d\nu = \int c(y,\eta)\eta(y)\zeta(y)[1 - \eta(x)][1 - \zeta(x)]d\nu$$

$$- \int c(x,\eta)\eta(x)\zeta(x)[1 - \eta(y)][1 - \zeta(y)]d\nu .$$

By assumption (4.1.1), $c(x,\eta) \leq c(x,\zeta)$ a.s. with respect to ν, so the result follows.

Theorem 4.2.8. If $\nu \in \bar{\mathcal{J}}$, then $\nu\{(\eta,\zeta) : \eta \leq \zeta \text{ or } \eta \geq \zeta\} = 1$.

Proof. Let $\alpha(x,y) = \int [c(x,\zeta) + c(y,\eta)]\zeta(x)\eta(y)[1 - \zeta(y)][1 - \eta(x)]d\nu$ and

$$\beta(x,y) = -\int [\eta(x) + \eta(y) - 1]c(x,\zeta)\zeta(x)[1 - \zeta(y)]d\nu$$

$$- \int [\zeta(x) + \zeta(y) - 1]c(x,\eta)\eta(x)[1 - \eta(y)]d\nu$$

$$+ \int |c(x,\eta) - c(x,\zeta)|\eta(x)\zeta(x)[1 - \eta(y)][1 - \zeta(y)]d\nu .$$

By (4.2.3) and $\int \bar{\Omega} f_y d\nu = 0$, we have for $n \geq 1$

$$(4.2.9) \quad \sum_{|y|\leq n} \sum_x q(y,x)[\alpha(x,y) + \alpha(y,x)] = \sum_{|y|\leq n} \sum_x q(y,x)[\beta(x,y) - \beta(y,x)]$$

$$= \sum_{|y|\leq n} \sum_x q(y,x)\beta(x,y) - \sum_{|x|\leq n} \sum_y q(y,x)\beta(x,y)$$

$$= \sum_{|y|\leq n} \sum_{|x|>n} q(y,x)\beta(x,y) - \sum_{|x|\leq n} \sum_{|y|>n} q(y,x)\beta(x,y)$$

$$= \sum_{|y|\leq n} \sum_{|x|>n} q(y,x)[\beta(x,y) - \beta(y,x)] .$$

Any weak limit $\bar{\nu}$ of Cesaro averages of successive translates of ν is in $\bar{\mathcal{J}} \cap \bar{\mathcal{S}}$. Therefore $\bar{\nu}$ concentrates on $\{\eta \leq \zeta$ or $\eta \geq \zeta\}$ by theorem 4.2.1, and so is a convex combination of two measures, each of which is in $\bar{\mathcal{J}} \cap \bar{\mathcal{S}}$, one of which concentrates on $\{\eta \leq \zeta\}$ and the other of which concentrates on $\{\eta \geq \zeta\}$. Therefore by corollary 4.2.6 and lemma 4.2.7,

$$\lim_{|n| \to \infty} \frac{1}{n} \sum_{k=0}^{n} [\beta(x+n, y+n) - \beta(y+n, x+n)] = 0 \ .$$

By (4.2.9), it then follows that $\Sigma_{x,y}\, q(x,y)[\alpha(x,y) + \alpha(y,x)] = 0$. Since $c(x,\eta)$ is bounded away from zero, this gives

$$(4.2.10) \qquad \nu\{(\eta,\zeta) : \eta(x) = \zeta(y) \neq \eta(y) = \zeta(x)\} = 0$$

for all x,y such that $q(x,y) > 0$. Using the irreducibility of q and the fact that $\nu \in \bar{\mathcal{J}}$ again, it then follows that (4.2.10) holds for all $x,y \in S$.

Corollary 4.2.11. $\mathcal{J} = \tilde{\mathcal{G}}$.

Proof. Suppose $\mu \in \mathcal{J}_e$ and $\rho \in [0,1]$. Then by theorem 4.2.8 and lemma 3.1.2 applied to this context, there exists $\alpha,\beta \geq 0$ with $\alpha + \beta = 1$ and $\mu_1, \mu_2 \in \mathcal{J}$ so that $\mu_1 \leq \mu \leq \mu_2$ and $\gamma_\rho = \alpha\mu_1 + \beta\mu_2$. Suppose $\alpha,\beta > 0$. Then μ_1 and μ_2 are absolutely continuous with respect to γ_ρ, and therefore

$$\mu_i\left\{\eta : \lim_{n\to\infty} \frac{1}{n} \sum_{x=1}^{n} \eta(x) = \rho\right\} = 1$$

for each i by the ergodic theorem applied to γ_ρ. Since $\mu_1 \leq \mu \leq \mu_2$, it follows that

$$\mu\left\{\eta : \lim_{n\to\infty} \frac{1}{n} \sum_{x=1}^{n} \eta(x) = \rho\right\} = 1 \ .$$

For a given $\mu \in \mathcal{J}_e$, this can be the case for at most one $\rho \in [0,1]$, so α and β are both positive for at most one $\rho \in [0,1]$. We conclude then that there is a $\rho_0 \in [0,1]$ so that $\mu \leq \nu_\rho$ for $\rho > \rho_0$ and $\mu \geq \nu_\rho$ for $\rho < \rho_0$. Since γ_ρ is weakly continuous in ρ, it follows that $\mu = \gamma_{\rho_0}$. Therefore $\mathcal{J}_e \subset \tilde{\mathcal{G}}$, and the desired result follows from $\tilde{\mathcal{G}} \subset \mathcal{J}$

BIBLIOGRAPHIE

1. A. B. Bortz, M. H. Kalos, J. L. Lebowitz, and J. Marro (1975). Time evolution of a quenched binary alloy II. Computer simulation of a three dimensional model system, Physical Review B, 12, 2000-2011.

2. A. B. Bortz, M. H. Kalos, J. L. Lebowitz, and M. A. Zendejas (1974). Time evolution of a quenched binary alloy. Computer simulation of a two dimensional model system, Physical Review B, 10, 535-541.

3. J. Chover (1975). Convergence of a local lattice process, Stochastic Processes and their Applications, 3, 115-135.

4. D. Dawson (1974). Information flow in discrete Markov systems, Journal of Applied Probability, 11, 594-600.

5. D. Dawson (1975). Synchronous and asynchronous reversible Markov systems, Canadian Mathematical Bulletin, 17, 633-649.

6. R. L. Dobrushin (1971). Markov processes with a large number of locally interacting components, Problems of Information Transmission, 7, 149-164, and 235-241.

7. R. Glauber (1963). The statistics of the stochastic Ising model, Journal of Mathematical Physics, 4, 294-307.

8. H. O. Georgii (1975). Canonical Gibbs states, their relation to Gibbs states, and applications to two-valued Markov chains, Z. Wahrscheinlichkeitstheorie verw. Geb., 32, 277-300.

9. H. O. Georgii (1976). On canonical Gibbs states and symmetric and tail events, Z. Wahrscheinlichkeitstheorie verw. Geb., 33, 331-341.

10. L. Gray and D. Griffeath (1976). On the uniqueness of certain interacting particle systems, Z. Wahrscheinlichkeitstheorie verw. Geb., 35, 75-86.

11. D. Griffeath (1975). Ergodic theorems for graph interactions, Advances in Applied Probability, 7, 179-194.

12. T. E. Harris (1972). Nearest-neighbor Markov interaction processes on multidimensional lattices, Advances in Mathematics, 9, 66-89.

13. T. E. Harris (1974). Contact interactions on a lattice, Annals of Probability, 2, 969-988.

14. T. E. Harris (1976). On a class of set-valued Markov Processes, Annals of Probability, 4, 175-194.

15. L. L. Helms (1974). Ergodic properties of several interacting Poisson particles, Advances in Mathematics, 12, 32-57.

16. Y. Higuchi ( ). A study on the canonical Gibbs states.

17. Y. Higuchi and T. Shiga (1975). Some results on Markov processes of infinite lattice spin systems, Journal of Mathematics of Kyoto University, 15, 211-229.

18. R. Holley (1970). A class of interactions in an infinite particle system, Advances in Mathematics, 5, 291-309.

19. R. Holley (1971). Free energy in a Markovian model of a lattice spin system, Communications in Mathematical Physics, 23, 87-99.

20. R. Holley (1971). Pressure and Helmholtz free energy in a dynamic model of a lattice gas, Proceedings of the Sixth Berkeley Symposium on Mathematical Statistics and Probability, III, 565-578.

21. R. Holley (1972). An ergodic theorem for interacting systems with attractive interactions, Z. Wahrscheinlichkeitstheorie verw. Geb., 24, 325-334.

22. R. Holley (1972). Markovian interaction processes with finite range interactions, The Annals of Mathematical Statistics, 43, 1961-1967.

23. R. Holley (1974). Recent results on the stochastic Ising model, Rocky Mountain Journal of Mathematics, 4, 479-496.

24. R. Holley and T. M. Liggett (1975). Ergodic theorems for weakly interacting infinite systems and the voter model, The Annals of Probability, 3, 643-663.

25. R. Holley and D. Stroock (1976). A martingale approach to infinite systems of interacting processes, The Annals of Probability, 4, 195-228.

26. R. Holley and D. Stroock (1976). Applications of the stochastic Ising model to the Gibbs states, Communications in Mathematical Physics, 48, 249-265.

27. R. Holley and D. Stroock ( ). Dual processes and their application to infinite interacting systems, Advances in Mathematics.

28. R. Holley and D. Stroock (1976). L_2 theory for the stochastic Ising model, Z. Wahrscheinlichkeitstheorie verw. Geb., 35, 87-101.

29. J. Kemeny, J. Snell, and A. Knapp (1966). Denumerable Markov Chains, Van Nostrand, Princeton, N. J.

30. R. Kinderman and J. Snell (1976). Markov random fields.

31. T. Kurtz (1969). Extensions of Trotter's operator semigroup approximation theorems, Journal of Functional Analysis, 3, 354-375.

32. T. M. Liggett (1972). Existence theorems for infinite particle systems, Transactions of the A.M.S., 165,471-481.

33. T. M. Liggett (1973). An infinite particle system with zero range interactions, The Annals of Probability, 1, 240-253.

34. T. M. Liggett (1973). A characterization of the invariant measures for an infinite particle system with interactions, Transactions of the A.M.S., 179, 433-453.

35. T. M. Liggett (1974). A characterization of the invariant measures for an infinite particle system with interactions II, Transactions of the A.M.S., 198,201-213.

36. T. M. Liggett (1974). Convergence to total occupancy in an infinite particle system with interactions, The Annals of Probability, 2, 989-998.

37. T. M. Liggett (1975). Ergodic theorems for the asymmetric simple exclusion process, Transactions of the A.M.S., 213, 237-261.

38. T. M. Liggett (1976). Coupling the simple exclusion process, The Annals of Probability, 4, 339-356.

39. K. Logan (1974). Time reversible evolutions in statistical mechanics, Cornell University, Ph.D. dissertation.

40. N. Matloff ( ). Ergodicity conditions for a dissonant voting model, The Annals of Probability.

41. C. J. Preston (1974). Gibbs states on countable sets, Cambridge University Press.

42. D. Ruelle (1969). Statistical mechanics, W. A. Benjamin.

43. D. Ruelle (1968). Statistical mechanics of a one-dimensional lattice gas, Communications in Mathematical Physics, 9, 267-278.

44. D. Schwartz (1976). Ergodic theorems for an infinite particle system with births and deaths, Annals of Probability, 4

45. F. Spitzer (1970). Interaction of Markov processes, Advances in Mathematics, 5, 246-290.

46. F. Spitzer (1971). Random fields and interacting particle systems, M. A. A. Summer Seminar Notes, Williamstown, Mass.

47. F. Spitzer (1974). Introduction aux processus de Markov a parametres dans Z_ν, Lecture Notes in Mathematics 390, Springer-Verlag.

48. F. Spitzer (1974). Recurrent random walk of an infinite particle system, Transactions of the A.M.S., 198, 191-199.

49. F. Spitzer (1975). Random time evolution of infinite particle systems, Advances in Mathematics, 16, 139-143.

50. O. N. Stavskaya and I. I. Pyatetskii-Shapiro (1971). On homogeneous nets of spontaneously active elements, Systems Theory Res., 20, 75-88.

51. W. G. Sullivan (1973). Potentials for almost Markovian random fields, Communications in Mathematical Physics, 33, 61-74.

52. W. G. Sullivan (1974). A unified existence and ergodic theorem for Markov evolution of random fields, Z. Wahrscheinlichkeitstheorie verw. Geb., 31, 47-56.

53. W. G. Sullivan (1976). Processes with infinitely many jumping particles, Proceedings of the A.M.S., 54, 326-330.

54. W. G. Sullivan (1975). Mean square relaxation times for evolution of random fields, Communications in Mathematical Physics, 40, 249-258.

55. W. G. Sullivan (1975). Markov processes for random fields, Communications of the Dublin Institute for Advanced Studies, series A, number 23.

56. W. G. Sullivan ( ). Specific information gain for interacting Markov processes.

57. L. N. Vasershtein (1969). Processes over denumerable products of spaces, describing large systems, Problems of Information Transmission, 3, 47-52.

58. D. Schwartz ( ). Applications of duality to a class of Markov processes,

59. D. Griffeath ( ). An ergodic theorem for a class of spin systems.

60. R. Lang ( ). Unendlich-dimensionale Wienerprozesse mit wechselwirkung I: Existenz.

61. L. N. Vasershtein and A. M. Leontovich (1970). Invariant measures of certain Markov operators describing a homogeneous random medium, Problems of Information Transmission, 6, 61-69.

62. C. Cocozza and C. Kipnis ( ). Existence de processus Markoviens pour des systems infinis de particules.

63. L. Gray and D. Griffeath ( ). On the nonuniqueness of proximity processes.

64. R. Holley and D. Stroock ( ). Nearest neighbor birth and death processes on the real line.

TOPICS IN PROPAGATION OF CHAOS

Alain-Sol SZNITMAN

The preparation of these notes was supported in part by NSF Grant No. DMS-8903858, and by an Alfred P. Sloan Foundation research fellowship.

Originally published in: *Ecole d'Eté de Probabilités de Saint-Flour XIX – 1989*, Lecture Notes in Mathematics, Vol. **1464**, 165–251, DOI: 10.1007/BFb0085169, © Springer-Verlag Berlin Heidelberg 1991, Reprint by Springer-Verlag Berlin Heidelberg 2012

TABLE DES MATIERES

A. S. SZNITMAN : "TOPICS IN PROPAGATION OF CHAOS"

0. INTRODUCTION

The terminology propagation of chaos comes from Kac. The initial motivation for the subject was to try to investigate the connection between a detailed and a reduced description of particles' evolution. For instance on the one hand one has Liouville's equations:

$$\partial_t u + \sum_1^N v_i \partial_{x_i} u + \sum_{j \neq i} -\nabla V_N(x_i - x_j)\partial_{v_j} u = 0 \,, \tag{0.1}$$

where $u(t, x_1, v_1, \ldots, x_n, v_N)$ is the density of presence at time t, assumed to be symmetric of N particles, with position x_i and velocity v_i, and pairwise potential interaction $V_N(\cdot)$.

On the other hand one also has for instance the Boltzmann equation for a dilute gas of hard spheres

$$\partial_t u + v.\nabla_x u = \int_{R^3 \times S_2} (u(x,\tilde{v})u(x,\tilde{v}') - u(x,v)u(x,v'))\,|(v'-v)\cdot n|\; dv'dn \tag{0.2}$$

where $\tilde{v}, \tilde{v}'$ are obtained from v, v' by exchanging the respective components of v and v' in direction n, that is:

$$\begin{aligned} \tilde{v} &= v + (v'-v)\cdot n\; n \\ \tilde{v}' &= v' + (v - v')\cdot n\; n \,, \end{aligned}$$

and $u(t, x, v)$ is the density of presence at time t of particles at location x with velocity v.

One question was to understand the nature of the connection between (0.1) and (0.2). There are now works directly attacking the problem (see Lanford [22], and more recently Uchiyama [54]). However at the time, Kac proposed to get insight into the problem, by studying simpler Markovian models of particles. The forward equations for the Markovian evolutions of N-particle systems come as a substitute for the Liouville equations. They are called master equations. Kac [15] traces the origin of these masters equations back to the forties with works of Norsdieck-Lamb-Uhlenbeck on a problem of cosmic rays, and of Siegert.

In the case of the Boltzmann equation, one can for instance forget about positions of particles and take as master equations:

(0.3)

$$\begin{aligned} &\partial_t u_t^N(v_1, \ldots, v_N) \\ &= \frac{1}{(N-1)} \sum_{1 \le i < j \le n} \int_{S_2} (u_t^N(v_1,..,\tilde{v}_i,..,\tilde{v}_j,..,v_N) - u_t^N(v_1,..,v_N))|(v_i - v_j)\cdot n|)dn \end{aligned}$$

One now tries to connect these master equations with the spatially homogeneous Boltzmann equations for hard spheres:

$$(0.4)\qquad \partial_t u = \int_{R^3\times S_2} (u(\tilde{v},t)u(\tilde{v}',t) - u(v,t)u(v',t))|(v'-v)\cdot n|dv'dn \ .$$

The idea proposed by Kac, which motivates the terminology "propagation of chaos" is the following. If one picks a "chaotic" initial distribution of particles: $u_N(0, v_1, \dots, v_N) = u_0(v_1)\dots u_0(v_N)$, for fixed N the evolution due to the master equations will in general destroy the independence property of $v_1, \dots, v_N$ at time t. However if one focuses on the reduced distribution at time t of the first k components, it should approximately be given when N is large by $u_t(v_1)\dots u_t(v_k)$, if $u_t(v)$ is the solution of equation (0.3) with initial condition $u_0(\cdot)$. So in this sense independence (or chaos) still propagates and eequation (0.4) emerges.

"Propagation of chaos" deals with symmetric evolution of particles, and this is not an innocent assumption. Among other things it tells us that the probability distribution of the first k particles $u^N(dv_1, \dots, dv_k)$ is the normalized k-particle correlation measure, that is the intensity of the random measure $\frac{1}{N(N-1)\cdots(N-k-1)} \sum_{i_\ell \text{distinct}} \delta_{(v_{i_1},\dots,v_{i_k})}$. The consequence is that the study of one individual gives information on the behavior of the group. We have so far presented the N-particle system and the nonlinear equation. There is a third actor in the play the "nonlinear process". It describes the limit behavior of the trajectory of one individual. It is sometimes called the tagged particle process, however we refrained from using this terminology for it can have different meanings, (see for instance the end of Chapter I). The time marginals of the nonlinear process will evolve according to the nonlinear equation under study, and this motivates the name of "nonlinear process", although of course in some examples, where interactions vanish all can be very linear in fact.

We will present in these notes a selection of topics on "propagation of chaos" which by no means covers the huge literature on the subject.

Let us now close the introduction with an informal discussion, along the lines of the "BBGKY-hierarchical method" for a model of interacting diffusions due to McKean [27]. This will motivate why propagation of chaos should hold in this case.

One looks at N particles on R^d, with initial "chaotic" distribution $u_0^{\otimes N}$, satisfying the S.D.E.:

$$(0.5)\qquad dx_t^i = dB_t^i + \frac{1}{N}\sum_1^N b(x_t^i, x_t^j)\ dt\ , \quad 1 \le i \le N\ ,$$

where $B^i_.$ are independent Brownian motions, and $b(\cdot,\cdot)$ is for instance smooth compactly supported. We are now going to explain in an informal way how the nonlinear equation

$$\partial_t u = \frac{1}{2}\Delta u - \operatorname{div}(\int b(\cdot,y)\, u(t,y)\, dy\, u) \tag{0.6}$$
$$u_{t=0} = u_0\,,$$

arises in the propagation of chaos effect.

Let us retinterpret equation (0.6). If P_t^0 denotes the Brownian transition density, we have the perturbation formula (obtained for instance by differentiating in s, $u_s P_{t-s}^0$):

$$u_t(x) - u_0\ P_t^0(x) = \int_0^t ds_1 \int dx_1 dx_2 u_{s_1}(x_1) u_{s_1}(x_2)\ b(x_1,x_2)\ \nabla_{x_1} P_{t-s_1}^0(x_1,x)\,.$$

Continuing the development of $u_{s_1}(x_1)u_{s_1}(x_2),\ldots$, by induction we find:

(0.7)

$$u_t = u_0 P_t^0 + \sum_{k=1}^m \int_{0<s_k<\cdots<s_1<t} ds_k..ds_1\ u_0^{\otimes k+1} P_{s_k}^0 B\cdot\nabla P_{s_{k-1}-s_k}^0 .. B\cdot\nabla P_{t-s_1}^0 + R_m\,,$$

$$\text{where } R_m = \int_{0<s_{m+1}<s_m<\cdots<s_1<t} ds_{m+1}..ds_1\ u_{s_{m+1}}^{\otimes m+2}\ B\cdot\nabla P_{s_m - s_{m+1}}^0 .. \nabla P_{t-s_1}^0\,.$$

Here we have adopted convenient notations, and P_t^0 acts naturally in a tensorial way on functions of an arbitrary number of variables, and $B\cdot\nabla$ maps functions of k variables into functions of $(k+1)$ variables, $k\geq 1$, by the formula:

$$[B\cdot\nabla]f(x_1,\ldots,x_{k+1}) = \sum_1^k b(x_i,x_{i+1})\ \nabla_i f(x_1,\ldots,x_k)\,. \tag{0.8}$$

If we use a similar perturbation method for the time marginals $u_{N,t}(x_1,\ldots,x_N)$ of the N-particle system, using the forward equation corresponding to (0.5), we now find:

(0.9)

$$u_{N,t} = u_0^{\otimes N} P_t^0 + \sum_{k=1}^m \int_{0<s_k<\cdots<s_1<t} u_0^{\otimes N}\ P_{s_k}^0\ B^N\ \nabla P_{s_k - s_{k-1}}^0 \cdots B^N \nabla P_{t-s_1}^0 + R_m^N\,,$$

$$\text{where } R_m^N = \int_{0<s_{m+1}<\cdots<s_1<t} u_{s_{m+1}}^N B^N \nabla P_{s_m - s_{m+1}}^0 \cdots B^N \nabla P_{t-s}^0\,, \text{ and}$$

$$[B^N\cdot\nabla]f(x_1,\ldots,x_N) = \frac{1}{N}\sum_{i,j=1}^N b(x_i,x_j)\ \nabla_i f(x_1,\ldots,x_N)\,. \tag{0.10}$$

Let us now see what happens as N goes to infinity when we consider $\langle u_{N,t}, f\rangle$ for a function $f(x_1)$ depending only on the first variable.

In (0.9), the first term

$$a_0^N = u_0^{\otimes N} P_t^0 f = u_0 P_t^0 f \ ,$$

coincides with the first term a_0 of (0.7).

The second term of (0.9) is

$$a_1^N = \int_0^t ds \ u_0^{\otimes N} \ P_{s_1}^0 B^N \cdot \nabla P_{t-s_1}^0 \ f \ ds_1 \ .$$

Since $P_{t-s_1}^0 f$ just depends on the x_1 variable

$$a_1^N = \int_0^t ds_1 \ u_0^{\otimes N} P_{s_1}^0 \left(\frac{1}{N} \sum b(x_1, x_j) \ \nabla_1 (P_{t-s_1}^0 f)(x_1) \right) ds_1$$

But $u_0^{\otimes N} P_{s_1}^0$ is symmetric. For $j \neq 1$, one can pick a permutation of $[1, N]$, leaving 1 invariant and mapping j on 2, so

$$a_1^N = \frac{N-1}{N} \int_0^t u_0^{\otimes N} P_{s_1}^0 \{ b(x_1, x_2) (\nabla_1 P_{t-s_1}^0 f)(x_1) \} \ ds_1 + o(N) \ ,$$

and in fact $u_0^{\otimes N}$ can be replaced by $u_0 \otimes u_0$ in the last expression so that a_1^N converges to the first term a_1 of (0.7).

The third term a_2^N of (0.9) is

$$a_2^N = \int_{0<s_2<s_1<t} ds_2 \ ds_1 \ u_0^{\otimes N} P_{s_2}^0 B^N \ \nabla P_{s_1-s_2}^0 \ B^N \nabla P_{t-s_1}^0 f \ ds_1 ds_2$$

Denote by ϕ the symmetric distribution $u_0^{\otimes N} P_{s_2}^0$ and by ψ the expression $u_0^{\otimes N} P_{s_2}^0 B^N \cdot \nabla P_{s_1-s_2}^0$. To obtain ψ one lets $P_{s_1-s_2}^0$ act on the right on:

$$- \sum_i \operatorname{div}_i (\phi(x_1, \ldots, x_N) \frac{1}{N} \sum_{j=1}^N b(x_i, x_j)) \ ,$$

which is also symmetric. Applying the same trick as before, we find:

$$\begin{aligned} a_2^N &= \frac{N-1}{N} \int_{0<s_2<s_1<t} ds_2 ds_1 \langle \psi \ , \ b(x_1, x_2) (\nabla_1 P_{t-s_1}^0 f)(x_1) \rangle + o(N) \\ &= \frac{(N-1)(N-2)}{N^2} \int_{0<s_2<s_1<t} ds_2 ds_1 \langle u_0^{\otimes N} P_{s_2}^0, (b(x_1, x_3) \nabla_1 + b(x_2, x_3) \cdot \nabla_2) \\ &\quad \cdot P_{s_1-s_2}^0 (b(x_1, x_2) \nabla_1 (P_{t-s}^0 f)(x_1) \rangle \\ &\quad + o(N) \ , \end{aligned}$$

and one sees that a_2^N converges to a_2 the third term in (0.7).

In the same way, one sees that there is a term by term convergence of a_k^N to a_k for each k. However, we cannot transform this into a bona fide propagation of chaos result, since we do not have a good control on our series.

Indeed, we have an estimate of the type

$$\|b(x_i, x_j)\nabla_i P_r^0\|_{L^\infty \to L^\infty} \leq \frac{c}{\sqrt{r}} \,,$$

and the generic term a_k in (0.7) for instance is naturally estimated in terms of

$$k! c^k \int_{0<s_2<\cdots<s_1<t} ds_k \ldots ds_1 [(t-s_1)\ldots(s_{k-1}-s_k)]^{-1/2} \,,$$

but this quantity tends to infinity.

Of course this convergence problem comes from the fact that it is unreasonable to use a perturbation series of P_t^0 to take care of the drift term $b(\cdot,\cdot)\cdot\nabla$. Nevertheless this gives a flavor of the propagation of chaos result. In the next section, we will provide a proof by a probabilistic approach.

I. Generalities and first examples.

1) A laboratory example.

We now come back to McKean's example of interacting diffusions, and we are going to use a probabilistic method to attack the problem.

We suppose $b(\cdot,\cdot)$ bounded Lipschitz $R^d \times R^d \to R^d$, and construct on $(R^d \times C_0(R_+, R^d))^{N^*}$, with product measure $(u_0 \otimes W)^{\otimes N^*}$, ($u_0$ probability on R^d, W standard R^d-Wiener measure) the $X^{i,N}_{\cdot}$, $i = 1, \dots, N$, satisfying

$$
(1.1) \qquad \begin{aligned}
dX_t^{i,N} &= dw_t^i + \frac{1}{N}\sum_{j=1}^{N} b(X_t^{i,N}, X_t^{j,N})dt\ , \quad i = 1,\dots,N \\
X_0^{i,N} &= x_0^i\ ,
\end{aligned}
$$

here x_0^i, $(w_{\cdot}^i)$, $i \geq 1$, are the canonical coordinates on the product space $(R^d \times C_0)^{N^*}$.

We are going to show that when N goes to infinity each $X^{i,N}_{\cdot}$, has a natural limit $\overline{X}^i$. Each $\overline{X}^i$ will be an independent copy of a new object: "the nonlinear process".

Nonlinear process:

On a filtered probability space $(\Omega, F, F_t, (B_t)_{t\geq 0}, X_0, P)$, endowed with an R^d-valued Brownian motion $(B_t)_{t\geq 0}$, and an u_0-distributed, F_0 measurable R^d-valued random variable X_0, we look at the equation:

$$
(1.2) \qquad \begin{aligned}
dX_t &= dB_t + \int b(X_t, y)\ u_t(dy)\ dt \\
X_{t=0} &= X_0\ , \quad u_t(dy) \text{ is the law of } X_t
\end{aligned}
$$

Theorem 1.1. *There is existence and uniqueness, trajectorial and in law for the solutions of* (1.2).

Remark 1.2. Let us notice that the nonlinear process has time marginals which satisfy in a weak sense the nonlinear equation

$$
(1.3) \qquad \partial_t u = \frac{1}{2}\Delta u - \operatorname{div}\Big(\int b(\cdot, y)\ u_t(dy) u\Big)\ .
$$

Indeed, for $f \in C_b^2(R^d)$, applying Ito's formula:

$$
f(X_t) = f(X_0) + \int_0^t f'(X_s)dB_s + \int_0^t \Big(\frac{1}{2}\Delta f + \int_{R^d} b(X_s, y)\ u_s(dy)\ \nabla f(X_s)\Big)\ ds\ ,
$$

after integration this yields a weak version of (1.3).

□

Proof: Let us now turn to the proof of Theorem 1.1. We introduce the Kantorovitch-Rubinstein or Vaserstein metric on the set $M(C)$ of probability measures on $C = C([0,T], R^d)$, defined by

$$D_T(m_1, m_2) = \inf\{ \int (\sup_{s \le T} |X_s(\omega_1) - X_s(\omega_2)| \wedge 1)\, dm(\omega_1, \omega_2)\ , \quad m \in M(C \times C)\ , \quad p_1 \circ m = m_1\ , \quad p_2 \circ m = m_2\} . \tag{1.4}$$

Here X_s is simply the canonical process on C.

Formula (1.4) defines a complete metric on $M(C)$, which gives to $M(C)$ the topology of weak convergence. The proof of this fact can be found in Dobrushin [8].

Take now $T > 0$, and define Φ the map which associates to $m \in M(C([0,T], R^d))$ the law of the solution of

$$X_t = X_0 + B_t + \int_0^t \Big(\int_C b(X_s, w_s)\, dm(w) \Big)\, ds\ , \quad t \le T . \tag{1.5}$$

Observe that this law does not depend on the specific choice of the space Ω, we use.

Next observe that if $X_t, t \le T$, is a solution of (1.2), then its law on $C([0,T], R^d)$, is a fixed point of Φ, and conversely if m is such a fixed point of Φ, (1.5) defines a solution of (1.2), up to time T. So now our problem is translated into a fixed point problem for Φ, and one has the contraction lemma:

Lemma 1.3. *For $t \le T$,*

$$D_t(\Phi(m_1), \Phi(m_2)) \le c_T \int_0^t D_u(m_1, m_2)\, du\ , \quad m_1, m_2 \in M(C_T)\ ,$$

c_T is a constant and, $D_u(m_1, m_2)$ ($\le D_T(m_1, m_2)$) is the distance between the images of m_1, and m_2 on $C([0,u], R^d)$.

Proof:

$$X_t^1 = X_0 + B_t + \int_0^t \Big(\int b(X_s^1, w_s) dm_1(w) \Big) ds\ , \quad t \le T\ ,$$

$$X_t^2 = X_0 + B_t + \int_0^t \Big(\int b(X_s^2, w_s) dm_2(w) \Big) ds\ , \quad t \le T\ .$$

So we find:

$$\sup_{s \le t} |X_s^1 - X_s^2| \le \int_0^t ds \Big| \int b(X_s^1(\omega), w_s) dm_1(w) - \int b(X_s^2(\omega), w_s) dm_2(w) \Big| .$$

But

$$\Big| \int b(x, w_s) dm_1(w) - \int b(y, w_s) dm_2(w) \Big| \le K[|x-y| \wedge 1 + \int |w_s^1 - w_s^2| \wedge 1\, dm(w^1, w^2)]$$

where m is any coupling of m_1, m_2 on $C([0,s], \mathbb{R}^d)$. From this

$$\sup_{s \le t} |X_s^1 - X_s^2| \le K \int_0^t ds |X_s^1(\omega) - X_s^2(\omega)| \wedge 1 + K \int_0^t D_s(m_1, m_2) ds .$$

Using Gronwall's lemma:

$$\sup_{s \le t} |X_s^1 - X_s^2| \wedge 1 \le K\, e^{KT} \int_0^t D_s(m_1, m_2) ds ,$$

from which the lemma follows.

□

From the lemma, we can immediately deduce weak and strong uniqueness for the solutions of (1.2). The existence part also follows now from a standard contraction argument.

Namely for $T > 0$, and $m \in M(C_T)$, iterating the lemma:

$$D_T(\Phi^{k+1}(m), \Phi^k(m)) \le c_T^k \frac{T^k}{k!} D_T(\Phi(m), m) . \tag{1.6}$$

So $\Phi^k(m)$ is a Cauchy sequence, and converges to a fixed point of $\Phi : P_T$. Now if $T' < T$, the image of P_T on $C([0,T'], \mathbb{R}^d)$ is still a fixed point, so the P_T are a consistent family, yielding a P on $C([0,\infty), \mathbb{R}^d)$. This provides the required solution.

□

Using Theorem 1.1, we now introduce on $(\mathbb{R}^d \times C_0)^{\mathbb{N}^*}$, where we have constructed in (1.1) our interacting diffusions $X_\cdot^{i,N}$, $i = 1, \dots, N$, the processes $\overline{X}_\cdot^i$, $i \ge 1$, solution of:

$$\begin{aligned} &\overline{X}_t^i = x_0^i + w_t^i + \int_0^t \int b(\overline{X}_s^i, y)\, u_s(dy)\, ds , \\ &u_s(dy) = \text{law}(\overline{X}_s^i) . \end{aligned} \tag{1.7}$$

Theorem 1.4. *For any $i \ge 1$, $T > 0$:*

$$\sup_N \sqrt{N} E[\sup_{t \le T} |X_t^{i,N} - \overline{X}_t^i|] < \infty . \tag{1.8}$$

Proof: Dropping for notational simplicity the superscript N, we have:

$$
\begin{aligned}
X_t^i - \overline{X}_t^i &= \int_0^t \Big(\frac{1}{N}\sum_{j=1}^N b(X_s^i, X_s^j) - \int b(\overline{X}_s^i, y) u_s(dy)\Big)\, ds \\
&= \int_0^t ds \frac{1}{N}\sum_{j=1}^N \{(b(X_s^i, X_s^j) - b(\overline{X}_s^i, X_s^j)) + (b(\overline{X}_s^i, X_s^j) - b(\overline{X}_s^i, \overline{X}_s^j)) \\
&\quad + (b(\overline{X}_s^i, \overline{X}_s^j) - \int b(\overline{X}_s^i, y)\, u_s(dy))\} .
\end{aligned}
$$

Writing $b_s(x,x') = b(x,x') - \int b(x,y)\, u_s(dy)$, we see that:

$$
E[|X^i - \overline{X}^i|_T^*] \le K \int_0^T ds \Big(E[|X_s^i - \overline{X}_s^i|] + \frac{1}{N}\sum_1^N E[|X_s^j - \overline{X}_s^j|] + E[|\frac{1}{N}\sum_{j=1}^N b_s(\overline{X}_s^i, \overline{X}_s^j)|]\Big)
$$

Summing the previous inequality over i, and using symmetry, we find:

$$
\begin{aligned}
N E[|X^1 - \overline{X}^1|_T^*] &= \sum_1^N E[|X^i - \overline{X}^i|_T^*] \\
&\le K' \int_0^T \sum_{i=1}^N \Big(E[|X_s^i - \overline{X}_s^i|] + E[|\frac{1}{N}\sum_{j=1}^N b_s(\overline{X}_s^i, \overline{X}_s^j)|]\Big)\, ds .
\end{aligned}
$$

Applying Gronwall's lemma, and symmetry, we find:

$$
E[|X^i - \overline{X}^i|_T^*] \le K(T) \int_0^T ds E[|\frac{1}{N}\sum_{j=1}^N b_s(\overline{X}_s^i, \overline{X}_s^j)|] .
$$

Our claim will follow provided we can show that:

$$
E[|\frac{1}{N}\sum_{j=1}^N b_s(\overline{X}_s^i, \overline{X}_s^j)|] \le \frac{C(T)}{\sqrt{N}} .
$$

But

$$
E[(\frac{1}{N}\sum_{j=1}^N b_s(X_s^i, X_s^j))^2] = \frac{1}{N^2} E[\sum_{j,k} b_s(\overline{X}_s^i, \overline{X}_s^j) b_s(\overline{X}_s^i, \overline{X}_s^k)] ,
$$

and because of the centering of $b_s(x,y)$ with respect to its second variable, when $j \neq k$

$$
E[b_s(\overline{X}_s^i, \overline{X}_s^j) b_s(\overline{X}_s^i, \overline{X}_s^k)] = 0 .
$$

The previous sum is then less than const.$/N$. Our claim follows.

□

Let us end this section with some comments.

— We have presented here a simple enough example. It is possible to let basically the same method work in the case of an additional interaction through the diffusion coefficient, see McKean [27], or [41]. The use in this context of the Vasershtein metric, can be found in Dobrushin [9]. Of course one can as well obtain higher moment estimates in (1.8). Theorem 1.4 suggests the possibility of a fluctuation theorem, which indeed exists, see for instance, Tanaka [51], Shiga-Tanaka [38], Kusuoka-Tamura [20], or [40], [41].

— As a result of the probabilistic proof we just described, we see that the introduction of a "nonlinear process", and not only of a nonlinear equation is fairly natural. For each t and k, the joint distribution of $(X_t^{1,N},\dots,X_t^{k,N})$ is converging to $u_t^{\otimes k}$, but in fact we have convergence at the level of processes.

— Let us also mention some interesting examples which arise in possibly singular cases, of the nonlinear equation

$$\partial_t u = \frac{\sigma^2}{2}\Delta u - \operatorname{div}\Big(\int b(\cdot,y)\ u_t(dy) u\Big)\ , \quad (\sigma = \text{constant}) \tag{1.9}$$

a) When $\sigma = 0$, and $\boldsymbol{R}^d = \boldsymbol{R}^3 \times \boldsymbol{R}^3 : (x,v)$, with $b((x,v),(x',v')) = (v, F(x-x'))$, one finds:

$$\partial_t u + v.\nabla_x u + \int F(x-x')\ u(dx',dv') \cdot \nabla_v u = 0$$

This is Vlasov's equation (see Dobrushin [9]).

b) $\boldsymbol{R}^d = \boldsymbol{R}$, $\sigma = 1$, $b(x,y) = c\delta(x-y)$, we have

$$\partial_t u = \frac{1}{2}u'' - c(u^2)'\ ,$$

this is Burger's equation, we will come back to this (singular) example in Chapter II.

c) $\boldsymbol{R}^d = \boldsymbol{R}^2$, $b(x,y) = b(x-y)$ is the Biot-Savart kernel, with $b(z) = \frac{1}{2\pi|z|^2}(-z_2, z_1)$, one now finds:

$$\partial_t u + (b\ *\ u)\nabla u = \frac{\sigma^2}{2}\Delta u\ ,$$

where we use $\operatorname{div}(b\ *\ u) = 0$. If one sets $v = (b\ *\ u)$, curl $v = \partial_2 v_1 - \partial_1 v_2 = u$, and u satisfies the vorticity equation for the Navier-Stokes equation in dimension 2. For this example, see Goodman [12], Marchioro-Pulvirenti [29], Osada [35].

d) Scheutzow [37], gives an example of an equation like (1.3), with polynomial coefficients, in $\boldsymbol{R}^2$, which admits some genuinely periodic solutions.

2) Some generalities

We will now give some definitions, that we will use for the study of propagation of chaos. $M(E)$ denotes here the set of probability measures on E.

Definition 2.1. E a separable metric space, u_N a sequence of symmetric probabilities on E^N. We say that u_N is u-chaotic, u probability on E, if for $\phi_1, \dots, \phi_k \in C_b(E)$, $k \geq 1$,

$$\lim_{N\to\infty} \langle u_N, \phi_1 \otimes \dots \otimes \phi_k \otimes 1 \cdots \otimes 1 \rangle = \prod_1^k \langle u, \phi_i \rangle \ . \tag{2.1}$$

As mentioned in the introductory chapter, the symmetry assumption on the laws u_N is less innocent than one may think at first. As the next (easy) proposition shows, the notion of u-chaotic means that the empirical measures of the coordinate variables of E^N, under u_N tend to concentrate near u. This is a type of law of large numbers. Condition (2.1) can also be restated as the convergence of the projection of u_N as E^k to $u^{\otimes k}$ when N goes to infinity, see [2], p. 20. In the coming proposition, we suppose u_N symmetric.

Proposition 2.2.

i) *u_N is u-chaotic is equivalent to $\overline{X}_N = \frac{1}{N}\sum_1^N \delta_{X_i}$ ($M(E)$-valued random variables on (E^N, v_N), X_i canonical coordinates on E^N) converge in law to the constant random variable u.* It is also equivalent to condition (2.1), with $k = 2$.

ii) *When E is a Polish space, the $M(E)$-valued variables $\overline{X}_N$ are tight if and only if the laws on E of X_1 under u_N are tight.*

Proof:

i) First suppose u_N satisfies (2.1) with $k = 2$, and consequently with $k = 1$ as well. Take ϕ in $C_b(E)$,

$$E_N[\langle \overline{X}_N - u, \phi\rangle^2] = \frac{1}{N^2}\sum_{i,j=1}^N E_N[\phi(X_i)\phi(X_j)] - \frac{2}{N}\sum_1^N E_N[\phi(X_i)]\langle u,\phi\rangle + \langle u,\phi\rangle^2 \ .$$

Using symmetry we find:

$$\frac{1}{N}E_N[\phi(X_1)^2] + \frac{(N-1)}{N}E_N[\phi(X_1)\phi(X_2)] - 2\langle u,\phi\rangle E_N[\phi(X_1)] + \langle u,\phi\rangle^2$$

which tends to zero by (2.1), with $k = 1, 2$. This implies that $\overline{X}_N$ converges in law to the constant random variable equal to u.

Conversely, suppose $\overline{X}_N$ converge in law to the constant u,

$$(2.2)\qquad \begin{aligned} &|\langle u_N, \phi_1 \otimes \ldots \otimes \phi_k \otimes 1 \ldots \otimes 1\rangle - \prod_1^k \langle u, \phi_i\rangle| \\ &\leq |\langle u_N, \phi_1 \otimes \ldots \otimes \phi_k \otimes 1 \cdots \otimes 1\rangle - \langle u_N, \prod_1^k \langle \overline{X}_N, \phi_i\rangle\rangle| \\ &+ |\langle u_N, \prod_1^k \langle \overline{X}_N, \phi_i\rangle\rangle - \prod_1^k \langle u_0, \phi_i\rangle| \,. \end{aligned}$$

The second term in the right member of (2.2) goes to zero. The first term using symmetry can be written as:

$$|\langle u_N, \frac{1}{N!} \sum_{\sigma \in S_N} \phi_1(X_{\sigma(1)}) \cdots \phi_k(X_{\sigma(k)}) - \prod_1^k \langle \overline{X}_N, \phi_i\rangle\rangle| \,.$$

Observe now that if $M \geq \|\phi_i\|_\infty$, $1 \leq i \leq k$.

$$(2.3)\qquad \begin{aligned} &\sup_{E^N} |\frac{1}{N!} \sum_{\sigma \in \rho_N} \phi_1(X_{\sigma(1)}) \ldots \phi_k(X_{\sigma(k)}) - \prod_1^k \langle \overline{X}_N, \phi_i\rangle| \\ &\leq M^k [(\frac{(N-k)!}{N!} - \frac{1}{N^k}) \cdot \frac{N!}{(N-k)!} + \frac{1}{N^k}(N^k - \frac{N!}{(N-k)!})] \\ &= 2M^k(1 - \frac{N!}{N^k(N-k)!}) \to 0 \,. \end{aligned}$$

Here we simply used that there are $N!/(N-k)!$ injections from $[1,k]$ into $[1,N]$, each of them has weight $(N-k)!/N!$ in the first sum of (2.3) and $1/N^k$ in the second sum, and in the second sum there are also $N^k - N!/(N-k)!$ terms where repetitions of coordinates occur. So we see that the first term of (2.2) goes to zero, and this proves i).

ii) For a probability $Q(dm)$ on $M(E)$, define the intensity $I(Q)$ as the probability measure

$$(2.4)\qquad \langle I(Q), f\rangle = \int_{M(E)} \langle m, f\rangle \, dQ(m) \,,$$

for $f \in bB(E)$.

We will in fact show the more general fact:

(2.5) tightness for a family of measure Q on $M(E)$ is equivalent to the tightness of their intensity measures $I(Q)$ on E.

Now in our situation of ii), by symmetry the intensity measure of the law of $\overline{X}_N$ is just the law of X_1, under E_N, so that our claim ii) follows from (2.5). Now the map

$Q \to I(Q)$ is clearly continuous for the respective weak convergence topologies. So, (2.5) will follow if we prove that whenever $I_n = I(Q_n)$ is tight, then Q_n is tight.

For each $\epsilon > 0$, denote by K_ϵ a compact subset of E, with $I_n(K_\epsilon^c) \leq \epsilon$, for every n. Now for $\epsilon, \eta > 0$, and any n, $Q_n(\{m, m(K_{\epsilon\eta}^c) \geq \eta) \leq \frac{1}{\eta} I_n(K_{\epsilon\eta}^c) \leq \epsilon$.

It follows that:

$$Q_n\big(\bigcup_{k\geq 1}\{m, m(K^c_{\epsilon 2^{-k}/k}) \geq 1/k\}\big) \leq \sum_{k\geq 1}^{\infty} \epsilon 2^{-k} = \epsilon\,,$$

this means that Q_n puts a mass greater or equal to $1 - \epsilon$ on the compact subset of $M(E)$: $\cap_{k\geq 1}\{m, m(K^c_{\epsilon 2^{-k}/k}) \leq 1/k\}$. This proves the Q_n are tight, and yields (2.5).

□

Remark 2.3.

1) It is clear from the proof of Proposition 2.2, that thanks to symmetry, the distribution under u_N of k particles chosen with the empirical distribution $\overline{X}_N$ is approximately the same as the law of $(X_1, \ldots, X_k)$ under u_N, when N is large. In fact this last distribution is the intensity of the empirical distribution of distinct k-uples: $\overline{X}_{N,k} = \frac{1}{N(N-1)\cdots(N-k+1)} \sum_{i_\ell \text{distinct}} \delta_{(X_{i_1},\ldots,X_{i_k})}$.

2) Suppose the empirical measures $\overline{X}_N = \frac{1}{N}\sum_1^N \delta_{X_i}$ converge in law to a constant $u \in M(E)$, with an underlying distribution v_N on E^N non necessarily symmetric. Then the symmetrized distribution u_N on E^N, preserves $\overline{X}_N$, and u_N is then u-chaotic in the sense of Definition 2.1. This remark applies to the case of deterministic sequences X_i, with convergent empirical distributions for example.

□

We will now give a result which will be helpful when transporting results from a space E to a space F. E and F are separable metric spaces, and ϕ is a measurable map from E to F. One also has the natural diagonal or tensor map $\phi^{\otimes N}$ from E^N into F^N. If u_N is a (non necessarily symmetric) probability on E^N, we will write $v_N = \phi^{\otimes N} \circ u_N$.

Proposition 2.4.

(2.6) If u_N is u-chaotic, and the continuity points C_ϕ of ϕ have full u-measure, v_N is v-chaotic if $v = \phi \circ u$.

(2.7) If $\overline{u}_\infty$ is a limit point of the laws of the empirical distributions $\overline{X}_N$ as N tends to infinity, C_ϕ has full measure under the intensity measure $I(\overline{u}_\infty)$ of $\overline{u}_\infty$, and v_N is v-chaotic, then for $\overline{u}_\infty$-a.e. m in $M(E)$, $\phi \circ m = v$.

(2.8) If $F = R$, $Q \in M(M(E))$ and ϕ is bounded and C_ϕ has full $I(Q)$ measure, then for any Q_n converging weakly to Q and any continuous real function $h(\cdot)$,

$$E_Q[h(\langle m, \phi \rangle)] = \lim_n E_{Q_n}[h(\langle m, \phi, \rangle)].$$

Proof: It is not difficult to see that (2.6), (2.7), (2.8) follow from the remark: if $Q \in M(M(E))$ and $I(C_\phi) = 1$, then for any Q-convergent sequence Q_n, $\Psi \circ Q_n$ converges weakly to $\Psi \circ Q$, if $\Psi : M(E) \to M(F)$ is the map: $m \to \phi \circ m$. The proof of this last statement is that $I(C_\phi) = 1$ implies $Q[\{m, \langle, m, C_\phi \rangle = 1\}] = 1$, and when $\langle m, C_\phi \rangle = 1$, then m is a continuity point of Ψ (see Billingsley [2] p. 30). So the continuity points of Ψ have full Q-measure which yield the remark by a second application of the quoted result.

□

We will now give some comments on how the material of this section will be used in a propagation of chaos context.

In several cases one has a symmetric law P_N, on $C(R_+, R^d)^N$ or $D(R_+, R^d)^N$ for instance, describing the interacting particle system. The initial conditions are supposed to be u_0-chaotic, $u_0 \in M(R^d)$, and one tries to prove that the P_N are P-chaotic, for some suitable P on $C(R_+, R^d)$ or $D(R_+, R^d)$ with initial condition u_0, by (2.7).

One way to prove such a statement is to show that the laws $\overline{P}_N$ of the empirical distributions $\overline{X}_N$ converge weakly to δ_P. This can be performed by checking tightness of the $\overline{P}_N$, that is tightness of the laws of X^1 on C or D under P_N, and identifying all possible limit points $\overline{P}_\infty$ as being concentrated on P.

This last step will involve finding some denumerable collection of functionals on probabilities on C or D, $G(m)$ having some continuity property (see (2.8)), for which P is the only common zero. Then we will show that $\overline{P}_\infty$ is concentrated on the zeros of each functional.

The reason for using this method is that in many cases, $G(m)$ will be a quadratic function of m, and we will end up calculating $E_{\overline{P}_\infty}[G(m)^2]$, as $\lim_\ell E_{N_\ell}[G(\overline{X}_{N_\ell})^2]$, with the help of (2.8), for a suitable subsequence N_ℓ. But the study of this last limit when G is quadratic in m, by symmetry basically involves only four particles. In many examples, it yields a line of proof which is more pleasant than working directly with (2.1), even with $k = 2$.

3) Examples.

a) Our laboratory example:

$$dX_t^i = dB_t^i + \frac{1}{N}\sum_1^N b(X_t^i, X_t^j)\, dt\ , \quad i = 1, \dots, N\ ,$$

$$X_0^i : \text{independent } u\text{-distributed}$$

here the X^i have symmetric laws P_N on $C(R_+, R^d)^N$, which thanks to Theorem 1.4 are P-chaotic, where P is the law of the "nonlinear process" solution of

$$dX_t = dB_t + \int b(X_t, y)\ u_t(dy)\ dt$$

$$X_0 : u - \text{distributed}, \ u_t(dy) = \text{law of } X_t\ .$$

(A number of assumptions: independence of X_0^i, Lipschitz character of b can in fact be relaxed, see references at the end of section 1).)

b) Uniform measure on the sphere of radius $\sqrt{n}$ in R^n.

Proposition 3.1. *The uniform distribution $ds_n(x)$ on the sphere of radius $\sqrt{n}$ in R^n is u-chaotic, if $u = 1/\sqrt{2\pi}\exp\{-x^2/2\}dx$.*

Proof: s_n is clearly symmetric. We will show directly that the projection of s_n on the first k components of R^n converges weakly to $u^{\otimes k}$.

Call $\mu_n(dr)$ the law of the radius under $u^{\otimes n}$, and $s_{n,r}(dx)$ the uniform distribution on the sphere of radius r in R^n.

Take now f continuous with compact support on R^k. By the law of large numbers for $(x_1^2 + \cdots + x_n^2)/n$, under $u^{\otimes n}$, we know that for $0 < a < 1 < b$,

$$\text{(3.1)} \qquad \lim_{n\to\infty} \Big| \int_{R^k} f(x) u^{\otimes k}(dx) - \int_{[a\sqrt{n}, b\sqrt{n}]} d\mu_n(r) \int ds_{n,r}(x)\ f(x_1, \dots, x_k) \Big| = 0\ .$$

But one also has:

$$\int_{a\sqrt{n}}^{b\sqrt{n}} \mu_n(dr) \int ds_{n,r}(x)\ f(x_1, \dots, x_k) = \int_{a\sqrt{n}}^{b\sqrt{n}} \mu_n(dr) \int ds_n(x)\ f(x_1 \frac{r}{\sqrt{n}}, \dots, x_k \frac{r}{\sqrt{n}})\ .$$

Now $r/\sqrt{n} \in [a, b]$ in the previous integral, and f is compactly supported and continuous, so that:

$$\text{(3.2)} \quad \lim_{a\to 1, b\to 1} \sup_{n\geq k} \Big| \int_{a\sqrt{n}}^{b\sqrt{n}} \mu_n(dr) \int ds_{n,r}(x) f - \mu_n([a\sqrt{n}, b\sqrt{n}]) \langle s_n, f \otimes 1 \cdots \otimes 1 \rangle \Big| = 0\ .$$

Now for each $a < 1 < b$, $\mu_n([a\sqrt{n}, b\sqrt{n}]) \to 1$ as n goes to infinity. By (3.1), (3.2), we see that when n goes to infinity $\langle s_n, f \otimes 1 \ldots \otimes 1\rangle$ tends to $\langle u^{\otimes k}, f\rangle$. This finishes the proof of Proposition 3.1.

c) Variation on a theme.

We will now present an example closely related to the previous one, which explains why one had the previous result. In fact one could modify it to include the previous example, however as explained before we are not seeking the maximum generality here.

Take $x_1, \ldots, x_n$, which are iid, with law $\mu(dx) = f(x)\,dx$, on R^d, where $f > 0$, is C^1 and such that

$$\int (f(x) + |\nabla f(x)|)\, e^{\lambda |x|} dx < \infty\,, \qquad \text{for any } \lambda\,. \tag{3.3}$$

Then it is known that for any $a \in R^d$, there is a unique $\lambda = \lambda_a$ such that $\frac{1}{Z_\lambda} \cdot e^{\lambda x} \mu(dz)$ (Z_λ normalization factor) has mean a. In fact if $I(\cdot)$ is the convex conjugate of the logarithm of the Laplace transform of μ:

$$I(x) = \sup_\lambda (\lambda \cdot x - \log(\int e^{\lambda \cdot y} \mu(dy)))\,,$$

we have: $\nabla I(a) = \lambda_a$.

Consider $s_n(dx) = \mu^{\otimes n}[(x_1, \ldots, x_n) \in dx / \frac{x_1 + \cdots + x_n}{n} = a]$, the conditional distribution of $(x_1, \ldots, x_n)$ given the mean $x_1 + \cdots + x_n/n = a$.

Proposition 3.2. *$s_n(dx)$ is ν-chaotic, where $\nu = \frac{1}{Z_\lambda} e^{\lambda \cdot x}\mu$ with $\lambda = \lambda_a$ determined by $\int x \cdot d\nu(x) = a$.*

Proof: We have, with obvious notations

(3.4)

$$\begin{aligned} s_n(dx) &= \frac{1}{z_n} f(x_1) \ldots f(x_{n-1})\, f(an - x_1 - \cdots - x_{n-1}) dx_1 \ldots dx_{n-1} \\ &= \frac{1}{e^{\lambda \cdot an} z_n} (e^{\lambda \cdot} f)(x_1) \ldots (e^{\lambda \cdot} f)(x_{n-1}) (e^{\lambda \cdot} f)(an - x_1 \ldots - x_{n-1})\, dx_1 \ldots dx_{n-1} \\ &= \nu^{\otimes n}[(x_1, \ldots, x_n) \in dx / x_1 + \cdots x_n = an]\,. \end{aligned}$$

Now looking at $x_i - a = y_i$, we see that we can assume that μ is centered and $a = 0$, and (3.3) holds.

It is now enough to show that for $\phi_1, \ldots, \phi_k \in C_K(R^d)$,

$$\langle s_n, \phi_1 \otimes \ldots \otimes \phi_k \otimes 1 \ldots \otimes 1\rangle \longrightarrow \prod_1^k \langle \mu, \phi_i\rangle\,, \qquad \text{as } N \text{ goes to infinity.} \tag{3.5}$$

Now from (3.4), the expression under study is

$$\begin{aligned}&(\phi_1 f) \, * \, \cdots \, * \, (\phi_k f) \, * \, f^{(n-k)*}(0) \; / \; f^{n*}(0) \\ &= \int \widehat{\phi_1 f}(\xi) \ldots \widehat{\phi_k f}(\xi) \; \hat{f}^{n-k}(\xi) \; d\xi \; / \; \int \hat{f}^n(\xi) \; d\xi \, ,\end{aligned}$$

using Fourier transforms.

Observe that $\int \hat{f}^n(\xi) d\xi = \frac{1}{n^{d/2}} \int \hat{f}^n(\frac{\xi}{\sqrt{n}}) \; d\xi$, now $\hat{f}^n(\frac{\xi}{\sqrt{n}}) \overset{n\to\infty}{\longrightarrow} \exp\{-\frac{1}{2} \, {}^t\xi A \xi\}$, for each ξ. On the other hand (3.3) ensures that we have the domination:

$$|\hat{f}(\xi)| \leq [1 + |\xi|^2]^{-\delta} \, , \quad \xi \in R^d \, ,$$

for a suitable δ. So we have $|\hat{f}^n(\frac{\xi}{\sqrt{n}})| \leq [(1 + \frac{|\xi|^2}{n})^n]^{-\delta} \leq (1 + \frac{|\xi|^2}{n_0})^{-n_0 \delta}$, when $n \geq n_0$, which is integrable if n_0 is large enough. So we see that as n tends to infinity

$$\int \hat{f}^n(\xi) d\xi \sim (\frac{n}{2\pi})^{-d/2} (\det A)^{-1/2} \text{ and } \int_{|\xi| \geq A} n^{d/2} |\hat{f}^{n-k}(\xi)| d\xi \to 0 \, , \quad \text{for } A > 0 \, .$$

So $\langle s_n, \phi_1 \otimes \cdots \otimes \phi_k \otimes 1 \ldots \otimes 1 \rangle$ converges to $\widehat{\phi_1 f}(0) \ldots \widehat{\phi_k f}(0) = \langle \mu, \phi_1 \rangle \ldots \langle \mu, \phi_k \rangle$, which yields our claim.

□

d) Symmetric self exclusion process.

In this example there will be a microscopic scale at which particles interact (exclusion condition) and a macroscopic scale. The interaction will in fact disappear in the limit.

Our N particles move on $\frac{1}{N}\mathbb{Z} \subset R$. We look at the simple exclusion process for these N particles from the point of view of N particles which evolve in a random medium given as follows: with the bonds $(i, i + \frac{1}{N})$, $i \in \frac{1}{N}\mathbb{Z}$, we associate a collection of independent Poisson counting processes $N_t^{i,i+1/N}$, with intensity $\frac{N^2}{2} dt$. The motion of any particle in this random medium is given by the rule:

If a particle is at time t at the location i it remains there until the first of the two Poisson processes $N_\cdot^{i,i+1/N}$ or $N_\cdot^{i-1/N,i}$ has a jump, and then performs a jump across the corresponding bond.

As a consequence of this rule if we consider N particles with N distinct initial conditions, these particles will never collide (self exclusion). Here we pick an initial distribution u_N on $(\frac{1}{N}\mathbb{Z})^N \subset R^N$, which is u_0-chaotic, where $u_0 \in M(R)$. Using Remark 2.3, we can for instance take the deterministic locations $(\frac{1}{N}, \frac{2}{N}, \ldots, 1)$ and symmetrize them, in this case $u_0(dx) = 1_{[0,1]} dx$. Given the initial conditions and the random medium, this provides us with N trajectories $(X_\cdot^1, \ldots, X_\cdot^N)$ and, we denote by

P_N, the symmetric law on $D(R_+,R)^N$, one obtains, when the initial conditions are picked independent of the medium, and u_N distributed.

Theorem 3.3. *P_N is P-chaotic, if P is the law of Brownian motion with initial distribution u_0.*

Corollary 3.4. *For each $s \geq 0$, the law of $(X_s^1,\dots,X_s^N)$ is $u_0 * p_s$-chaotic if $p_s = (2\pi s)^{-1/2} \exp -\frac{x^2}{2s}$.*

Proof: This is an immediate application of Proposition 3.4, where the map ϕ, is the coordinate at time s on $D(R_+,R)$.

□

Before giving the proof of Theorem 3.3, let us give some words of explanation about the point of view we have adopted here. As mentioned before, for the propagation of chaos result one works with symmetric probabilities. This is why we have constructed the self-exclusion process, using the trajectories of particles interacting with a given random medium. If one instead is just interested in "density profiles", that is, the evolution of:

$$\eta_t^N = \frac{1}{N}\sum_1^N \delta_{X_t^i} \in M(R)\ ,$$

one can directly build it as a Markov process evolving on the space of simple point measures, but one loses the notion of particle trajectories, especially when two neighboring particles perform a jump (see Liggett [24], DeMasi-Ianiro-Pellegrinotti-Presutti [7]). Notice however that Corollary 3.4 can be restated purely in terms of η_t^N. Indeed by Proposition 2.2 i), it means that when N tends to infinity η_t^N converges in law to the constant probability $u_0 * p_t$, that is belongs with vanishing probability to the complement of any neighborhood of $u_0 * p_t$ for the weak topology.

So we can derive a corollary of Theorem 3.3 purely in terms of "profile measures" η_t^N, stating that if η_0^N is a sum of N distinct atoms of mass $1/N$, and converges in law to the constant probability u_0 on R, then the random probabilities η_t^N converge in law to $u_0 * p_s$. Of course Theorem 3.3 has more in it. From a "profile point of view", we have introduced the symmetric variables through the trajectories $(X_.^1,\dots,X_.^N)$.

Proof of Theorem 3.3: We consider the empirical measures $\overline{X}_N = \frac{1}{N}\sum_1^N \delta_{X_.^i} \in M(D)$, where $D = D(R_+,R)$. We will show the $\overline{X}_N$ concentrate their mass around P. We will first show that the laws of the $\overline{X}_N$ are tight, and then we will identify any possible limit point as being δ_P the Dirac mass at P. This will precisely mean the $\overline{X}_N$ converge in law to the constant P.

Tightness:

By Proposition 2.2 ii), it boils down to checking tightness for $X^1_\cdot$, under P_N. But here $X^1_\cdot$ is distributed as a simple random walk on $\frac{1}{N}\mathbb{Z}$ in continuous time with jump intensity $= N^2 dt$, and initial distribution $x^1 \circ u_N$. Classically, the law of $X^1_\cdot$ under P_N converges weakly to P, and this yields tightness.

Identification of limit points:

Take $\overline{P}_\infty$ a limit point of the laws of the $\overline{X}_N$. We already know by the tightness step (see Proposition 2.2 ii) that $I(\overline{P}_\infty)$, the intensity of $\overline{P}_\infty$ is P. By Proposition 2.4, we know that $\overline{P}_\infty$ is concentrated on measures for which X_0 is u_0-distributed. If we introduce $F(m) = E_m[(e^{i\lambda(X_t - X_s) + (\lambda^2/2)(t-s)} - 1)\psi_s(X)]$, where $\lambda \in \mathbb{R}$, $t > s$, and $\psi_s(X) = \phi_1(X_{s_1})\dots\phi_k(X_{s_k})$, with $0 \le s_1 < \dots < s_k \le s$ and $\phi_1, \dots, \phi_k \in C_b(\mathbb{R})$, it is enough to show that $F(m) = 0$, $\overline{P}_\infty$-a.s., for then varying over countable families of λ, $t > s$, s_i, ϕ_i, we will find that $X_t - X_s$ is independent of $\sigma(X_u,\ u \le s)$, and $N(0, t-s)$ distributed, for $\overline{P}_\infty$-a.e. m, which implies $\overline{P}_\infty = \delta_P$.

Using (2.8), since $(e^{i\lambda(X_t - X_s) + \lambda^2(t-s)} - 1)$ has continuity points of full measure under $P = I(\overline{P}_\infty)$, we find

$$\begin{aligned} E_{\overline{P}_\infty}[|F(m)|^2] &= \lim_k E_{N_k}[|F(\overline{X}_{N_k})|^2] \\ &= \lim E_{N_k}[|\frac{1}{N}\sum_1^N (e^{i\lambda(X^i_t - X^i_s) + (\lambda^2/2)(t-s)} - 1)\ \psi_s(X^i)|^2]\ , \end{aligned}$$

using symmetry, the latter quantity equals:

(3.6)

$$\lim E_{N_k}[(e^{i\lambda(X^1_t - X^1_s) + (\lambda^2/2)(t-s)} - 1)(e^{-i\lambda(X^2_t - X^2_s) + (\lambda^2/2)(t-s)} - 1)\ \psi_s(X^1)\ \psi_s(X^2)]\ .$$

Observe that $E_{N_k}[e^{i\lambda(X^1_t - X^1_s) + (\lambda^2/2)(t-s)}\ \psi_s(X^1)\ \psi_s(X^2)] =$ $E_{N_k}[e^{i\lambda(X^1_t - X^1_s) + (\lambda^2/2)(t-s)}]\ \ E_{N_k}[\psi_s(X^1)\psi_s(X^2)]$, and the limit of the first term of the product is 1, by the weak convergence of the law of X^1 to P. In view of (3.6), to prove tha $F(m) = 0$, $\overline{P}_\infty$ a.s., it is enough to check that:

$$A_N = E_N[(\exp\{i\lambda[(X^1_t - X^2_t) - (X^1_s - X^2_s)] + \lambda^2(t-s)\} - 1)\ \psi_s(X^1)\psi_s(X^2)]$$

tends to zero when N goes to infinity. Observe now that (X^1_t, X^2_t) is a pure jump process on $(\frac{1}{N}\mathbb{Z})^2$, with bounded generator:

$$
\begin{aligned}
Lf(x_1,x_2) =& \frac{N^2}{2}(\Delta_N^1 + \Delta_N^2)\, f(x_1,x_2) \\
& - \frac{N^2}{2} 1\{x_1 - x_2 = \frac{1}{N}\} \cdot D_1 D_2 f(x_1 - \frac{1}{N}, x_2) \\
& - \frac{N^2}{2} 1\{x_2 - x_1 = \frac{1}{N}\} \cdot D_2 D_1\, f(x_1, x_2 - \frac{1}{N}) ,
\end{aligned}
\tag{3.7}
$$

here Δ^1, Δ^2, D_1, D_2, are the discrete Laplacian and the difference operators with respect to first and second variables. If we pick now $f(x_1,x_2) = \exp\{i\lambda(x_1 - x_2)\}$, we see that

$$
\left|\frac{Lf(x_1,x_2)}{f} - N^2(e^{i\lambda/N} + e^{-i\lambda/N} - 2)\right| \le \text{const } 1\{|x_1 - x_2| = \frac{1}{N}\} . \tag{3.8}
$$

Using now the fact that $f(X_t^1, X_t^2)\exp - \int_0^t \frac{Lf(X_u)}{f} du$ is a bounded martingale, we have

$A_N =$

$$
E_N[f(X_t^1,X_t^2)/f(X_s^1,X_s^2)\; e^{\lambda^2(t-s)}(1 - \exp - \int_s^t (\frac{Lf}{f} + \lambda^2)(X_u)du)\; \psi_s(X^1)\psi_s(X^2)] ,
$$

using (3.8) we see that:

$$
|A_N| \le o(N) + \text{const } E_N[\int_0^t 1\{|X_s^1 - X_s^2| = \frac{1}{N}\}ds]
$$

So Theorem 3.3 will follow from

Lemma 3.5.

$$
\lim_{N\to\infty} E_N[\int_0^\infty e^{-s} 1\{|X_s^1 - X_s^2| = \frac{1}{N}\}ds] = 0 .
$$

Proof: $Y_s = N|X_{s/N^2}^1 - X_{s/N^2}^2|$, as follows from (3.7) is a jump process on $\mathbb{N}^*$, with generator

$$
L'f(k) = 1\{k \ne 1\}\Delta f(k) + 1\{k = 1\}[f(2) - f(1)] ,
$$

and the quantity under the limit sign in Lemma 3.5 equals:

$$
E_N[\int_0^\infty e^{-s/N^2} 1\{Y_s = 1\}\frac{ds}{N^2}] \le \frac{C}{N^2}\sum_{k=0}^\infty Q_2[e^{-\tau/N^2}]^k = \frac{C}{N^2}\,\frac{1}{1 - Q_2[e^{-\tau/N^2}]} , \tag{3.9}
$$

where τ is the hitting time of 1, and Q_2 is the law of Y, starting from 2. If we now pick $a(N) < 1$, such that $a + a^{-1} - 2 = 1/N^2$, using the bounded martingale $a^{Y_{s\wedge\tau}}\, e^{-(s\wedge\tau)/N^2}$, we find

$$E_2[e^{-\tau/N^2}] = a(N) = 1 + \frac{1}{2N^2} - \left((1+\frac{1}{2N^2})^2 - 1\right)^{1/2} = 1 - \frac{1}{N} + o(\frac{1}{N}) .$$

It follows that $\frac{C}{N^2}(1-a)^{-1} \sim \frac{C}{N}$ which tends to zero, and in view of (3.9) proves the lemma.

□

e) Reordering of Brownian motions.

We consider N independent Brownian motions $X^1_\cdot, \dots, X^N_\cdot$ on $\mathbb{R}$, with initial law u_0, which we suppose atomless. We then introduce the increasing reorderings $Y^1_t \le \dots \le Y^N_t$, of the $X^1_\cdot, \dots, X^N_\cdot$, so that $Y^1_t = \inf_i\{X^i_t\}$, $Y^2_t = \sup_{|A|=N-1} \inf_{i\in A}\{X^i_t\}$, etc. Now the processes $(Y^1_\cdot, \dots, Y^N_\cdot)$ are reflected Brownian motions on the convex $\{y_1 \le y_2 \le \dots \le y_N\}$, but they are not symmetric any more, so we consider the symmetrized processes $(Z^1_\cdot, \dots, Z^N_\cdot)$ on the enlarged space $(\mathcal{S}_N \times C^N, d\nu_N \otimes P_{u_0}^{\otimes N})$, where $\mathcal{S}_N$ is the symmetric group on $[1,N]$, $d\nu_N$ the normalized counting measure, and $Z^i_t = Y^{\sigma(i)}_t$, σ being the $\mathcal{S}_N$ valued component on $\mathcal{S}^N \times C^N$.

The interest of this example comes from the fact that on the one hand the $(X^1_\cdot, \dots, X^N_\cdot)$ and $(Z^1_\cdot, \dots, Z^N_\cdot)$ have the same density profile:

$$\frac{1}{N}\sum_1^N \delta_{X^i_t} = \frac{1}{N}\sum_1^N \delta_{Z^i_t} , \tag{3.10}$$

but on the other hand we are going to prove that the $Z^i_\cdot$ are Q-chaotic, where Q is a different law from P_{u_0}. Of course the $X^i_\cdot$ are P_{u_0}-chaotic. As a result of (3.10), Q and P_{u_0} will share the same time marginals, namely $u_0 * (\frac{1}{\sqrt{2\pi s}}\exp\{-\frac{x^2}{2s}\})$. This emphasizes the fact that the limit behavior of the profile evolution is not enough to reconstruct the "nonlinear process".

Let us describe the law Q. The distribution function F_t of u_t, is strictly increasing for $t > 0$. If $C = \text{supp } u_0$, we can write the complement of C as a union of disjoint intervals (a_n, b_n). The points of $D = C \setminus \cup\{a_n, b_n\}$ are points of left and right increase of F_0, and D has full u_0 measure.

We define for $x \in D$,

$$\psi_t(x) = F_t^{-1} \circ F_0(x) , \tag{3.11}$$

one in fact has $\lim_{t\to 0} \psi_t(x) = x$, and $\psi : x \to (\psi_t(x))_{t\ge 0}$, defines a measurable map from D in $C(\mathbb{R}_+, \mathbb{R})$.

Theorem 3.6. *The laws of $(Z^1_\cdot, \dots, Z^N_\cdot)$ are Q-chaotic where $Q = \psi \circ u_0$.*

Proof: We first give a lemma, making precise the structure of $Y^1_\cdot, \dots, Y^N_\cdot$ as reflected Brownian motion.

Lemma 3.7. *There are N independent $\sigma(X_u,\ u \le t)$ Brownian motions on $C(\boldsymbol{R}_+, \boldsymbol{R})^N$, $\beta^1_\cdot, \dots, \beta^N_\cdot$, and $(N-1)$ continuous adpated increasing processes $\gamma^1_t, \dots, \gamma^{N-1}_t$ such that*

$$
\begin{aligned}
&\gamma^i_t = \int_0^t 1(Y^i_s = Y^{i-1}_s)\, d\gamma^i_s\,, \\
&Y^1_t = Y^1_0 + \beta^1_t - \frac{1}{2}\gamma^1_t\,, \\
&Y^k_t = Y^k_0 + \beta^k_t - \frac{1}{2}\gamma^k_t + \frac{1}{2}\gamma^{k-1}_t \\
&Y^N_t = Y^N_0 + \beta^N_t + \frac{1}{2}\gamma^{N-1}_t\,.
\end{aligned}
\tag{3.12}
$$

Proof: One uses induction, by stopping the processes at the successive times where distinct X^i, X^j meet, and applies Tanaka's formula. Since these stopping times tend to infinity, one then obtains the lemma. For details see [42].

Let us check tightness of the laws of $Z^1_\cdot$.

Remark 3.8. Before proving this point, let us mention here that the symmetrized sampling of $Y^1_\cdot, \dots, Y^N_\cdot$ by the random permutation σ is crucial for tightness. Suppose u_0 has support in $[0,1]$, if the X^i are constructed on the infinite product space, one knows that for $t > 0$, $\overline{\lim}_{N\to\infty} \frac{M^N_t}{\sqrt{2t\log N}} = 1$, a.s., with $M^N_\cdot = \sup(X^1_\cdot - X^1_0, \dots, X^N_\cdot - X^N_0)$. But $Y^N_t = \sup(X^1_t, \dots, X^N_t) \ge M^N_t$, so one cannot expect tightness for the law of $Y^N_\cdot$. □

By symmetry and (3.10), the law of Z^1_t is u_t. So to prove our claim it is enough to prove

Lemma 3.9.

$$
E\Big[\sum_1^N |Y^i_t - Y^i_s|^4\Big] \le 3N(t-s)^2\,.
$$

Proof: We apply Ito-Tanaka's formula to

$$
\begin{aligned}
\|Y_t - Y_s\|^2 &= \sum_1^N (Y^i_t - Y^i_s)^2 \\
&= 2\sum_i \int_s^t (Y^i_u - Y^i_s)\, d\beta^i_u \\
&\quad + \sum_1^{N-1} \int_s^t [(Y^{i+1}_u - Y^{i+1}_s) - (Y^i_u - Y^i_s)]\, d\gamma^i_u + N(t-s)\,.
\end{aligned}
$$

Now by (3.12), $Y_u^{i+1} = Y_u^i\ d\gamma_u^i$-a.s., and $Y_s^i \le Y_s^{i+1}$, so after taking expectations

$$E[\sum_1^N (Y_t^i - Y_s^i)^2] \le N(t-s)\,. \tag{3.13}$$

In the same way, we find:

$$\begin{aligned}\sum_1^N (Y_t^i - Y_s^i)^4 =& 4\sum_1^N \int_s^t (Y_u^i - Y_s^i)^3 d\beta_u^i \\ &+ 2\sum_1^{N-1} \int_s^t [(Y_u^{i+1} - Y_s^{i+1})^3 - (Y_u^i - Y_s^i)^3]\ d\gamma_u^i \\ &+ 6\sum_1^N \int_s^t (Y_u^i - Y_s^i)^2 du\ ,\end{aligned}$$

and again $(Y_u^{i+1} - Y_s^{i+1})^3 - (Y_u^i - Y_s^i)^3 \le 0$, $d\gamma_u^i$ a.s., so that taking expectation and (3.13),

$$E[\sum_1^N (Y_t^i - Y_s^i)^4] \le 6N \int_s^t (u-s)du = 3N(t-s)^2\ .$$

□

Let us show now that for fixed k, the law of $(Z_.^1, \dots, Z_.^k)$ converges to $Q^{\otimes k}$. Since for every t, $(Z_t^1, \dots, Z_t^N)$ has the same reordered sequence $(Y_t^1, \dots, Y_t^N)$ as $(X_t^1, \dots, X_t^N)$, symmetry immediately implies that for each t, $(Z_t^1, \dots, Z_t^N)$ has the same distribution as $(X_t^1, \dots, X_t^N)$, that is $u_t^{\otimes N}$. Using this fact for $t = 0$, one sees easily that it is enough to prove that

(3.14) for any limit point $\overline{P}$ of the laws of $Z_.^1$, and $s > 0$, $\overline{P}[Z_s = \psi_s(Z_0)] = 1$,

to deduce that any limit point of the laws of $(Z^1, \dots, Z^k)$ is in fact $Q^{\otimes k}$. To check this statement observe that

$$\begin{aligned}E_{\overline{P}}[|F_0(Z_0) - F_s(Z_s)|] &= \lim_{N_k} E_{N_k}[|F_0(Z_0^1) - F_s(Z_s^1)|] \\ &\le \overline{\lim}_k E_{N_k}[|F_0(Z_0^1) - \frac{1}{N}\sum_{i=1}^N 1(X_0^i \le Z_0^1)|] \\ &+ \overline{\lim}_k E_{N_k}[|F_s(Z_s^1) - \frac{1}{N}\sum_{i=1}^N 1(X_s^i \le Z_s^1)|] \\ &+ \overline{\lim}_k E_{N_k}[|\frac{1}{N}\sum_{i=1}^N 1(X_0^i \le Z_0^1) - 1(X_s^i \le Z_s^1)|]\ .\end{aligned}$$

The first two terms go to zero. We have in fact

$$\lim E_N[\sup_x |F_0(x) - \frac{1}{N}\sum_i 1(X_0^i \le x)|] = 0 ,$$

since u_0 is atomless, and a similar result at time s. On the other hand we know that since $Z_.^1 = Y_.^{\sigma(1)}$,

$$|\sum_i 1(X_0^i \le Z_0^1) - 1(X_s^i \le Z_s^1)| \le 1 ,$$

so the last term goes to zero as well.

□

If we set $x_t = \psi_t(x)$, x_t satisfies the differential equation for $t > 0$,

$$\frac{dx_t}{dt} = -\frac{\partial_t F}{\partial_x F}(t, x_t) = -\frac{1}{2}(\log u_t)'(x_t) , \tag{3.15}$$

as is seen from the implicit equation $F(t, x_t) = \text{const}$. So Q is the law of a deterministic evolution with initial random distribution u_0, and u_t is a solution of the corresponding forward equation

$$\begin{aligned} &\partial_t u - \partial_x(\frac{1}{2}(\log u_t)'u) = 0 \\ &v(t=0,\cdot) = u_0 . \end{aligned}$$

Let us mention finally that when x is a "bad point" in the convex hull of the support of C, that is $x \in [a_n, b_n]$, with $-\infty < a_n < b_n < \infty$, then one sees easily that $\lim_{t\to 0} \psi_t(x) = (a_n + b_n)/2$, on the equation:

$$0 = F_t(x_t) - F_0(x) = u_+ \ * \ p_t(-\infty, x_t] - u_- \ * \ p_t[x_t, \infty) , \quad t > 0 ,$$

with $u_- = 1_{(-\infty, a_n]} \cdot u_0$, $u_+ = 1_{[b_n, \infty)} \cdot u_0$.

f) Colored particles and nonlinear process.

Now we look at P_N on $C(R_+, R^d)^N$, which are P-chaotic. Let I_0 be a subset of R^d such that $\{X_0 \in I_0\}$ is a continuity set for P, for instance a product of intervals if $u_0 = X_0 \circ P$ has a density. We color in blue the particles which are in I_0 at time zero, and we are interested in the empirical measure at time t of the blue particles:

$$\nu_t^N(dx) = \frac{1}{N}\sum_i 1(X_0^i \in I_0)\delta_{X_t^i} . \tag{3.16}$$

Then Proposition 2.4, easily implies that

$$\lim_{N\to\infty} E_N[\nu_t^N \in U^c] = 0 ,$$

for any neighborhood for the weak topology on $M_{+,b}(R^d)$ of

$$\nu_t(dx) = P[X_0 \in I_0\ ,\quad X_t \in dx]\ . \tag{3.17}$$

The coloration of particles is one way to recover some trajectorial information, and gain some knowledge on P, if one uses profile measures.

For applications of coloring of particles, in a propagation of chaos context, to stochastic mechanics, we refer the reader to Nagasawa-Tanaka [31].

g) Loss of Markov property and local fluctuations.

The reader might be tempted to think that when the N-particle system follows a symmetric Markovian evolution, which is chaotic, the law P of the "nonlinear process" will inherit a Markov property. We will now give an example showing that this is not the case.

Heuristically, what happens, is that we have an N-particle system having local interactions, but there is no mechanism to "average out" the local fluctuations of the interaction. One interest of the example is that the presence of these local fluctuations does not prevent the chaotic behavior of the N-particle system (so there is propagation of chaos in this sense). The limit law P does not have the Markov property, and the limit of the density profile of particles which is governed by the time marginals of P, does not correspond now to a nonlinear forward equation.

We consider the N-particle system $(Z^1_.,\dots,Z^N_.) \in C(R_+,E)^N$, where $E = R/Z{\times}R$, and $Z^i_. = (Y^i_.,X^i_.)$. It follows a symmetric Markovian evolution E^N, given as follows:

- The $Y^i_.$ are constant in time, and the Y^i_0 are i.i.d. dx-distributed, on R/Z.
- $X^i_t = \sigma_i B^i_t$, $t \geq 0$, $1 \leq i \leq N$, where $B^i_.$, $1 \leq i \leq N$, are i.i.d. standard Brownian motions independent of the $(Y^i_.)$, $1 \leq i \leq N$, and $\sigma_i = (1 + \sum_{j\neq i} 1\{Y_j \in \Delta_N(Y_i)\})^{1/2}$,

$\Delta_N(x)$ denoting the only interval $[\frac{k}{N},\frac{k+1}{N})$, $0 \leq k < N$, containing $x \in R/Z$.

Let us now describe the law P on $C(R_+,E)$, which appears in the propagation of chaos result. Under P:

- $Y_t \equiv Y_0$ is dy distributed on R/Z.
- $X_.$ is independent of $Y_.$ and distributed as a mixture of Brownian motions starting from zero, with trajectorial variance $\sigma^2 = 1 + m$, m being distributed as a Poisson, mean one, variable.

Let us right away observe that the law P is not Markovian. Indeed if (F_t) denotes the natural filtration on $C(R_+, E)$, $\sigma^2 = \overline{\lim}_{t\to 0} \frac{X_t^2}{2t \log\log 1/t}$ is F_1-measurable, and we have:

$$E^P[X_2^2 \mid F_1] = X_1^2 + \sigma^2 \ ,$$

whereas:

$$\begin{aligned} E^P[X_2^2 \mid X_1] &= X_1^2 + E^P[\sigma^2 \mid X_1] \\ &= X_1^2 + (\sum_{m\geq 0} \frac{(1+m)}{m!} p_{1+m}(X_1)) \;/\; (\sum_{m\geq 0} \frac{1}{m!} p_{1+m}(X_1)) \ . \end{aligned}$$

Here $p_t(x)$ denotes $(2\pi t)^{-1/2} \exp\{-\frac{x^2}{2t}\}$.

We have

Theorem 3.10.

(3.18) The laws P_N of $(Z_\cdot^1, \dots, Z_\cdot^N)$ are P-chaotic.

(3.19) For $t \geq 0$, the random measures $\frac{1}{N}\sum_1^N \delta_{(Y^i, X_t^i)} \in M(R/Z \times R)$ ("density profiles") converge in law to the constant $dy \otimes (e^{-1} \sum_{m\geq 0} \frac{1}{m!} p_{t(1+m)}(x)\, dx)$.

Proof: (3.19) is an immediate consequence of (3.18). Let us prove (3.18). In view of Proposition 2.2, we will simply show that the law of $(Z_\cdot^1, Z_\cdot^2)$ converges weakly to $P^{\otimes 2}$. Anyway, the case of $(Z_\cdot^1, \dots, Z_\cdot^k)$ is also obvious from our proof. It is enough to show for $f_i \in C(R/Z)$, $g_i \in C_b(C(R_+, R))$, $i = 1, 2$, that:

$$\lim_N E_N[f_1(Y_\cdot^1) g_1(X_\cdot^1) f_2(Y_0^2) g_2(X_\cdot^2)] = \prod_{i=1}^2 E^P[f_i(Y_0) g_i(X_\cdot)] \ . \tag{3.20}$$

The left member of (3.20) equals

$$\begin{aligned} &\lim_N E_N[f_1(Y_0^1) g_1(\sigma_1 B_\cdot^1) f_2(Y_0^2) g_2(\sigma_2 B_\cdot^2)] \\ &\quad = \lim_N E_N[f_1(Y_0^1) \phi_1(\sigma_1) f_2(Y_0^2) \phi_2(\sigma_2)] \ , \end{aligned} \tag{3.21}$$

where $\phi_i(a) = E[g_i(aB_\cdot)]$, $i = 1, 2$ (Wiener expectation), are continuous bounded functions. The expression inside the limit in (3.21) involves an expectation on the Y^i variables, $1 \leq i \leq N$, alone. To prove (3.20), it suffices to show that conditionally on $Y^1 = y_1$, $Y^2 = y_2$, $y_1 \neq y_2$, the law of $m_i = \sum_{j\neq i} 1\{Y_j \in \Delta_N(Y_i)\}$, $i = 1, 2$, converges weakly to the law of two independent mean one Poisson variables. To check this last point observe that for large N, $\Delta_N(y_1) \cap \Delta_N(y_2)$ is empty. So for $a_1, a_2 > 0$,

$$\begin{aligned}
&\lim_N E_N[e^{-a_1 m_1 - a_2 m_2} / Y_1 = y_1 \ , Y_2 = y_2] \\
&= \lim \Big(\int_0^1 \exp\{-a_1 1_{\Delta_N(y_1)}(x) - a_2 1_{\Delta_N(y_2)}(x)\} \ dx \Big)^{N-2} \\
&= \lim_N \big(1 + \frac{1}{N}(e^{-a_1} - 1 + e^{-a_2} - 1)\big)^{N-2} = \exp\{(e^{-a_1} - 1) + (e^{-a_2} - 1)\} \ ,
\end{aligned}$$

from which our claim follows. □

An example of a situation with a loss of the Markov property can also be found in Uchiyama [54].

h) Tagged particle: a counterexample.

In the introduction, we mentioned that we refrained from calling the "nonlinear process", the "tagged particle process", because this expression has a variety of meanings. We will give here an easy example where the tagged particle is the trajectory of the particle with initial starting point nearest to a certain point. It will turn out that the law of the tagged particle will converge to a limit distinct from the natural law of the nonlinear process conditioned to start from this point.

We consider the N-particle system $(Z^1_., \dots, Z^N_.) \in C(\mathbb{R}_+, E)$, where $E = [0,1) \times \mathbb{R}$. The processes $Z^i_. = (Y^i_., X^i_.)$, $1 \le i \le N$, will be independent Markov processes, satisfying:

- The $Y^i_.$ are constant in time, and the Y^i_0 are i.i.d. dy-distributed, on $[0,1)$.
- $X^i_t = \sigma_N(Y^i_0) B^i_t$, $t \ge 0$, $1 \le i \le N$, where $B^i_.$, $1 \le i \le N$, are i.i.d. standard Brownian motions independent of the $(Y^i_.)$, $1 \le i \le N$, and

$$\sigma_N(y) = 1\{y \in \bigcup_{k \ even} [\frac{k}{N} \, , \, \frac{k+1}{N})\} + \sqrt{2} \, 1\{y \in \bigcup_{k \ odd} [\frac{k}{N} \, , \, \frac{k+1}{N})\} \, .$$

The tagged particle process $\overline{Z}^N_.$ will be defined by $\bar{Z}^N_. = (Y^i_., X^i_.)$ on $\{Y^i_0 = \min Y^j_0\}$, $1 \le i \le N$. This defines the tagged particle a.s., with no ambiguity, and since the starting point of $Z^i_.$ is $(Y^i_0, 0)$, the tagged particle corresponds to the particle with initial starting point closest to (0,0) in $[0,1) \times \mathbb{R}$.

If we still denote by dy the measure on trajectories $Y_.$, constant in $[0,1)$, with initial point dy distributed, and by $W^1_.$, $W^2_.$ Wiener measure with respective variance 1 and 2, it is easy to see that the laws of the $Z^i_.$, which are independent, are P-chaotic, if $P = dx \otimes (\frac{1}{2} W^1 + \frac{1}{2} W^2)$ on $C(\mathbb{R}_+, E)$.

Notice, by the way, that the nonlinear process has also lost its Markov property.

We are now going to show:

Proposition 3.11: $\overline{Z}^N_.$ *converges in law to* $Q = \delta_0 \otimes ((1+e^{-1})^{-1}W^1 + e^{-1}(1+e^{-1})^{-1}W^2)$ *(here* δ_0 *denotes the Dirac mass on the constant trajectory equal to* 0).

Proof: It is enough to show that for $f \in C_b([0,1))$, $g \in C_b(C(R_+,R))$,

$$(3.22)\quad \lim_N E[f(\overline{Y}_0^N)g(\bar{X}_.^N)] = f(0) \times \left((1+e^{-1})^{-1}E^{W^1}[g] + e^{-1}(1+e^{-1})E^{W^2}[g]\right).$$

Set $\tilde{Y}^N = N\bar{Y}^N$, we have

$$E[f(\bar{Y}_0^N)g(\bar{X}_.^N)] = \sum_{0\le k \text{ even } <N} E[f(\frac{\tilde{Y}^N}{N})\, 1\{k \le \tilde{Y}_N < k+1\}]\, E^{W^1}[g]$$

$$+ \sum_{0\le k \text{ odd } <N} E[f(\frac{\tilde{Y}^N}{N})\, 1\{k \le \tilde{Y}_N < k+1\}]\, E^{W^2}[g]\,.$$

Observe now that $\tilde{Y}^N$ converges in law to an exponential variable of parameter 1. Indeed, for $t > 0$, N large,

$$E[\tilde{Y}^N > t] = E[\bigcap_1^N (Y^i > \frac{t}{N}] = (1 - \frac{t}{N})^N\,,$$

which tends to e^{-t}.

From this it is easy to argue that the expression in (3.23) tends to:

$$\sum_{k \text{ even } \ge 0} (e^{-k} - e^{-(k+1)})\, f(0)\, E^{W^1}[g] + \sum_{1\le k \text{ odd}} (e^{-k} - e^{-(k+1)})\, f(0) E^{W^2}[g]\,,$$

which is equal to the right member of (3.22). This yields our claim. □

So we see that the limit law for $\bar{Z}^N$ is given by Q which is distinct from the natural conditioning of the nonlinear process to be zero at time zero corresponding to $\delta_0 \otimes (\frac{1}{2}W^1 + \frac{1}{2}W^2)$. We also refer the reader to Guo-Papanicolaou [13], where the limit behavior of a tagged particle process for a system of interacting Brownian motions is studied.

II. A local interaction leading to Burgers' equation

The object of this chapter is to present one model of local interactions, proposed by McKean [27], which leads to Burgers' equation, as the forward equation of the nonlinear process.

The "laboratory example" we discussed in Chapter I, section 1),

$$dX_t^i = dw_t^i + \frac{1}{N}\sum_1^N b(X_t^i, X_t^j)\, dt\,, \quad i = 1, \ldots, N\,.$$

with $b(\cdot,\cdot)$ bounded Lipschitz, has an interaction term $\frac{1}{N}\sum_1^N b(X_t^i, X_t^j)dt$. Such a function b, independent of N and regular, corresponds to an interaction at a macroscopic distance. In this chapter we will be dealing with the one dimensional situation, when $b(\cdot,\cdot)$ =const $\delta(x-y)$, and the interaction will be local in nature. We will start first in section 1) with a warm up calculation, of δ-like interaction terms, for independent particles.

1) A warm up calculation.

In our "laboratory example", as a result of the Lipschitz property of $b(\cdot,\cdot)$, and Theorem 1.4 of Chapter I, one sees easily that for each t,

$$\lim_{N\to\infty} E\Big[\frac{1}{N}\sum_{i=1}^N \Big(\frac{1}{N}\sum_{j=1}^N b(X_t^i, X_t^j) - \int b(X_t^i, y)u_t(dy)\Big)^2\Big] = 0\,.$$

In other words, the instantaneous drift term $\frac{1}{N}\sum_j b(X_t^i, X_t^j)$ seen by particle i, is getting close to the quantity $\int b(X_t^i, y)\; u_t(dy)$ which simply depends on X_t^i.

We are now going to analyze similar quantities when $b(\cdot,\cdot)$ is replaced by $\phi_{N,a}(x - y) = N^{ad}\phi(N^a(x-y))$, with $\phi(\cdot) \geq 0$, smooth, compactly supported, on $\mathbb{R}^d$, $\int\phi(x)dx = 1$. The $X_\cdot^i$ will be independent d-dimensional Brownian motion, with initial distribution $u_0(dx) = u_0(x)dx$ having smooth compactly supported density. The quantities $Z_i = \frac{1}{(N-1)}\sum_{j\neq i}\phi_{N,a}(X_t^i - X_t^j)$ will play the role of an instantaneous "pseudo drift" seen by particle i. We will denote by $p_s(x,y)$ the Gaussian transition density. Now for $0 < a < \infty$, we will look at the $N\to\infty$, behavior of

$$a_N = E\Big[\frac{1}{N}\sum_{i=1}^N \Big(\frac{1}{(N-1)}\sum_{j\neq i}\phi_{N,a}(X_t^i - X_t^j) - u_t(X_t^i)\Big)^2\Big]\,. \tag{1.1}$$

The interpretation of the parameter a, is that now the interaction range between particles is of order N^{-a}. Thanks to symmetry,

$$\begin{aligned} a_N &= E\big[\big(\frac{1}{N-1}\sum_{j=2}^{N}\phi_{N,a}(X_t^1 - X_t^j) - u_t(X_t^1)\big)^2\big] \\ &= E\big[\big(\frac{1}{(N-1)}\sum_{j=2}^{N}\phi_{N,a}(X_t^1 - X_t^j) - \phi_{N,a} * u_t(X_t^1)\big)^2\big] \\ &\quad + E\big[\big(\phi_{N,a} * u_t(X_t^1) - u_t(X_t^1)\big)^2\big] . \end{aligned}$$

The second term of the previous expression clearly tends to zero. Expanding the square in the first term we find

$$\begin{aligned} a_N &= \frac{1}{(N-1)}E\big[\big(\phi_{N,a}(X_t^1 - X_t^2) - \phi_{N,a} * u_t(X_t^1)\big)^2\big] + o(N) \\ &= \frac{1}{(N-1)}E\big[\phi_{N,a}^2(X_t^1 - X_t^2)\big] + o(N) \qquad (1.2) \\ &= \frac{1}{(N-1)}N^{ad}E\big[N^{ad}\phi^2(N^a(X_t^1 - X_t^2))\big] + o(N) . \end{aligned}$$

So we see that

$$\begin{aligned} &\text{— when} \quad 0 < a < 1/d\,, \quad \lim_N a_N = 0\,, \\ &\text{— when} \quad a = 1/d\,, \quad \lim_N a_N = \int \phi^2 dx \times \|u_t\|_{L^2}^2 > 0\,, \qquad (1.3) \\ &\text{— when} \quad a > 1/d\,, \quad \lim_N a_N = \infty\,. \end{aligned}$$

The case $0 < a < 1/d$ corresponds to "moderate interaction" (see Oelschläger [34]). In fact when $a = 1/d$, we are in a "Poisson approximation" regime, and conditionally on $X_t^1 = x$, the sum

$$\frac{1}{(N-1)}\sum_{j=2}^{N}\phi_{N,a}(x - X_t^j) = (1 + o(N)) \times \sum_{j=2}^{N}\phi(N^{1/d}(x - X_t^j))\,,$$

converges in law to the distribution of $\int_{\mathbb{R}^d} M(dy)\phi(y)$, with $M(dy)$ Poisson point process of intensity $u_t(x)dy$ (there is no misprint here). So conditioned on X_t^i there is a true fluctuation of the quantity $\frac{1}{(N-1)}\sum_{j\neq i}\phi_{N,1/d}(X_t^i - X_t^j)$, for each i.

When $a > 1/d$, conditionally on $X_t^1 = x$, the quantity $\frac{1}{(N-1)}\sum_{j=2}^{N}\phi_{N,a}(x - X_t^j)$ is zero with a probability going to 1 uniformly in x, but has conditional expectation approximately $u_t(x)$. So now we really have huge fluctuations.

Let us by the way mention that even in the presence of fluctuations a propagation of chaos result may hold. One can in fact see that the symmetric variables $Z_i =$

$\frac{1}{N}\sum_{j\neq i}\phi_{N,a}(X_t^i - X_t^j)$, $s \leq i \leq N$, are v_a-chaotic. Here v_a stands for the law of $u_t(X_t)$, when $a < 1/d$, the law of $\int_{R^d} M(dy)\,\phi(y)$, where conditionally on X_t, M is a Poisson point process with parameter $u_t(X_t)$, when $a = 1/d$, and trivially the Dirac mass in 0 when $a > 1/d$. The case $a = 1/d$ is somewhat comparable to example g) in Chapter I, section 3).

Since we are interested in interactions going very fast to δ (in fact being δ), if we hope to see our "pseudodrift" seen by particle i close for N large to a quantity just depending on particle i, some helping effect has to come to rescue us. This helping effect will be integration over time.

Integration over time as a smoothing effect: We are now going to replace the quantity $\frac{1}{(N-1)}\sum_{j\neq i}\phi_{N,a}(X_t^i - X_t^j)$, for each i, by $\frac{1}{(N-1)}\sum_{j\neq i}\int_0^t \phi_{N,a}(X_s^i - X_s^j)ds$. Correspondingly we are now interested in the limit behavior of:

$$b_N = E\Big[\frac{1}{N}\sum_{i=1}^{N}\Big(\frac{1}{(N-1)}\sum_{j\neq i}\int_0^t \phi_{N,a}(X_s^i - X_s^j)\,ds - \int_0^t u_s(X_s^i)ds\Big)^2\Big]\,. \tag{1.4}$$

An analogous calculation as before yields:

$$b_N = \frac{1}{(N-1)}E\Big[\Big(\int_0^t ds\phi_{N,a}(X_s^1 - X_s^2)ds\Big)^2\Big] + o(N)\,.$$

If we introduce $W_s = X_s^1 - X_s^2$, W_s is a Brownian motion with initial distribution $u_0 \,*\, \check{u}_0 = v_0$, and transition density $p_{2s}(x,y)$. We find that

(1.5)

$$b_N = \frac{2}{(N-1)}\int v_0(dx)\,dy\,dz\int_0^t du\;p_{2u}(x,y)\;\phi_{N,a}(y)\int_0^{t-u} dv\;p_{2v}(y,z)\;\phi_{N,a}(z){+}o(N)\,.$$

It is clear that in dimension $d = 1$, for any value of a, (and formally in(1.5), even if $\phi_{N,a} = \delta$), $\lim_N b_N = 0$. In dimension $d \geq 2$, taking as new variables $N^a y$, and $N^a z$, we find

$$b_N = \frac{2}{(N-1)}\int v_0(dx)dy\;dz\int_0^t du$$

$$p_{2u}(x,\frac{y}{N^a})\;\phi(y)\int_0^{(t-u)} dv\;p_{2v}(\frac{y}{N^a},\frac{z}{N^a})\;\phi(z) + o(N)\,.$$

In dimension $d = 2$, we see again from the logarithmic Green's function singularity appearing in the term $\int_0^{(t-u)} dv\;p_{2v}(y/N^a, z/N^a)$, that for any $a < \infty$, $\lim_N b_N = 0$, and in fact it is clear that b_N will not be vanishing unless the interaction range is exponentially small. In dimension $d \geq 3$, using the transition density scaling

$$b_N = \frac{2}{(N-1)} N^{a(d-2)} \int v_0(dx) dy\, dz \int_0^t du$$
$$p_{2u}(x, \frac{y}{N^a})\ \phi(y) \int_0^{(t-u)N^{2a}} dv\ p_{2v}(y,z)\ \phi(z) + o(N)\ .$$

It is now clear that

$$\text{(1.6)}\qquad \begin{aligned} &\text{for} \quad a < \frac{1}{d-2}\,, \quad \lim_N b_N = 0\,,\\ &\text{for} \quad a = \frac{1}{d-2}\,, \quad \lim b_N = \text{const} > 0\,.\\ &\text{for} \quad a > \frac{1}{d-2}\,, \quad \lim b_N = \infty\,. \end{aligned}$$

So we see that the integration over time has removed the existence of a critical exponent $1/d$, in dimension 1. In dimension 2 the new critical regime corresponds to an exponentially small range of interaction. In dimension $d \geq 3$, the critical exponent $a = 1/d$ is raised to $1/(d-2)$.

In Chapter III, we will see that there is a "Poissonian picture" corresponding to these critical regimes in dimension $d \geq 2$.

As for dimension 1, we have seen that $\lim b_N = 0$, for any a. In fact it is the consequence of an even stronger result, namely:

$$\lim E\Big[\frac{1}{N(N-1)} \sum_{i \neq j} \Big(\int_0^t \phi_{N,a}(X_s^i - X_s^j) ds - \frac{1}{2} L^0(X^i - X^j)_t \Big)^2\Big] = 0\,.$$

We will use these type of ideas in the next sections, in our approach to the propagation of chaos result. For other approaches, we refer the reader to Gutkin [14], Kotani-Osada [21].

2. The N-particle system and the nonlinear process.

The N-particle system will be given by the solution

$$\text{(2.1)}\qquad \begin{aligned} &dX_t^i = dB_t^i + \frac{c}{N} \sum_{j \neq i} dL^0(X^i - X^j)_t\,,\\ &(X_0^i)_{1 \leq i \leq N} \in \Delta_N\,, \quad u_N - \text{distributed}, \quad c > 0\,. \end{aligned}$$

Here $L^0(X^i - X^j)_t$ denotes the symmetric local term in 0 of $X^i_. - X^j_.$, B^i are independent 1-dimensional Brownian motions, Δ_N is the subset of $\boldsymbol{R}^N$ where no three coordinates are equal, and of course the initial conditions X_0^i are independent of the Brownian motions. Existence and uniqueness of solutions of (2.1) holds trajectorically and in law.

The solution is Δ_N valued, strong Markov. It has the Brownian like scaling property: $\lambda X^x_{\cdot/\lambda^2} \overset{\text{law}}{\sim} X^{\lambda x}_{\cdot}$, for an initial starting point $x \in \Delta_N$. The δ-interaction interpretation comes from the fact that the solution of

$$
(2.2) \qquad \begin{aligned} dX^{i,\epsilon}_t &= dB^i_t + \frac{2c}{N}\sum_{j\neq i}\phi_\epsilon(X^{i,\epsilon}_t - X^{j,\epsilon}_t)\, dt\ , \quad \phi_\epsilon(\cdot) = \frac{1}{\epsilon}\phi(\frac{\cdot}{\epsilon})\ , \\ X^{i,\epsilon}_0 &= X^i_0\ , \end{aligned}
$$

converge a.s. uniformly on compact intervals to the solution of (2.1), as ϵ goes to zero. Finally if u_N is symmetric the law of the $(X^i_\cdot)$ is symmetric as well. For these results we refer the reader to [42] and [46].

Let us now present the nonlinear process. One might be tempted to define the process as the solution of

$$
(2.3) \qquad \begin{aligned} &dX_t = dB_t + 2c\; u(t, X_t)\; dt \\ &X_0\ ,\ u_0 - \text{distributed,} \end{aligned}
$$

for $u(t,x)$ the density of X_t at time t, from which one would deduce that $u(t,x)$ is the solution of Burgers' equation:

$$
\begin{aligned} \partial_t u &= \frac{1}{2}\partial^2_x u - 2c\ \partial_x(u^2) \\ u_{t=0} &= u_0\ , \end{aligned}
$$

However the density of the law at time t of a process is an ill behaved function for the weak topology on $M(C(\check{R}_+, R^d))$, and such a characterization of the nonlinear process is not best suited to apply the strategy explained in Chapter I 2). From the previous section, we know that quantities integrated over time are better behaved, and we are going to characterize the law of the nonlinear process as the unique law of continuous semimartingales $X_\cdot$, on some filtered probability space endowed with a Brownian motion $B_\cdot$, solution of:

(2.4) $X_t = X_0 + B_t + A_t$, X_0, u_0-distributed, A_t continuous adapted, of integrable variation with $A_t = cE_Y[L^0(X-Y)_t]$. $Y_t = Y_0 + \overline{B}_t + \overline{C}_t$ defined on an independent filtered space endowed with a Brownian motion $\overline{B}_\cdot$ has the same law as $(X_\cdot)$, and $\overline{C}$ is a continuous adapted integrable variation process. Here $L^0(X-Y)_\cdot$ denotes the symmetric local time in zero of $X_\cdot - Y_\cdot$, defined on the product space.

For the moment we will be concerned with the uniqueness statement. Indeed the more important role of the nonlinear process, in the proof of "propagation of chaos" comes in the identification of limit points of laws of empirical measures. We will see in the course of the proof that solutions of (2.4) have indeed the structure of (2.3). We will start with some a priori estimates:

Proposition 2.1.

i) *If*

$$X_t = X_0 + B_t + A_t \text{ ,(}B.\text{ Brownian motion, } A.\text{ integrable variation),} \tag{2.5}$$

the image of $1_{[0,T]} ds\, dP$ *under the map* $(s,\omega) \to (s, X_s(\omega)) \in [0,T] \times \boldsymbol{R}$, *for* $T < \infty$, *has an* L^2 *density* $u(s,x)$, *with* $\|u\|_{L^2([0,T]\times \boldsymbol{R})} \le C(E[|A|_T])$.

ii) *If* $Y.$ *on some independent space has a decomposition as in* (2.5), *then*

$$E_Y[L^0(X-Y)_t] = 2\int_0^t u(s,X_s)\, ds \ , \tag{2.6}$$

and this defines a continuous increasing integrable process which only depends on the law of Y. *Here* u *is the density associated to* Y *in* i).

Proof:

i) Take $\phi(\cdot) \ge 0$, smooth, symmetric supported in $[-1/2, 1/2]$, $\int \phi = 1$. Define $\psi = \phi * \phi$ so that $\psi \ge 0$, is symmetric supported in $[-1,1]$ and $\int \psi = 1$. If we define $\psi_n(\cdot) = n\,\psi(n\cdot)$, $\phi_n(\cdot) = n\phi(n\cdot)$, we have $\psi_n(\cdot) = \phi_n * \phi_n$. We now use ψ_n to define our test function

$$F_n(x) = \int_{-\infty}^x \int_{-\infty}^u (\psi_n(v) - \psi_1(v))\, dv\, du \ , \tag{2.7}$$

so that $F_n'(x) = 0$, for x outside $[-1,+1]$, $|F_n'| \le 1$, and $|F_n| \le 2$.

Take $X_t(\omega')$ an independent copy of $X_t(\omega)$, and set on the product space $Z_t = X_t(\omega) - X_t(\omega')$. Applying Ito's formula to $F_n(Z_t)$ we find after integration:

$$\begin{aligned} E[F_n(Z_t)] =& E[F_n(Z_0)] + E[\int_0^t F_n'(Z_s)\, d(A_s(\omega) - A_s(\omega'))] \\ &+ E[\int_0^t (\psi_n(Z_s) - \psi_1(Z_s)) ds] \ . \end{aligned} \tag{2.8}$$

From this it follows that:

$$E[\int_0^T \psi_n(Z_s) ds] \le 4 + \|\psi\|_\infty T + 2E[|A|_T] \ . \tag{2.9}$$

But

$$\begin{aligned} E[\int_0^T \psi_n(Z_s) ds] &= \int_0^T \langle u_s \otimes u_s, \int \phi_n(y-z)\phi_n(y'-z) dz\rangle \\ &= \|u_n\|^2_{L^2([0,T]\times \boldsymbol{R})} \ , \end{aligned}$$

where u_s denotes the law of X_s at time s, and $u_n(s,x) = \int u_s(dy)\ \phi_n(x-y)$.

i) easily follows from (2.9) now.

ii) Denote by u and v the densities corresponding to Y and X, using i). The quantity $\int_0^T ds\ u(s,X_s)$ does not depend on the version of u which is used and

$$E[\int_0^T u(s,X_s)ds] = \langle u,v\rangle_{L^2([0,T]\times R)} < \infty\ .$$

In our case the corresponding uniform estimate to (2.9), shows that

$$E[L^0(X-Y)_T] < \infty\ ,$$

and in fact working with the limit $F_\infty(\ )$ of the functions F_n, applying to it Tanaka's formula, one finds easily by studying $F_\infty(Z_t) - F_n(Z_t)$ that

$$L^0(X-Y)_t = \lim_{n\to\infty} 2\int_0^t \psi_n(X_s - Y_s)\ ds \quad \text{in } L^1\ . \tag{2.10}$$

The same with approximations of δ supported in R_+ or R_- gives the corresponding result for the right and left continuous local times $L^{0,r}(X-Y)$, and $L^{0,\ell}(X-Y)$.. From (2.10), after integration over Y we find, u_n denoting now the ψ_n regularization of u,

$$\lim_n E[|E_Y[L^0(X-Y)_t] - 2\int_0^t u_n(s,X_s)ds|] = 0\ .$$

On the other hand

$$\begin{aligned} lim_n E[|\int_0^t u_n(s,X_s)ds - \int_0^t u(s,X_s)ds|] &\le \lim_n E[\int_0^t |u_n - u|(s,X_s)ds] \\ &\le \lim_n \|u-u_n\|_2 \|v\|_2 = 0\ . \end{aligned}$$

ii) easily follows. From our proof it is also clear that similarly

$$E_Y[L^{0,r}(X-Y)_t] = E_Y[L^{0,\ell}(X-Y)_t] = 2\int_0^t u(s,X_s)ds\ , \tag{2.11}$$

the approximating density from the right or from the left just as well converging in L^2 to u.

□

We now state our required uniqueness statement, in fact existence will be proved later.

Theorem 2.2. *Let $S(u_0)$ be the set of laws of solutions of (2.4). $S(u_0)$ has at most one element. If P is the law of a solution of (2.4), $u_t = X_t \circ P$, satisfies*

(2.12) $\exp\{-4cF_t(x)\} = \exp\{-4cF_0\} * p_t(x)$, F_t distribution function of u_t, (that is u_t satisfies Burgers' equation).

Proof: By Proposition 2.1, we know that $X_t = X_0 + B_t + 2c\int_0^t u(s, X_s)ds$, where $u \in L^2([0,T] \times \boldsymbol{R})$ is the density of the law of X_s for a.e. s. Applying Ito's formula to $f(T, X_T)$, with $f \in C_K^\infty((0,T) \times \boldsymbol{R})$, we see that

$$0 = E[\int_0^T (\partial_s f + \frac{1}{2}\partial_x^2 f + 2cu\partial_x f)(s, X_s)ds] ,$$

using the definition of u we deduce that:

(2.13) $-\partial_s u + \frac{1}{2}\partial_x^2 u - 2c\partial_x(u^2) = 0$, in the distribution sense on $(0,T) \times \boldsymbol{R}$.

If we now set: $F(t,x) = \int_{-\infty}^x u_t(dy)$, $-\partial_t F + \frac{1}{2}\partial_x^2 F - 2cu^2$ is a distribution invariant under spatial translations. The value of this distribution tested against $f \in C_K^\infty((0,T) \times \boldsymbol{R})$ is equal to

$$\int_{(0,T)\times\boldsymbol{R}} dt\ dx\ F(t, x-z)(\partial_t f + \frac{1}{2}\partial_x^2 f)(t,x) + 2cu^2(t, x-z)\partial_x f(t,x)$$

for any z in $\boldsymbol{R}$. Letting z tend to $+\infty$, we see this last expression goes to zero, so that

$$-\partial_t F + \frac{1}{2}\partial_x^2 F - 2cu^2 = 0 \text{ in the distribution sense.}$$

Now $w'/w = -4cu$ is a change of unknown function which linearizes Burger's equation into $\partial_t w = \frac{1}{2}\partial_x^2 w$. So we intoduce a regularization by convolution in space time F_λ in $(\epsilon, T-\epsilon)$, and set $w_\lambda(t,x) = \exp\{-4cF_\lambda(t,x)\}$. One now has: $\partial_t w_\lambda - \frac{1}{2}\partial_x^2 w_\lambda = 8c^2 w_\lambda[(u^2)_\lambda - (u_\lambda)^2]$, but as $\lambda \to 0$ $u_\lambda^2 \to u^2$ in $L^1((\epsilon, T-\epsilon), \boldsymbol{R})$, $u_\lambda \to u$ in $L^2((\epsilon, T-\epsilon) \times \boldsymbol{R})$. From this letting λ go to zero we see that $\partial_t w - \frac{1}{2}\partial_x^2 w = 0$, in the distribution sense, so that by hypoellipticity, w is in fact smooth, and bounded, and from this for $0 < s < t < T$, $w_t = w_s * p_{t-s}$. Letting now s go to zero we find (2.12). So u is in fact the solution of Burgers' equation. Now $u(v,x)$, $v \geq s > 0$, is Lipschitz and bounded, and

$$X_t = X_s + (B_t - B_s) + \int_s^t 2cu(v, X_v)\ dv\ , \quad t \geq s\ ,$$

X_s is $u_s(dx)$ distributed.

It follows that any two solutions of (2.4) generate the same law on $C([s,+\infty), \boldsymbol{R})$. Since s is arbitrary our claim follows.

□

Remark 2.3. When $u_0(dx) = u_0(x)\,dx$, with u_0 bounded measurable, one sees easily that the solution of Burgers' equation given by (2.12) is bounded measurable. Now one has trajectorial uniqueness for the equation

$$X_t = X_0 + B_t + 2c \int_0^t u(s, X_s)\, ds\ ,$$

see Zvonkin [58]. The proof of Theorem 2.2 now yields a trajectorial uniqueness for the solution of (2.4), when X_0, $(B_.)$ are given.

□

3) The Propagation of chaos result.

In this section we will prove the propagation of chaos result:

Theorem 3.1. *If u_N, supported on Δ_N, is u_0-chaotic, then $(X^1_., \dots, X^N_.)$ solutions of (2.1) are P_{u_0}-chaotic, where P_{u_0} is the unique element of $S(u_0)$.*

With no restriction of generality we will assume that $c > 0$.

We will use in this section tightness estimates, which will be proved in section 4), namely

Proposition 3.2. *There is a $K > 0$, such that for $N > 2c$, $1 \le i \ne j \le N$, $s \le t$,*

$$\begin{aligned} &E[|X^i_t - X^i_s|^4] \le K|t-s|^2 \\ &E[(L^0(X^i - X^j)^t_s)^4] \le K|t-s|^2\ . \end{aligned} \tag{3.1}$$

Theorem 3.1 is in fact the corollary of a stronger statement, that will be presented now. Let us first introduce some notation. We denote by $\tilde{H}$ the closed subset of $C(R_+, R) \times C_0(R_+, R)$:

$$\tilde{H} = \{(X_., B_.) \in C \times C_0 : X_. - X_0 - B_. \in C_0^+\}\ .$$

C_0 and C_0^+ are respectively the space of continuous and continuous increasing functions from R_+ to R, with value zero at time zero. In the course of the proof we will show that in fact $S(u_0)$ has indeed one element. $\tilde{P}_{u_0}$ will stand for the joint distribution of $(X_., B_.)$, if X is the nonlinar process and $B_.$ its driving Brownian motion. Precisely $\tilde{P}_{u_0} \in M(H)$ will be the measure on $\tilde{H}$ image of P_{u_0} under the map: $X_. \in C \to (X_., X_. - X_0 - 2c\int_0^. u(s, X_s)ds) \in \tilde{H}$, where u is the solution of Burgers' equation with initial condition u_0. Let us mention that $u(s,x) \le \text{const}\ s^{-1/2}$, so that the map is in fact continuous. Finally we will consider the law Q image of $\tilde{P}_{u_0} \otimes \tilde{P}_{u_0}$ on the space $H = \tilde{H} \times \tilde{H} \times C_0^+$, under the map:

$$(X^1_., B^1_.)(X^2_., B^2_.) \to (X^1_., B^1_., X^2_., B^2_., L^0(X^1 - X^2)_.)\ ,$$

which is clearly $\tilde{P}_{u_0} \otimes \tilde{P}_{u_0}$ a.s. defined.

Using Proposition 2.4 of Chapter I, Theorem 3.1 is an easy consequence of

Theorem 3.3. *The empirical measures*

$$\overline{Y}_N = \frac{1}{N(N-1)} \sum_{i \neq j} \delta_{(X^i_., B^i_., X^j_., B^j_., L^0(X^i - X^j)_.)} \in M(H) \, ,$$

converge in law to the constant Q.

Before embarking on the proof of Theorem 3.3, we are going to give some implications of a result such as Theorem 3.3, in terms of the quantities which were presented in section 1. In our present context the reader who is an afficionado of the "density profile" point of view is in fact interested merely in the behavior for large N of $\langle \eta^N_t, f \rangle = \frac{1}{N} \sum_i f(X^i_t)$, and using Ito's formula, only the bounded variation term $c \int_0^t \frac{1}{N^2} \sum_{i \neq j} f'(X^i_s) dL^0(X^i - X^j)_s$, is really problematic.

So we will give an implication of Theorem 3.3 in terms of this quantity.

Corollary 3.4.

$$\text{(3.2)} \qquad \lim_{N \to \infty} E\Big[\frac{1}{N} \sum_i \Big(\frac{1}{N} \sum_{j \neq i} L^0(X^i - X^j)_t - 2 \int_0^t u(s, X^i_s) ds\Big)^2\Big] = 0 \, , \; t \geq 0 \, .$$

More generally for f continuous bounded $\boldsymbol{R} \to \boldsymbol{R}$

$$\lim_{N \to \infty} E\Big[\frac{1}{N} \sum_i \Big(\frac{1}{N} \sum_{j \neq i} \int_0^t f(X^i_s) \, dL^0(X^i - X^j)_s - 2 \int_0^t f(X^i_s) u(s, X^i_s) ds\Big)^2\Big] = 0 \, ,$$

and

$$\text{(3.4)} \quad \lim_{N \to \infty} E\Big[\Big| \int_0^t \frac{1}{N^2} \sum_{i \neq j} f(X^i_s) dL^0(X^i - X^j)_s - 2 \int_0^t \int_{\boldsymbol{R}} u^2(s,x) \, f(x) \, ds \, dx \Big|\Big] = 0 \, .$$

Proof: First (3.2) is a special case of (3.3), and (3.4) is an immediate consequence of (3.3), and the fact that thanks to Theorem 3.1

$$\lim_N E\Big[\Big| \frac{1}{N} \sum_i \int_0^t f(X^i_s) u(s, X^i_s) ds - \int_0^t \int_{\boldsymbol{R}} u^2(s,x) \, f(x) \, ds \, dx \Big|\Big] = 0 \, .$$

Now to prove (3.3), notice that

$$(X_., B_.) \in \tilde{H} \to \int_0^t f(X_s) c^{-1} d(X_. - X_0 - B_.)_s \in \boldsymbol{R} \, ,$$

is a continuous map, and the expression under study is

$$E\Big[\frac{1}{N}\sum_i\Big(\int_0^t f(X_s^i)c^{-1}d(X_\cdot^i - X_0^i - B_\cdot^i)_s - 2\int_0^t u(s, X_s^i)ds\Big)^2\Big]\,. \tag{3.5}$$

If we now replace the square in (3.5) by $(\ )^2 \wedge A$, using Theorem 3.3, the limit as N goes to infinity of the new quantity is

$$E_{\bar{P}_{u_0}}\Big[\Big(\int_0^t f(X_s)c^{-1}d(X_\cdot - X_0 - B_\cdot)_s - 2\int_0^t f(X_s)u(s, X_s)ds\Big)^2 \wedge A\Big] = 0\,.$$

Using estimates (3.1), to remove the truncation, one easily proves (3.3).

□

Proof of Theorem 3.3: The tightness of the laws of the $\overline{Y}_N$,comes from (3.1), together with the fact that u_N is u_0-chaotic. We now have to prove that any limit point $\overline{Q}_\infty$ of the laws of the $\overline{Y}_N$, is in fact δ_Q. The proof of the following lemma is easy and we refer to [42], for more details. G_t will denote the natural σ-field on H.

Lemma 3.5. *For* $\overline{Q}_\infty$ a.e. $m \in M(H)$, $(X_\cdot^1, B_\cdot^1)$ and $(X_\cdot^2, B_\cdot^2)$ *are* m *independent identically distributed,* $(B_\cdot^1, B_\cdot^2)$ *is a two dimensional* G_t*-Brownian motion, and the law of* X_0^1 *(or* X_0^2*) under* m *is* u_0.

Let us first introduce some notation. We define the following functions on H:

$A_t^i = X_t^i - X_0^i - B_t^i\ ,\ \ i = 1,2\ , \in C_0^+$ (continuous increasing process),

A_t is the C_0^+ valued component of H ,

$$H_t = |X_t^1 - X_t^2| - |X_0^1 - X_0^2| - \int_0^t \mathrm{sign}^+(X_s^1 - X_s^2)\, dA_s^1$$
$$- \int_0^t \mathrm{sign}^+(X_s^2 - X_s^1) dA_s^2 - A_t\,,$$
$$D_t = |X_t^1 - X_t^2| - |X_0^1 - X_0^2| - \int_0^t \mathrm{sign}^-(X_s^1 - X_s^2) dA_s^1$$
$$- \int_0^t \mathrm{sign}^-(X_s^2 - X_s^1) dA_s^2 - A_t\,.$$

Lemma 3.6. *For* $\overline{Q}_\infty$ a.e. $m \in M(H)$,

$$A_t^i = cE_m[A_t/\sigma(X_\cdot^i, B_\cdot^i)]\,,\quad i = 1,2, \tag{3.6}$$

$$H_t \textit{ is a continuous } G_t - \textit{supermartingale}, \tag{3.7}$$

$$D_t \textit{ is a continuous } G_t - \textit{submartingale}. \tag{3.8}$$

Proof: Let us first prove (3.6). Set $F(m) = \langle m, (cA_t - A_t^1)\ g(X_\cdot^1, B^1)\rangle$, where g is continuous bounded, and define $F_\alpha(m)$ by the same expression, replacing cA_t by $(cA_t) \wedge \alpha$ and A_t^1 by $A_t^1 \wedge \alpha$. It is enough to show that $F(m) = 0$, $\overline{Q}_\infty$-a.s. Observe now that by (3.1), for $k \leq \infty$,

$$\begin{aligned} E_{\overline{Q}_{N_k}}[|F(m) - F_\alpha(m)|] &\leq \text{const } E_{\overline{Q}_{N_k}}[\langle m, (cA_t - \alpha)_+ + (A_t^1 - \alpha)_+\rangle] \\ &\leq \text{const } \alpha^{-1} . \end{aligned}$$

It follows that

$$\begin{aligned} &E_{\overline{Q}_\infty}[|F(m)|] \leq \underline{\lim}_k E_{\overline{Q}_{N_k}}[|F(m)|] \\ &= c\ \underline{\lim}_k E_{N_k}[|\frac{1}{N(N-1)} \sum_{i \neq j} \{(L^0(X^i - X^j)_t - \frac{1}{N} \sum_{k \neq i} L^0(X^i - X^k)_t)\ g(X^i, B^i)\}|] \\ &= c\ \underline{\lim}_k E_{N_k}[|\frac{1}{N} \sum_i (\frac{1}{N(N-1)} \sum_{j \neq i} L^0(X^i - X^j)_t)\ g(X^i, B^i)|] = 0 . \end{aligned}$$

Let us now prove (3.7), (3.8) being proved in a similar fashion.

We now introduce for $t > s$

$$F(m) = \langle m, (H_t - H_s) \cdot g_s \rangle ,$$

where g_s is a nonnegative G_s-measurable continuous bounded function. Take now $K(m) \geq 0$, a continuous bounded function, it is enough to show that

$$E_{\overline{Q}_\infty}[F(m)\ K(m)] \leq 0 ,$$

to be able to conclude that (3.7) holds. Observe now that the functional ($\alpha > 0$), on H,

$$-\Big(\int_0^t \text{sign}^+(X_s^1 - X_s^2)\ d(A^1 \wedge \alpha)_s + \int_0^t \text{sign}^+(X_s^2 - X_s^1)\ d(A^2 \wedge \alpha)_s\Big)$$

is bounded lower semicontinuous. Using very similar truncation arguments, we see that

$$\begin{aligned} &E_{\overline{Q}_\infty}[F(m)\ K(m)] \\ &\leq \underline{\lim}_k E_{N_k}[K(\overline{Y}_N)(\frac{1}{N(N-1)} \sum_{i \neq j} \{(|X_t^i - X_t^j| - |X_s^i - X_s^j| \\ &- c \int_s^t \text{sign}^+(X_u^i - X_u^j) \times \frac{1}{N} \sum_{k \neq i} dL^0(X^i - X^k)_u \\ &- c \int_s^t \text{sign}^+(X_u^j - X_u^i) \times \frac{1}{N} \sum_{h \neq j} dL^0(X^j - X^h)_u \\ &- L^0(X^i - X^j)_t)\ g_s(X_\cdot^i, B_\cdot^i, X_\cdot^j, B_\cdot^j, L^0(X^i - X^j)_\cdot)\}\})] \end{aligned}$$

Since the process $(X^1_\cdot, \ldots, X^N_\cdot)$ is Δ_N valued, we can in fact replace sign^+ by sign in the previous expression, and find using Tanaka's formula:

$$\underline{\lim}_k E_{N_k}[K(\overline{Y}_N) \times \frac{1}{N(N-1)} \sum_{i \neq j} \int_s^t \text{sign}(X^i_u - X^j_u)\, d(B^i_u - B^j_u) \cdot g^{ij}_s] \,,$$

with obvious notations. This is less than:

$$\text{const} \cdot \underline{\lim}_k E_{N_k}[\{\frac{1}{N(N-1)} \sum_{i \neq j} \int_s^t \text{sign}(X^i_u - X^j_u)\, d(B^i_u - B^j_u) \cdot g^{ij}_s\}^2]^{1/2} \,,$$

which is easily seen to be zero after expanding the square and using the orthogonality of terms (i,j), (k,ℓ) with $\{i,j\} \cap \{k,\ell\} = \emptyset$.

□

Let us now continue the proof of Theorem 3.3. For an m satisfying the properties of Lemmas 3.5, 3.6, we know that

$$D_t = \int_0^t \text{sign}^+(X^1_s - X^2_s)\, d(B^1 - B^2)_s + L^{0,\ell}(X^1 - X^2)_t - A_t + 2\int_0^t 1(X^1_s = X^2_s)\, dA^1_s \,,$$

is a G_t-submartingale. From this we deduce that the bounded variation process

$$K^+_t = L^{0,\ell}(X^1 - X^2)_t - A_t + 2\int_0^t 1(X^1_s = X^2_s)\, dA^1_s \,, \tag{3.9}$$

is continuous increasing. Similarly, we see that:

$$K^-_t = L^{0,\ell}(X^1 - X^2)_t - A_t - 2\int_0^t 1(X^1_s = X^2_s)\, dA^2_s \,, \tag{3.10}$$

is a continuous decreasing process. From section 2 (2.11), and the independence under m of $(X^1_\cdot, B^1_\cdot)$, $(X^2_\cdot, B^2_\cdot)$, we know that:

$$E_m[L^{0,\ell}(X^1 - X^2)_t / (X^1, B^1)] = 2\int_0^t u(s, X^1_s)\, ds \,.$$

Conditioning (3.9) with respect to (X^1, B^1), we see that

$$\frac{1}{c} A^1_t + C^1_t = 2\int_0^t u(s, X^1_s)\, ds + 2\int_0^t p(s, X^1_s)\, dA^1_s \,, \tag{3.11}$$

where C^1_t is a continuous increasing process depending on (X^1, B^1), and $p(s,x) = \int 1(X^i_s = x) dm$, for $i = 1,2$. We can write

$$dA^1_t = 1(p(t, X^1_t) < \frac{1}{4c})\, dA^1_t + 1(p(t, X^1_t) \geq \frac{1}{4c})\, dA^1_t \,,$$

and from equation (3.11), we already know that

$$(\frac{1}{c} - 2p(t, X_t^1))\ 1(p(t, X_t^1) < \frac{1}{4c})\ dA_t^1$$

is absolutely continuous with respect to Lebesgue measure. Let us now study the measure $1(p(t, X_t^1) \geq \frac{1}{4c})dA_t^1$. It is supported by the closed set

$$F = \{t \geq 0\ ,\ \exists x \in R\ ,\ p(t, x) \geq \frac{1}{4c}\}\ ,$$

which has measure zero, since the law of X_t^1 (or X_t^2) has a density with respect to dx for almost every t.

Let us show that $F \subset \{0\}$. If not there is $I = (a, b) \subset F^c$, with $b < \infty$, $b \in F$. On I, $dA_t^1 = 1(p(t, X_t^1) < \frac{1}{4c})dA_t^1$, so that $1_I dA_t^1 << dt$. From this it immediately follows that $\int_I 1(X_s^2 = X_s^1)dA_s^1 = \int_I 1(X_s^2 = X_s^1)dA_s^2 = 0$, and by (3.9), (3.10) we get:

$$1_I \cdot dA_t = 1_I \cdot dL^{0,\ell}(X^1 - X^2)_t = 1_I dL^0(X^1 - X^2)_t\ . \tag{3.12}$$

Now the process $\overline{X}_t^1 = X_{t+a^1}^1$, for $a^1 \in I$, and $0 \leq t < b - a^1$, satisfies with obvous notations:

$$\overline{X}_t^1 = \overline{X}_0^1 + \overline{B}_t^1 + c \int_{\tilde{H}} L^0(\overline{X}^1 - \overline{X}^2)_t\ d\tilde{m}(\overline{X}^2, \overline{B}^2)\ .$$

From section 2), Theorem 2.2, this implies that

$$\exp\{-4c\overline{F}_t(x)\} = \exp\{-4c\overline{F}_0\}\ *\ p_t(x)\ ,\quad 0 \leq t < b - a^1\ .$$

So for t near $b - a^1$, $\exp\{-4c\overline{F}_t\}$ is uniformly continuous, bounded above and away from zero,so that $b \notin F$. This shows that $F \subset \{0\}$. Now the same reasoning we just made shows that $A_t = L^{\ell,0}(X^1 - X^2)_t = L^0(X^1 - X^2)_t$, and

$$X_t^1 = X_0^1 + B_t^1 + cE_{\tilde{m}}^2[L^0(X^1 - X^2)_t]\ , \tag{3.13}$$

where $\tilde{m}$ is the law of (X^1, B^1) or $(X_\cdot^2, B_\cdot^2)$ under m, and a similar equation for $X_\cdot^2$. So $S(u_0)$ is not empty and $m = Q$. This proves Theorem 3.3.

□

Let us mention that for any $u_0 \in M(R)$ we can find a sequence u_N u_0-chaotic and concentrated in Δ_N, so that we have indeed $S(u_0) = \{P_{u_0}\}$, by any u_0.

From convergence in law to trajectorial convergence:

In the case where $u_0(dx)$ has a bounded density with respect to Lebesgue measure, we can in fact consider the trajectorial solutions

$$(3.14) \qquad \overline{X}^i_t = X^i_0 + B^i_t + 2c \int_0^t u(s, \overline{X}^i_s)\, ds\ , \quad 1 \le i \le N\ ,$$

where u is the solution of Burgers' equation, with initial condition u_0. As already mentioned in Remark 2.3 we have pathwise uniqueness for the solution of (3.14). We can in fact obtain a trajectorial convergence in the fashion of Theorem 1.4 of Chapter I.

Theorem 3.7. *Suppose the (X^i_0) are independent u_0-distributed, for any i, T,*

$$(3.15) \qquad \lim_{N\to\infty} E[\sup_{t\le T} |X^{i,N}_t - \overline{X}^i_t|] = 0$$

Proof: If one now defines

$$\overline{Z}_N = \frac{1}{N(N-1)} \sum_{i\neq j} \delta_{(X^i_\cdot, B^i_\cdot, \overline{X}^i_\cdot, X^j_\cdot, B^j_\cdot, \overline{X}^j_\cdot, L^0(X^i - X^j)_\cdot)} \in M(\tilde{H} \times C \times \tilde{H} \times C \times C^+_0)\ ,$$

by the same proof as before, one sees that $\overline{Z}_N$ converges in law to the Dirac mass on the law of $(\overline{X}^1_\cdot, B^1_\cdot, \overline{X}^1_\cdot, \overline{X}^2_\cdot, B^2_\cdot, \overline{X}^2_\cdot, L^0(\overline{X}^1 - \overline{X}^2)_\cdot)$. Taking $F_\alpha(m) = \langle m, \sup_{s\le T} |X^1_s - \overline{X}^1_s| \wedge \alpha\rangle$, we see that $\lim_N E[\sup_{s\le T} |X^1_s - \overline{X}^1_s| \wedge \alpha] = 0$. Estimates (3.1) then allow us to prove (3.15).

□

4) Tightness estimates.

We are now going to explain how one derives the estimates (3.1) of Proposition 3.2. To obtain these estimates, we are going to use the increasing reordering $Y^1_t \le \ldots \le Y^N_t$ of the processes $(X^1_t, \ldots, X^N_t)$, (see Chapter I section 3) example e)). On the one hand, we will be able to use techniques of reflected processes to derive estimates on $Y^1_\cdot, \ldots, Y^N_\cdot$ and on the other hand the identity $Y^1_t + \cdots + Y^N_t = X^1_t + \cdots + X^N_t$ will yield a piece of information on the bounded variation term of $X^1_t + \cdots + X^N_t$.

In a first step, let us explain how an exponential control on $X^1_t + \cdots + X^N_t$ gives an individual exponential control on the X^i_t, $1 \le i \le N$, and yields (3.1).

Proposition 4.1. *Suppose there exist $d_1, d_2 > 0$, such that for $N > 2c$, any $x \in \Delta_N$, and $t > 0$,*

$$(4.1) \qquad E_x[\exp\{\frac{d_1}{\sqrt{t}} \sum_{i=1}^N (X^i_t - x^i)\}] \le d_2^N\ ,$$

then there exists $d, \bar{d} > 0$, such that for $N > 2c$, $i \in [1, N]$, $t > 0$,

$$(4.2) \qquad E_x[\exp\{\frac{d}{\sqrt{t}}(X_t^i - x^i)\}] \leq \bar{d}\,,$$

and (3.1) *holds.*

Proof: Let us first prove (4.2). Using scaling we can assume $t = 1$. From the estimate for $i \in [1, N]$:

$$E[\exp\{d(B_1^i + \frac{c}{N}\sum_{j\neq i} L^0(X^i - X^j)_1)\}]$$
$$\leq \exp\{d^2\} \times E[\exp\{\frac{2dc}{N}\sum_{j\neq i} L^0(X^i - X^j)_1\}]^{1/2}\,,$$

we see it is enough to focus on $E[\exp\{\frac{2d}{N}\sum_{j\neq i} L^0(X^i - X^j)_1\}]$. (4.2) will follow from a link between the individual terms $\frac{1}{N}\sum_{j\neq i} L^0(X^i - X^j)_1$ for $i \in [1, N]$ and the sums $\frac{1}{N}\sum_{1\leq i\neq j\leq N} L^0(X^i - X^j)_1$. We introduce to this end: f such that f, f' are bounded, $f'' = \delta_0 +$ bounded function:

$$f(y) = [\arctan(y)]_+\,,$$
$$f'(y) = 0, \quad y < 0, 1/2\,, \quad y = 0, (1 + y^2)^{-1}, \quad y > 0\,.$$
$$g(y) = -\frac{2y_+}{(1 + y^2)^2}\,, \text{ the regular part of } f''\,.$$

For simplicity, let us pick $i = 1$ as the individual term to be estimated. Set

(4.3)

$$S_t = \frac{-1}{N}\sum_{j\neq 1}\{f(X_t^1 - X_t^j) - f(X_0^1 - X_0^j)\} + \frac{c}{N^2}\sum_{\substack{j\neq 1\\ k\neq 1}}\int_0^t f'(X_s^1 - X_s^j)\,dL^0(X^1 - X^k)_s$$
$$- \frac{c}{N^2}\sum_{\substack{j\neq 1\\ k\neq j}}\int_0^t f'(X_s^1 - X_s^j)\,dL^0(X^j - X^k)_s + \frac{1}{N}\sum_{j\neq 1}\int_0^t g(X_s^1 - X_s^j)\,ds\,.$$

From Tanaka's formula, $S_t + \frac{1}{2N}\sum_{j\neq 1} L^0(X^1 - X^j)_t$ is a martingale with increasing process

$$(4.4) \quad U_t = \frac{1}{N^2}\sum_{j\neq 1}\int_0^t (f')^2(X_s^1 - X_s^j)\,2\,ds + \frac{1}{N^2}\sum_{j\neq k}\int_0^t f'(X_s^1 - X_s^j)\,f'(X_s^1 - X_s^k)\,ds\,.$$

We now have for $\lambda > 0$,

(4.4)

$$E[\exp\{\frac{\lambda}{2N}\sum_{j\neq 1} L^0(X^1 - X^j)_1\}]$$

$$= E[\exp\{\frac{\lambda}{2N}\sum_{j\neq 1} L^0(X^1 - X^j)_1 + \lambda S_1 - \lambda^2 U_1 + \lambda^2 U_1 - \lambda S_1\}]$$

$$\leq E[\exp\{\frac{\lambda}{N}\sum_{j\neq 1} L^0(X^1 - X^j)_1 + 2\lambda S_1 - 2\lambda^2 U_1\}]^{1/2} E[\exp\{2\lambda^2 U_1 - 2\lambda S_1\}]^{1/2} .$$

Now by the exponential martingale property the first term of the last expression is smaller than 1. So we have:

$$E[\exp\{\frac{\lambda}{2N}\sum_{j\neq 1} L^0(X^1 - X^j)_1\}] \leq E[\exp\{2\lambda^2 U_1 - 2\lambda S_1\}]^{1/2} . \tag{4.5}$$

Observe now that U_1 is bounded, and that the only two dangerous terms of S_1 are the second and third in (4.3). However the second term only comes with a negative sign in $-2\lambda S_1$, so that we only have to worry about the third:

$$\frac{2\lambda c}{N^2}\sum_{\substack{j\neq 1\\ k\neq j}} \int_0^1 f'(X_s^1 - X_s^j)\, dL^0(X^j - X^k)_s \leq 2\lambda c \frac{\|f'\|_\infty}{N^2} \sum_{1\leq j\neq k\leq N} L^0(X^j - X^k)_1 .$$

But from the Cauchy-Schwarz inequality:

$$E[\exp\{2\lambda\|f'\|_\infty \frac{c}{N^2} \sum_{1\leq j\neq k\leq N} L^0(X^j - X^k)_1\}]$$

$$\leq E[\exp\{4\lambda\|f'\|_\infty \frac{1}{N}\sum_{i=1}^N (X_1^i - x^i)\}]^{1/2} E[\exp\{-4\lambda\|f'\|_\infty \frac{1}{N}\sum_i B_1^i\}]^{1/2}$$

$$\leq E[\exp\{4\lambda\|f'\|_\infty \sum_{i=1}^N (X_1^i - x^i)\}]^{1/2N} \exp\{4\lambda^2\|f'\|_\infty^2/N\} .$$

Picking λ small enough we see that (4.2) follows from (4.1).

Using scaling and the Cauchy Schwarz inequality, one easily deduces from (4.2), the first estimate $E[|X_t^i - X_s^i|^4] \leq K|t-s|^2$, in (3.1). On Tanaka's formula:

$$L^0(X^i - X^j)_s^t = |X_t^i - X_t^j| - |X_s^i - X_s^j| - \int_s^t \operatorname{sign}(X_u^i - X_u^j)\, d(X_u^i - X_u^j) ,$$

it is easy to obtain the second estimate of (3.1)

$$E[\{L^0(X^i - X^j)_s^t\}^4] \leq K(t-s)^2 .$$

□

So we have reduced the proof of (3.1) to that of (4.1). Using scaling as before and the Cauchy-Schwarz inequality, it is enough to check that for some $d_1, d_2 > 0$:

$$E[\exp\{\frac{d_1}{N}\sum_{i\neq j} L^0(X^i - X^j)_1\}] \leq d_2^N \,, \tag{4.6}$$

for any initial point $x \in \Delta_N$ and $N > 2c$. As we mentioned before we are now going to use the increasing reordering $Y_t^1 \leq \ldots \leq Y_t^N$ of the processes $X_t^1, \ldots, X_t^N$. The semimartingale structure of the $(Y_\cdot^i)$ processes is given by the following lemma whose proof is very similar to that of Lemma 3.7, once one knows the $(X_\cdot)$ process is Δ_N valued (see [42] for details):

We suppose our process $X_\cdot$ is constructed on some filtered, probability space $(\Omega, F, F_t, (B_t^i), P)$, endowed with F_t-Brownian motions $B_\cdot^i$, $1 \leq i \leq N$.

Lemma 4.2. *There are N independent F_t-Brownian motions $W_t^1, \ldots, W_t^N$, and $(N-1)$ continuous increasing processes $\gamma_t^1, \ldots, \gamma_t^{N-1}$ such that:*

$$\begin{aligned}
Y_t^1 &= Y_0^1 + W_t^1 - \frac{1}{2}a\gamma_t^1 \,, \\
Y_t^k &= Y_0^k + W_t^k - \frac{1}{2}a\gamma_t^k + \frac{1}{2}b\gamma^{k-1} \,, \quad 2 \leq k \leq N-1 \,, \\
Y_t^N &= Y_0^N + W_t^N \qquad + \frac{1}{2}b\gamma_t^{N-1} \,,
\end{aligned} \tag{4.7}$$

where $a = 1 - 2c/N$, $b = 1 + 2c/N$, and

$$\gamma_t^i = \int_0^t 1(Y_s^i = Y_s^{i+1})\, d\gamma_s^i \,. \tag{4.8}$$

□

The identity $Y_t^1 + \cdots + Y_t^N = X_t^1 + \cdots X_t^N$, now yields

$$\frac{2}{N}(\gamma_t^1 + \cdots + \gamma_t^{N-1}) = \frac{1}{N}\sum_{i\neq j} L^0(X^i - X^j)_t \,. \tag{4.9}$$

So our estimate (4.6) can be rephrased in terms of the $(\gamma_\cdot^i)$ processes. Let us introduce some more convenient processes, namely:

$$\begin{aligned}
&D_t^k = b^{-(k-1)}(Y_t^{k+1} - Y_t^k)\,, \quad H_t^k = b^{-(k-1)}(W_t^{k+1} - W_t^k)\,, \\
&C_t^k = b^{-(k-1)}\gamma_t^k\,, \quad 1 \leq k \leq N-1\,.
\end{aligned} \tag{4.10}$$

One can see easily that:

$$D_t^1 = D_0^1 + H_t^1 + C_t^1 - \frac{\alpha}{2} C_t^2$$
$$D_t^k = D_0^k + H_t^k + C_t^k - \frac{\alpha}{2} C_t^{k+1} - \frac{1}{2} C_t^{k-1}\,, \quad 2 \le k \le N-2$$
$$D_t^{N-1} = D_0^{N-1} + H_t^{N-1} + C_t^{N-1} \qquad - \frac{1}{2} C_t^{N-2}\,, \quad \text{with } \alpha = ab = 1 - \frac{4c^2}{N^2}\,.$$

Moreover $C_t^i = \int_0^t 1(D_s^i = 0) dC_s^i$. It now follows from the solution to the Skorohod problem that: $(C_\cdot^k)_{1 \le k \le N-1} = F((D_0^k + H_\cdot^k), (C_\cdot^k))$, where $F : (C \times C_0^+)^{N-1} \to (C_0^+)^{N-1}$, is the map:

$$F(v,c)_t^1 = \sup_{s \le t} (-v_\cdot^1 + \frac{1}{2} \alpha c_\cdot^2)_+\,,$$
$$F(v,c)_t^k = \sup_{s \le t} (-v_\cdot^k + \frac{1}{2} \alpha c_\cdot^{k+1} + \frac{1}{2} c_\cdot^{k-1})_+$$
$$F(v,c)_t^{N-1} = \sup_{s \le t} (-v_\cdot^k \qquad + \frac{1}{2} c_\cdot^{N-2})_+$$

Set $|w|_t = \sum_{i=1}^{N-1} \sup_{s \le t} |w_s^i|$, for $w \in C^{N-1}$. It is not difficult to prove (see [42]) that:

Lemma 4.3.

$$|F(v_\cdot, c_\cdot) - F(v_\cdot, c_\cdot')|_t \le \frac{1}{2}(1 + |\alpha|)|c_\cdot - c_\cdot'|_t\,. \tag{4.12}$$

If $c = F(v_\cdot, c_\cdot)$, $\bar{c}_\cdot = F(\bar{v}_\cdot, \bar{c}_\cdot)$, *then*

$$|c_\cdot - \bar{c}_\cdot|_t \le \frac{2}{1 - |\alpha|} |v_\cdot - \bar{v}_\cdot|_t\,, \quad t \ge 0\,, \tag{4.13}$$

and

$$v_\cdot \le \bar{v}_\cdot \Rightarrow \bar{c}_\cdot \le c_\cdot\,. \tag{4.14}$$

□

Because of (4.12), when $N > 2c$, we can consider the fixed point solution $c_\cdot = F(v_\cdot, c_\cdot)$ for any $v_\cdot \in C^{N-1}$, which is obtained by iteration.

Now because of (4.14), if one replaces in (4.11) D_0^k by 0, this increases the corresponding $c_\cdot$ processes. If one then replaces D_0^k by $\frac{b}{N}^{-(k-1)}$ we see from (4.13) that the corresponding fixed point $\bar{c}_\cdot$, satisfies

$$\frac{1}{N}(c_1^1 + \cdots + c_1^{N-1}) \le \frac{1}{N}(\bar{c}_1^1 + \cdots + \bar{c}_1^{N-1}) + \frac{1}{N}\frac{2}{1-\alpha} \times \sum_{k=1}^{N-1} \frac{b}{N}^{-(k-1)}$$
$$\le \frac{1}{N}(\bar{c}_1^1 + \cdots + \bar{c}_1^{N-1}) + \frac{N}{2c^2}\,.$$

The constant b^k, for $k \in [1,N]$ satisfy: $1 \le b^k \le e^{2c}$, and from the previous inequality we simply have to prove

$$E[\exp\{\frac{d}{N}(\gamma_1^1 + \cdots + \gamma_1^{N-1})\}] \le \bar{d}^N \ , \quad \text{for some } d, \bar{d} > 0 \ , \tag{4.15}$$

when the initial point is now $x = (0, \frac{1}{N}, \ldots, \frac{N-1}{N})$. Set $\rho = (1-2c/N \ / \ 1+2c/N) = a/b$, one easily sees that for $i \in [1,N]$, $0 < \gamma \le \rho^i \le 1$, where γ is independent of N. By Ito's formula

$$\begin{aligned}\sum_1^N \rho^i (Y_t^i - Y_0^i)^2 &= 2\sum_{i=1}^N \int_0^t \rho^i (Y_u^i - Y_0^i)\, dW_u^i \\ &+ \sum_{i=1}^{n-1} \int_0^t [\rho^{i+1} b(Y_u^{i+1} - Y_0^{i+1}) - \rho^i a(Y_u^i - Y_0^i)]\, d\gamma_u^i \\ &+ t\sum_1^N \rho^i \ .\end{aligned}$$

Using now the fact that $\rho^{i+1} b = \rho^i a$, $Y_u^{i+1} = Y_u^i$ $d\gamma_u^i$-a.s., and $Y_0^{i+1} - Y_0^i = \frac{1}{N}$, we find:

$$\sum_1^N \rho^i (Y_t^i - Y_0^i)^2 + \frac{1}{N}\sum_1^{N-1} \rho^i a \gamma_t^i \le 2\sum_{i=1}^N \int_0^t \rho^i (Y_u^i - Y_0^i)\, dW_u^i + Nt \ . \tag{4.16}$$

So to obtain a control such as (4.15), it is enough to prove

$$E[\exp\{d\sum_1^N \int_0^1 \rho^i (Y_u^i - Y_0^i) dW_u^i\}] \le \bar{d}^N \ , \quad \text{for some } d, \bar{d} > 0 \ .$$

Using an exponential martingale and Cauchy Schwarz inequality, as we did in (4.5), and then the convexity of the exponential, it is easily seen that it is enough to show that for some $d, \bar{d} > 0$,

$$\int_0^1 E[\exp\{d\sum_1^N (Y_t^i - Y_0^i)^2\}]\, dt \le \bar{d}^N \tag{4.17}$$

Let us set $|Y_t - Y_0|^2 = \sum_1^N \rho^i (Y_t^i - Y_0^i)^2$, and $U_t = \exp\{\frac{\lambda}{t+1}|Y_t - Y_0|^2\}$. Applying Ito's formula, we find:

$$\begin{aligned}U_t =&1+\sum_1^N\int_0^t 2\lambda\rho^i\frac{(Y_s^i-Y_0)}{s+1}U_s\, dW_s^i\\&+\lambda\sum_{i=1}^{N-1}\int_0^t\{\rho^{i+1}b(Y_s^{i+1}-Y_s^i)-\rho^i a(Y_s^i-Y_0^i)\}\ (s+1)^{-1}\ U_s\ d\gamma_s^i\\&+\frac{1}{2}\int_0^t U_s(\frac{4\lambda^2}{(s+1)^2}\sum_i\rho^{2i}(Y_s^i-Y_0^i)^2-2\lambda\frac{|Y_s-Y_0|^2}{(s+1)^2})\ ds\\&+\sum_1^N\lambda\rho^i\int_0^t(s+1)^{-1}U_s\ ds\ .\end{aligned}$$

Now by the same reason as in (4.16), the third term of the last expression is nonpositive. For $\lambda\le\frac{1}{2}$ since $\rho\le 1$, the fourth term is as well nonpositive. It then follows that for $\lambda\le 1/2$:

$$U_t\le 1+\lambda N\int_0^t U_s\ ds+\ \text{local martingale.}$$

It then follows using a familiar stopping time argument and Gronwall's inequality, that:

$$E[\exp\{\frac{1}{2(t+1)}\sum_1^N\rho^i(Y_t^i-Y_0^i)^2\}]\le\exp\{\frac{N}{2}t\}\ .$$

From this (4.17) follow, and we get our claim (4.1).

□

5) Reordering of the interacting particle system.

We look now at the same problem in our present context as we did for independent Brownian motions, in section 3 example e) of Chapter I. As we will see a very similar result holds in our case as well. We again suppose u_0 atomless, and u_N u_0-chaotic,Δ_N supported. We introduce the symmetrized, reordered processes $(Z_\cdot^1,\dots,Z_\cdot^N)$ defined by:

(5.1) $$Z_t^i=Y_t^{\sigma(i)}\ ,\quad 1\le i\le N\ ,$$

where σ is a uniformly distributed permutation of $[1,N]$, independent of the space where the $(X_\cdot^1,\dots,X_\cdot^N)$ process is constructed.

We will see that the $(Z^i_\cdot)$ are R-chaotic where R is the law of a "deterministic" type evolution. Let us describe R. more precisely. We define for $x \in D = \text{supp}\, u_0 \backslash \cup_n \{a_n, b_n\}$, where $(\text{supp}\, u_0)^c = \cup_n (a_n, b_n)$, (a_n, b_n) disjoint intervals,

$$\psi_t(x) = F_t^{-1} \circ F_0(x) , \tag{5.2}$$

where F_t is the distribution function of the solution at time t of Burgers' equation, with initial u_0, that is;

$$\exp\{-4cF_t\} = \exp\{-4cF_0\} \, * \, p_t \, .$$

D has full measure under u_0, and for $x \in D$, $\lim_{t\to 0} \psi_t(x) = x$. The probability R is defined as $\psi \circ u_0$. We have

Theorem 5.1. *The laws* $(Z^1_\cdot, \ldots, Z^N_\cdot)$ *are* R*-chaotic.*

Proof: The proof is a repetition of the proof of Theorem 3.6 of Chapter I. The only point to explain is the tightness estimate. Recall that $p = \frac{a}{b} = \frac{1-2c/N}{1+2c/N}$, with the notations of section 4. Our required tightness estimates follow from

Lemma 5.2.

$$E[\sum_1^N \rho^i (Y_t^i - Y_s^i)^4] \le 3N(t-s)^2 \, . \tag{5.3}$$

Proof: Basically by the same argument as in (4.16), we know that

$$E[\sum_1^N \rho^i (Y_t^i - Y_s^i)^2] \le N(t-s) \, . \tag{5.4}$$

Applying then Ito's formula to $\sum_1^N \rho^i (Y_t^i - Y_s^i)^4$, we find;

$$\begin{aligned} \sum_1^N \rho^i (Y_t^i - Y_s^i)^4 = & 4 \int_s^t \sum_i \rho^i (Y_u^i - Y_s^i)^3 \, dW_u^i \\ & + 2 \sum_1^{N-1} \int_s^t [\rho^{i+1} b (Y_u^{i+1} - Y_s^{i+1})^3 - \rho^i a (Y_u^i - Y_s^i)^3] \, d\gamma_u^i \\ & + 3 \int_s^t \sum_1^N \rho^i (Y_u^i - Y_s^i)^2 \, du \, . \end{aligned}$$

By a now familiar argument the second term of the last expression is nonpositive, and since $\rho \le 1$, from (5.4) after integration we easily find our claim (5.3).

□

The proof of Theorem 5.1 is then basically a repetition of that of Theorem 3.6, of Chapter I.

□

Before closing this section let us mention that now for $t > 0$, $x_t = \psi_t(x)$ satisfies the O.D.E.

$$\dot{x}_t = -\frac{\partial_t F}{\partial_x F}(t, x_t) = -\frac{1}{u}\left(\frac{1}{2}\partial_x^2 F - 2cu^2\right) = \left(-\frac{1}{2}(\log u)' + 2cu\right)(t, x_t)\,. \tag{5.5}$$

So R is the law of the deterministic solution of O.D.E. (5.5) with initial random condition u_0 distributed.

III. The constant mean free travel time regime

In the warm up calculation of chapter II, section 1, we have seen that the interaction range $N^{-1/d-2}$, in dimension $d \geq 3$, and $\exp\{-\text{const.}N\}$ in dimension $d=2$, is critical in the study of the quantity:

$$b_N = E\left[\frac{1}{N}\sum_{i=1}^{N}\left(\frac{1}{(N-1)}\sum_{j\neq i}\int_0^t \phi_{N,a}(X_s^i - X_s^j)\,ds - \int_0^t u_s(X_s^i)\,ds\right)^2\right],$$

for independent d-dimensional Brownian motions X^i, whereas there is no critical regime in dimension 1.

The object of this chapter is to study in more detail this critical regime, which we call "constant mean free travel time regime". The reason for this name is that in the limit regime, a "typical particle" does not feel any influence from the other particles before a positive time. We will see that there is a very algebraic Poissonian picture which governs the limiting regime.

In section 1 we will start with a study of annihilated Brownian spheres, motivated by the work of Lang-Nguyen [23]. However we will follow a different route from the hierarchy method they use (see introduction). Although the results of section 1 answer the "propagation of chaos" question, they somehow miss the deeper Poissonian limit picture which is then explained in sections 2 and 3.

1) Annihilating Brownian spheres

We consider $X^1_\cdot, \dots, X^N_\cdot$, independent Brownian motions with initial distribution u_0, in $\mathbb{R}^d$, $d \geq 2$, which are centers of "Brownian soap bubbles" of radius $\frac{1}{2}s_N$. This means that when the centers of two such spheres which are still intact come to a distance smaller than s_N, then both spheres are destroyed. We precisely pick

$$(1.1) \qquad \begin{aligned} s_N &= N^{-1/d-2}\,, \quad d \geq 3\,, \\ &= \exp\{-N\}\,, \quad d = 2\,, \end{aligned}$$

and we assume that u_0 has a bounded density V. We will denote by τ_i the death time of the ith sphere. We are going to study the chaotic behavior of the laws of the symmetric variables $(X^1_\cdot, \tau_1), \dots, (X^N_\cdot, \tau_N)$ on $(C(\mathbb{R}_+, \mathbb{R}^d) \times [0, +\infty])^N$. The result we will obtain will already motivate the terminology of "constant mean free travel time regime".

Theorem 1.1. *The laws of the* $((X^i, \tau_i))_{1\leq i\leq N}$, *are* P*-chaotic, where* P *on* $C \times [0, +\infty]$ *is defined by:*

- *$X_\cdot$ is a Brownian motion with initial distribution u_0.*

- $P[\tau > t/X.] = \exp\{-c_d \int_0^t u(s, X_s)ds\}$, *where u is the unique bounded solution of*

$$\partial_t u = \frac{1}{2}\Delta u - c_d u^2 \quad (1.2)$$
$$u_{t=0} = u_0$$

and

$$c_d = (d-2)\mathrm{vol}(S_d) \ , \quad d \geq 3 \ , \qquad = 2\pi, \qquad d = 2 \ .$$

Remark 1.2. Let us first give some comments,
1) the nonlinear equation (1.2) is understood in the integral form:

$$u_t = u_0 P_t - c_d \int_0^t u_s^2 \, P_{t-s} \, ds \ ,$$

where P_t is the Brownian semigroup. We let P_t act on the right in the previous integral formula to indicate that we are in fact dealing with a forward equation, although this is somehow obscured here by the self-adjointness of P_t.
2) An application of Ito's formula to $f(t, X_t)\exp\{-c_d \int_0^t u(s, X_s)ds\}$, when $f(s,x) = P_{t-s}\phi(x)$, $\phi \in C_k^\infty(R^d)$, easily shows that the subprobability distribution of the alive particle under P is u_t, that is:

$$E_P[\phi(X_t) \ 1(\tau > t)] = \int \phi(x) \ u_t(x) \ dx \ .$$

3) In fact one can get various reinforcements of the basic weak convergence result corresponding to Theorem 1.1. For instance if f_i, $i = 1, \ldots, k$ belong to $L^1(\mu)$ (μ: Wiener measure with initial distribution in u_0), and ϕ_i are bounded functions on $[0, +\infty]$, with a set of discontinuities included in $(0, \infty)$ having zero Lebesgue measure, then:

$$\lim_N E_N[f_1(X_\cdot^1)\phi_1(\tau_1) \ldots f_k(X_\cdot^k)\phi_k(\tau_k)] = \prod_1^k E_p[f_i(X)\phi_i(\tau)] \ .$$

Also if one applies Theorem 2.2 of Baxter-Chacon-Jain [1], to $\tau_1 \wedge \cdots \wedge \tau_k$, one can also see that the density of presence of the first k particles, when they are all alive at time t, (that is $\tau_1 \wedge \ldots \wedge \tau_k > t$), converges in variation norm to $u(t, x_1) \ldots u(t, x_k) \ dx_1 \ldots dx_k$.
4) The tightness of the laws of $\overline{X}_N = \frac{1}{N}\sum_i \delta_{(X_\cdot^i, \tau^i)}$ is clearly immediate, by Proposition 2.2 of Chapter I. This is an indication that this point will not be very helpful in the proof. One of the difficulties of the problem is that annihilation induces a fairly complicated and long range dependence structure. The proof has somehow to rely on the treatment of terms like $1\{\tau_i \leq t\}$.

One may be tempted to express an event like $\{\tau_1 \leq t\}$ in terms of the basic independent variables which are the processes $(X_\cdot^1, \ldots, X_\cdot^N)$, for instance see Lang-Nguyen [23], p. 244.

However we will refrain from doing this. Loosely speaking we will stop at the first step of the unraveling, where the killer of particle 1 is considered: $P_N[\tau_1 \leq t] \simeq (N - 1)P[X^2$ had a collision with X^1 before time t, both were alive at that time]. Such an equality somehow bootstraps the law of (X^2, τ_2) in the law of $(X_\cdot^2, \tau_1)$. The idea is that this should force limit points of empirical laws to be concentrated on measures m satisfying a self-consistent property which will be characteristic of the law P.
5) The collision region between two particles corresponds to the set $|x_1 - x_2| \leq s_N$, in R^{2d}.
It roughly corresponds to the region where the two-particle 1-potential generated by Lebesgue measure sitting on the diagonal of $R^d \times R^d$,

$$h(x_1, x_2) = \int_0^\infty e^{-s} ds \int_{R^d} p_s(x_1, z)\, p_s(x_2, z)\, dz$$

is larger than $c_d^{-1} N$. This notion of level set of potentials is appropriate to find the "right" collision sets in non Brownian situations, see [45]. One can see that if $T_{1,2}$ is the entrance time in the region $|x_1 - x_2| \leq s_N$, then as N goes to infinity for $x_1 \neq x_2 \quad N E_{x_1,x_2}[T_{1,2} \in dt,\ (X^1_{T_{1,2}}, X^2_{T_{1,2}}) \in dx]$ converges vaguely to the measure $c_d p_s(x_1, z)\, p_s(x_2, z)\, dz\, ds$ on $(0, \infty) \times (R^d)^2$ (see [45]). Here dz stands for Lebesgue measure on the diagonal of $(R^d)^2$. This also makes plausible that we are dealing with a kind of Poisson limmit theorem.

□

We are going to prove Theorem 1.1, in a number of steps. The first step will be to give another characterization of the law P, given in Theorem 1.1, getting us closer to the actual form we will use to identify limit points of laws of empirical measures.

Lemma 1.2. *P is the only probability m on $C \times [0, \infty]$ such that*
- X is a Brownian motion with initial distribution u_0,

$$\text{(1.3)} \qquad \text{- for } t \geq 0\,, \quad E_m[1(\tau \leq t) - \int_0^t c_d\, 1(\tau > s)\, u(s, X_s)\, ds / X_\cdot] = 0\,;$$

where $u(s, x) \in L^\infty(ds\, dx)$ is the density of the image of the measure $1(\tau \geq s)\, ds\, dm$ under (s, X_s).

Sketch of Proof: In view of Remark 1.2 2), which identifies the solution of non-linear equation (1.2), with the density of presence of the not yet destroyed particle, for P, one checks readily that P satisfies the required conditions.

Uniqueness:

Let $\nu(X_., dt)$ be the conditional distribution on $[0,\infty]$ of τ given $X_.$ which is μ-distributed, where μ is Wiener measure with initial condition u_0. Condition (1.3) implies that μ-a.s., for any t:

$$\nu_{X_.}([0,t]) - \int_0^t c_d \, \nu_{X_.}((s,\infty)) \, u(s,X_s) \, ds = 0 ,$$

which implies that $\nu_{X_.}((t,\infty)) = \exp\{-c_d \int_0^t u(s,X_s)ds\}$. It then follows that for $\phi \in C_K^\infty$, and a.e. t:

$$\langle u_t, \phi\rangle = E[\phi(X_t) \, 1(\tau \geq t)] = \langle u_0, P_t\phi\rangle - c_d \int_0^t \langle u_s^2, P_{t-s}\phi\rangle \, ds ,$$

so that u is the unique bounded solution of the integral equation corresponding to (1.2). This yields that $m = P$.

□

The present characterization of P has the advantage for us that u does not refer to the solution of (1.2) any more, but can be directly measured on m. We also got rid of the exponential term $\exp\{-c_d \int_0^t u(s,X_s)ds\}$, and deal instead with $c_d \int_0^t 1(\tau > s) \, u(s,X_s)ds$. This will reduce the complexity of computations. Even so $u(s,x)$ is a (very) ill behaved function of m under weak convergence. So later we are in fact going to reinterpret this quantity $u(s,x)$, using a priori knowledge on our possible limit points of empirical measures to obtain continuous enough functionals characterizing P.

Before that we introduce the idea of chain reactions, in order to restore some independence between particles. We set for $1 \leq i \neq j \leq N$

$$T_{i,j} = \inf\{s \geq 0 \, , \, |X_s^i - X_s^j| \leq s_N\} . \tag{1.4}$$

The set $A_{N,t}^{i,j}$ will represent the occurrence between the Brownian trajectories $X_.^1, \ldots, X_.^N$, and forgetting about any destruction, of a chain reaction leading from i to j before time t. Precisely (for $i = 2$, $j = 1$):

$$\begin{aligned} A_{N,t}^{2,1} = \{&T_{2,1} \leq t \text{ or } \exists k_1, \ldots, k_p \text{ distinct in } [3,N] \text{ such that} \\ &S_1 = T_{2,k_1} \leq S_2 = T_{k_1,k_2} \leq \ldots \leq S_{p+1} = T_{k_p,1} \leq t\} . \end{aligned} \tag{1.5}$$

One now introduces τ_j^i, the death time of particle j if one does not take into account the trajectory X^i for the determination of collisions. The interest for us of the chain reaction sets $A_{N,t}^{i,j}$ comes from

Lemma 1.3. *For x, y in R^d, $t > 0$:*

$$P_{x,y} - \text{ a.s. } \{t \wedge \tau_j \neq t \wedge \tau_j^i\} \subseteq A_{N,t}^{i,j} .$$

($P_{x,y}$ is the probability on C^N for which the X^ℓ are independent Brownian motions, with initial distribution u_0 for $\ell \neq i,j$, δ_x for $\ell = i$, and δ_y for $\ell = j$).

For the proof of Lemma 1.3, we refer to [43].

Since clearly τ_j^i is independent of X^i, we are now interested in showing that $P_N[A_{N,t}^{i,j}] \to 0$, when N goes to infinity. This will provide us with a tool to restore independence.

Proposition 1.4. *For $\eta > 0$, $\lim_N \sup_{|x-y|>\eta} P_{x,y}[A_{N,t}^{2,1}] = 0$.*

$$\lim_N P_N[A_{N,t}^{2,1}] = 0 . \tag{1.7}$$

Proof: The second statement is an immediate consequence of the first statement since u_0 has a bounded density. Let us prove the first statement. Using symmetry we have

$$P_{x,y}[A_{N,t}^{2,1}] \le \sum_{p=0}^{N-2} N^p E_{x,y}[S_1 \le \ldots \le S_{p+1} \le t] , \tag{1.8}$$

where now $k_1 = 3, \ldots, k_p = 2 + p \le N$. For $\lambda > 0$, we set

$$h_N^\lambda(z) = E_{z,0}[\exp\{-\lambda T_{1,2}\}] = (g_\lambda(|z|)/g_\lambda(s_N)) \wedge 1 ,$$

where $g_\lambda(|x-y|)$ is the λ-Green's function for Brownian motion with covariance $2Id$. As $u \to 0$,

$$\begin{aligned} g_\lambda(u) &\sim c_d^{-1} u^{2-d} , \quad d \ge 3 , \\ &\sim (2\pi)^{-1} \log(u^{-1}) , \quad d = 2 , \end{aligned}$$

It follows that for $N \ge N_0(\lambda)$:

$$N h_N^\lambda(z) \le 2c_d \, g_\lambda(|z|) . \tag{1.9}$$

Set now $a_p^\lambda = N^p E_{x,y}[S_1 \le \ldots \le S_p \; e^{-\lambda S_p}]$. We have for $N \ge N_0(\lambda)$:

$$\begin{aligned} a_p^\lambda &\le N^p \; E_{x,y}[(S_1 \le \ldots \le S_{p-1}) \; e^{-\lambda S_{p-1}} \; h_N^\lambda(X_{S_{p-1}}^{p+2} - X_{S_{p-1}}^{p+1})] \\ &\le N^{p-1} \; E_{x,y}[(S_1 \le \ldots \le S_{p-1}) \; e^{-\lambda S_{p-1}} \; 2c_d \; g_\lambda(|X_{S_{p-1}}^{p+2} - X_{S_{p-1}}^{p+1}|)] . \end{aligned}$$

If we integrate over $X^{p+2}_{\cdot}$ in the last expression, since $\int g_\lambda(|y|)dy = \lambda^{-1}$, we find:

$$a_p^\lambda \le N^{p-1} E_{x,y}[S_1 \le \cdots \le S_{p-1}\ e^{-\lambda S_{p-1}}]\ \frac{2c_d\|V\|_\infty}{\lambda}\ ,$$

and now by induction:

$$\text{(1.10)} \qquad \text{for } N \ge N_0(\lambda)\ ,\quad a_p^\lambda \le \left(\frac{2c_d\|V\|_\infty}{\lambda}\right)^p .$$

Denote by a_p the pth term of the series in (1.8). Set $\lambda = 4c_d\|V\|_\infty$. For $N \ge N_0(\lambda)$:

$$a_0 \le e^{\lambda t}\ E_{x,y}[e^{-\lambda T_{2,1}}] \le e^{\lambda t}\frac{2c_d}{N}\ g_\lambda(|x-y|)\ .$$

Consider now $p \ge 1$. Pick $\epsilon \in (0,1)$.

For $N \ge N_0\left(\frac{2c_d\|V\|_\infty}{\epsilon}\right)$, by (1.10):

$$N^p\ E_{x,y}[S_1 \le \cdots \le S_p \le \frac{\epsilon}{2c_d\|V\|_\infty}] \le e\epsilon^p\ .$$

It follows that for $N \ge N_0(\lambda) \vee N_0(2c_d\|V\|_\infty/\epsilon)$:

$$\begin{aligned} &N^p E_{x,y}[S_1 \le \ldots \le S_p \le S_{p+1} \le t] \\ &\quad \le N^p\ E_{x,y}[S_1 \le \ldots \le S_p \le \frac{\epsilon}{2c_d\|V\|_\infty}] \\ &\qquad + e^{\lambda t} N^p E_{x,y}[1(S_1 \le \ldots \le S_p \le S_{p+1})\ 1(S_p \ge \frac{\epsilon}{2c_d\|V\|_\infty})\ e^{-\lambda S_{p+1}}]\ . \end{aligned}$$

The first term of the last expression is smaller than $e\ \epsilon^p$, and the second is smaller than:

$$\text{(1.11)} \quad e^{\lambda t} N^{p-1} E_{x,y}[(S_1 \le \ldots \le S_p)\ 1(S_p \ge \frac{\epsilon}{2c_d\|V\|_\infty})\ e^{-\lambda S_p} 2c_d g_\lambda(|X^1_{S_p} - X^{2+p}_{S_p}|)]$$

The distribution of $X^1_{S_p}$ conditionally on $(X^2_{\cdot}, \ldots, X^N_{\cdot})$, is Gaussian with covariance $S_p I_d$, and mean x. It has a density which is uniformly bounded when $S_p \ge \frac{\epsilon}{2c_d\|V\|_\infty}$ by $K = (\frac{c_d\|V\|_\infty}{\pi\epsilon})^{d/2}$.

Integrating in (1.11) over $X^1_{\cdot}$, we find the upper bound $K\ e^{\lambda t}\ N^{p-1}\ E_{x,y}[(S_1 \le \ldots \le S_p)\ e^{-\lambda S_p}] \le \frac{K e^{\lambda t}}{N}(\frac{1}{2})^p$, by (1.10).

It follows that for $p \ge 1$, $\epsilon > 0$, and $N \ge N_0(\lambda) \vee N_0(2c_d\|V\|_\infty/\epsilon)$:

$$N^p E_{x,y}[S_1 \le \ldots \le S_p \le S_{p+1} \le t] \le e^{\lambda t}\frac{K}{N}(\frac{1}{2})^p + e\ \epsilon^p\ .$$

It now follows that:

$$P_{x,y}[A^{2,1}_{N,t}] \le e^{\lambda t}(\frac{2c_d}{N} g_\lambda(|x-y|) + \frac{K}{N}) + e\ \frac{\epsilon}{1-\epsilon}\ .$$

From this our claim follows.

□

As we mentioned already the tightness of the laws of the $\overline{X}_N = \frac{1}{N}\sum_i \delta_{(X^i,\tau_i)} \in M(C \times [0,\infty])$ is immediate. Let us now identify the possible limit points. In order to be able to identify such a limit point $\overline{P}_\infty$ as δ_P, we will use the following idea. The density $u(s,x)$ which appears in (1.3), is ill behaved for the weak convergence topology. However we know that for $\overline{P}_\infty$-a.e. m, the $X_.$ component will be μ-distributed (that is Brownian motion with initial distribution u_0). So now for such an m:

$$\begin{aligned}\langle u_s, f\rangle &= \langle m, f(X_s)\rangle - \langle m, f(X_s)\, 1(\tau < s)\rangle \\ &= \langle V_s, f\rangle - \langle m, f(X_s)\, 1(\tau < s)\rangle \;,\end{aligned}$$

where $V(s,x) = u_0 \;*\; p_s(x)$. If now we anticipate the fact that for $\overline{P}_\infty$-a.e. m there should be a "Markov property" at time τ, then for such an m, the density $u(s,x)$ should be given by:

$$u(s,x) = V(s,x) - \langle m, p_{s-\tau}(X_\tau - x)\, 1(s > \tau)\rangle \;.$$

So we will interpret u, in (1.3), in terms of this last formula, and then we will check that for $\overline{P}_\infty$-a.e. m this expression defines the density of the image of $1(\tau \geq s)ds\; dm$ under (s, X_s). So the quantity $\int_0^t 1(\tau > s)\; u(s,X_s)ds$ is now replaced by $\int_0^{t\wedge\tau} V(s,X_s)ds - \tilde{E}_m[\int_0^{(t\wedge\tau-\tilde{\tau})_+} p_s(X_{\tilde{\tau}+s} - \tilde{X}_{\tilde{\tau}})ds]$, which has nicer continuity properties. In view of these comments it is natural to try now to obtain

Proposition 1.5. *For* $\overline{P}_\infty$-a.e. m, for $t \geq 0$:

$$(1.12)\;\; E_m[1(\tau \leq t) - c_d \int_0^{t\wedge\tau} V(s,X_s)ds + c_d\tilde{E}_m[\int_0^{(t\wedge\tau-\tilde{\tau})_+} p_s(X_{\tilde{\tau}+s} - \tilde{X}_{\tilde{\tau}})ds/X_.] = 0$$

Proof: We call D the at most denumerable set of t such that $I(\overline{P}_\infty)\;[\{\tau = t\}] \neq 0$. We then define for $h(\cdot) \in C_b(C)$, $t \notin D$:

$$G(m) = \langle m, \{1(\tau \leq t) - c_d \int_0^{t\wedge\tau} V(s,X_s)ds + \tilde{E}_m[c_d \int_0^{(t\wedge\tau-\tilde{\tau})_+} p_s(X_{\tilde{\tau}+s} - \tilde{X}_{\tilde{\tau}})ds]\}\; h(X)\rangle$$

and we define the smoothed $G_\epsilon(m)$, $\epsilon > 0$, where the last integral in the previous expression is replaced by $\int_0^{(t\wedge\tau-\tilde{\tau}-\epsilon)_+} p_{s+\epsilon}(X_{\tilde{\tau}+s+\epsilon} - \tilde{X}_{\tilde{\tau}})ds$. Since for $\overline{P}_\infty$-a.e. m, the $X_.$ component is μ-distributed under m, one easily sees that the expression in the conditional expectation in (1.12) is integrable (and meaningful) and that

$$E_{\overline{P}_\infty}[|G(m)|] \leq \text{const.}\|V\|_\infty \epsilon + \lim_k E_{N_k}[|G_\epsilon(\overline{X}_{N_k})|] \;.$$

Now by considering for $1 \le i \neq j \le N$, separately $X^i_\cdot$, $X^j_\cdot$, and the remaining group of $(N-2)$ particles one sees that

$$1(\tau_i \le t) - \sum_{j \neq i} 1(T_{i,j} \le t)\ 1(\tau^i_j > T_{i,j})\ 1(\tau^j_i > T_{i,j})$$

equals zero except maybe when $\tau_i = 0$, but anyway the quantity remains bounded in absolute value by $\sum_{j\neq i} 1(T_{i,j} = 0)$. Since $N\ s^d_N$ converges to zero, the expectation of this quantity is easily seen to go to zero. From this remark, applying now the Cauchy Schwarz inequality and symmetry, we find:

$$\begin{aligned}
&(\lim_k E_{N_k}[|G_\epsilon(\overline{X}_{N_k})|])^2 \le \\
&\overline{\lim}_N E_N[(N 1(T_{1,2} \le t)\ 1(\tau^1_2 > T_{1,2})\ 1(\tau^2_1 > T_{1,2}) - c_d \int_0^{t\wedge\tau_1} V(s, X^1_s)\ ds \\
&\quad + c_d \int_0^{(t\wedge\tau_1-\tau_2-\epsilon)_+} p_{s+\epsilon}(X^1_{\tau_2+s+\epsilon} - X^2_{\tau_2}) ds)\ h(X^1) \\
&\quad \text{(same expression with particles 3 and 4) } h(X^3)] \\
&+ \overline{\lim}_N O(\frac{1}{N})\ E_N[N^2 1(T_{1,2} \le t)\ 1(T_{2,3} \le t) + N\ 1(T_{1,2} \le t) + 1] \\
&+ \overline{\lim} O(\frac{1}{N^2})\ E_N[N^2\ 1(T_{1,2} \le t) + N 1(T_{1,2} \le t) + 1]\ .
\end{aligned} \tag{1.13}$$

Now the last two terms are in fact zero. For instance integrating over X^1_0 and X^3_0, we have:

$$\begin{aligned}
&E_N[N^2\ 1(T_{1,2} \le t)\ 1(T_{2,3} \le t)] \le \\
&\|V\|^2_\infty E_N[N|W_{N,t}(X^2_0 + B^2_\cdot - B^1_\cdot)| \times N|W_{N,t}(X^2_0 + B^2_\cdot - B^3_\cdot)|]\ ,
\end{aligned}$$

where $W_{N,t}$ denotes the Wiener sausage of radius s_N, in time t of the process inside the brackets, and $|\cdot|$ Lebesgue volume. But $N|W_{N,t}(X^2_0 + B^2_\cdot - B^1_\cdot)| = N|W_{N,t}(B^2_\cdot - B^1_\cdot)|$ is bounded in any L^p, $p < \infty$, uniformly in N, using usual estimates on the volume of Wiener sausage. It follows that $\frac{1}{N} E_N[N^2 1(T_{1,2} \le t)\ 1(T_{2,3} \le t)]$ converges to zero. The other terms are treated similarly.

Introduce now $\overline{\tau}_i$ the destruction time of particle i, $1 \le i \le 4$, if one replaces the set $[1, N]$, by $\{i\} \cup [5, N]$, when defining the collisions. Observe now that for instance

$$\{t \wedge \tau^2_1 \neq t \wedge \overline{\tau}_1\} \subseteq A^{3,1}_{N-2,t} \cup A^{4,1}_{N-2,t}\ , \tag{1.14}$$

where the subscript $(N-2)$ refers to the fact that the k_i in definition (1.5) are now supposed to belong to $[5, N]$. Indeed one simply uses the last occurrence of 3, 4, in any chain reaction $\{T_{3,k_1} < \ldots < T_{k_p,1} \le t\}$, or $\{T_{4,k_1} < \ldots < T_{k_p,1} \le t\}$, since on the set where $\{t \wedge \tau_1^2 \neq t \wedge \bar\tau_1\}$, one such chain reaction occurs. Similarly

$$\{t \wedge \tau_1 \neq t \wedge \bar\tau_1\} \subseteq A^{2,1}_{N-2,t} \cup A^{3,1}_{N-2,t} \cup A^{4,1}_{N-2,t} \,.$$

Let us now consider the quantity obtained by replacing in the first expression in the right member of inequality (1.13), the τ_i^j and τ_i by $\bar\tau_i$, i,j in [1,4]. The claim is that once "$\overline{\lim}_N$" is performed one does not change anything. There is a somewhat tricky point about this, that we will describe now. For the full details, however we refer to [43]. The point is that when getting bounds on the difference of the two expressions one obtains terms like:

$$E_N[N^2 1(T_{1,2} \le t)\, 1(T_{3,4} \le t)\, 1_{A^{3,1}_{N-2,t}}]\,,\; E_N[N\, 1(T_{1,2} \le t)\, 1_{A^{3,1}_{N-2,t}}]\,,$$
$$\text{or } E_N[N\, 1(T_{3,4} \le t)\, 1_{A^{1,2}_{N-2,t}}]\,.$$

These terms go to zero with N because one can force volume of Wiener sausage in these terms. For instance, we can integrate over X_0^2 and X_0^4 in the first term and get an upper bound by $\|V\|^2_\infty E_N[N|W_{N,t}(X_0^1 + B^1_\cdot - B^2_\cdot)\; N|W_{N,t}(X_0^3 + B^3_\cdot - B^4_\cdot)|\; 1_{A^{1,2}_{N-2,t}}]$, which is easily seen to go to zero,thanks to the usual estimates on the volume of the Wiener sausage and the fact that $E_N[A^{1,2}_{N-2,t}] \to 0$. The point is that in the difference one does not need to generate terms like:

$$E[N^2 1(T_{1,2} \le t)\, 1(T_{3,4} \le t)\, 1_{A^{4,3}_{N-2,t}}]\,,\text{ or}$$
$$E[N 1(T_{1,2} \le t)\, 1_{A^{2,1}_{N-2,t}}]\,,$$

for which we cannot use our previous reduction to an estimate on the volume of Wiener sausage. Indeed $\{T_{1,2} \le t\} \subset A^{2,1}_{N-2,t}$, and the last term for instance does not go to zero.

Our last reduction step is to observe (see [43]) that $N\, 1(T_{1,2} \le t)\, 1(\bar\tau_1 > T_{1,2})\, 1(\bar\tau_2 > T_{1,2}) = N\, 1(T_{1,2} \le t \wedge \bar\tau_1) - N\, 1(\bar\tau_2 \le T_{1,2} \le t \wedge \bar\tau_1)$ a.s. on the set where $\{T_{1,2} > 0\}$, and a similar equality holds for particles 3, 4, on $\{T_{3,4} > 0\}$. With this, one easily concludes that the first expression on the right of inequality (1.13) in fact equals:

$$\overline{\lim}_N E_N[(E_{\mu\otimes\mu}[\tilde I])^2]\,,$$

$$\text{where } \tilde I = \Big(N 1(T_{1,2} \le t \wedge \bar\tau_1) - c_d \int_0^{t\wedge\bar\tau_1} V(s, X_s^1)ds\Big)\, h(X^1) \tag{1.15}$$
$$- \Big(N 1(\bar\tau_2 \le T_{1,2} \le t \wedge \bar\tau_1) - c_d \int_0^{(t\wedge\bar\tau_1 - \bar\tau_2 - \epsilon)_+} p_{s+\epsilon}(X^1_{\bar\tau_2+s+\epsilon} - X^2_{\bar\tau_2})ds\Big)\, h(X^1_\cdot)$$

In order to study expression (1.15), we now introduce the collision intensity for $w^1 \in C$ and $v \in M(R^d)$, $t \geq 0$:

$$C_t^N(w^1, v) = E_v^2[N\ 1\{T_{1,2} \leq t\}]\ . \tag{1.16}$$

The expectation E_v^2 in (1.16) is performed with respect to the variable w^2, with Wiener measure having initial condition v.

Lemma 1.6. *Let $f(\cdot)$ belong to $C_0(R^d)$, for $T > 0$:*

$$\lim_N \sup_{u \in M(R^d)} E_u^1[\sup_{x \in R^d, t \leq T} |C_t^N(w^1, f(y-x)dy) - c_d \int_0^t f(s, X_s(w^1) - x)ds|] = 0\ ,$$

where $f(s,x) = f * p_s(x)$.

Let us mention, although we will not really develop this point here, that one interest of the quantity $C_t^N(\cdot, v)$ is that it has nice limit properties for a wide range of processes, and does not really rely for its definition on the additive structure of R^d. For more details see [45].

Proof: Since C_t^N is nondecreasing in t and the limit $c_d \int_0^t f(s, X_s(w^1) - x)ds$ is, uniformly in x, continuous in t, by a well known technique (as in Dini's second theorem), it is enough to show that for fixed t

$$\lim_N \sup_u E_u^1[\sup_x |C_t^N(w^1, f(y-x)dy) - c_d \int_0^t f(s, X_s(w^1) - x)ds|] = 0\ .$$

Now for each N and u the quantity we consider is smaller than:

$$E_u^1 \otimes E_{\delta_0}^2[\sup_x |N \int f(y-x)dy\ 1\{\inf_{0 \leq s \leq t} |X_0^1 + B_s^1 - B_s^2 - y| \leq s_N\}$$
$$- c_d \int_0^t f(X_0^1 + B_s^1 - B_s^2 - x)ds|]$$

where $B_\cdot^i = X_\cdot^i - X_0^i$, $i = 1, 2$. But the last expression equals

$$E_{\delta_0}^1 \otimes E_{\delta_0}^2[\sup_x |N \int f(y-x)dy\ 1\{\inf_{0 \leq s \leq t} |B_s^1 - B_s^2 - y| \leq s_N\}$$
$$- c_d \int_0^t f(B_s^1 - B_s^2 - x)ds|]\ .$$

This latter quantity converges to zero by a refinement on the usual limit result for the Wiener sausage of small radius, for details see [43], Lemma 3.4.

□

Observe now that for any $(X^5_\cdot,\dots,X^N_\cdot)$:

$$
\begin{aligned}
&|E_{\mu\otimes\mu}[(C^N_{t\wedge\bar\tau_1}(X^1,u_0) - c_d\int_0^{t\wedge\bar\tau_1} V(s,X^1_s)ds)\ h(X^1)]| \\
&= |E_\mu[(C^N_{t\wedge\bar\tau_1}(X^1,u_0) - c_d\int_0^{t\wedge\bar\tau_1} V(s,X^1_s)ds)\ h(X^1)]| \\
&\le E_\mu[\sup_{s\le t}|C^N_s(X^1,u_0) - c_d\int_0^s V(u,X^1_u)du|]\ \|h\|_\infty
\end{aligned}
$$

which converges to zero as N goes to infinity by a slight variant of Lemma 1.6 (here u_0 does not necessarily have a density in $C_0(R^d)$).

So we see that we can replace $E_{\mu\otimes\mu}[\tilde I]$ in (1.15) by

$$
(1.17)\quad E_{\mu\otimes\mu}[(N1(\bar\tau_2\le T_{1,2}\le t\wedge\bar\tau_1) - c_d\int_0^{(t\wedge\bar\tau_1-\bar\tau_2-\epsilon)_+} p_{s+\epsilon}(X^1_{\bar\tau_2+s+\epsilon}-X^2_{\bar\tau_2})ds)\ h(X^1)]
$$

without changing the limit result. The study of this last term is naturally more delicate. We cannot directly integrate over particle 2, and see a "collision intensity" term appear. The study of this last term will require some "surgery" on trajectories. Of course now $(X^5_\cdot,\dots,X^N_\cdot)$ are held fixed and represent an outside random medium, and we are going to derive uniform estimates over this random medium. We start with the identity:

$$
\begin{aligned}
1\{\bar\tau_2\le T_{1,2}\le t\wedge\bar\tau_1\} =& 1\{\bar\tau_2\le T_{1,2}\le t\wedge\bar\tau_1\}\times 1\{T_{1,2}<\bar\tau_2+\epsilon\} \\
&+ 1\{\bar\tau_2+\epsilon\le T_{1,2}\le t\wedge\bar\tau_1\}
\end{aligned}
$$

Moreoever we can write:

$$
\begin{aligned}
1\{\bar\tau_2+\epsilon\le T_{1,2}\le t\wedge\bar\tau_1\} =& (1-1\{\bar\tau_2+\epsilon>t\wedge\bar\tau_1\}) \\
&\times(1-1\{t\wedge\bar\tau_2<T_{1,2}<t\wedge\bar\tau_2+\epsilon\} - 1\{T_{1,2}\le t\wedge\bar\tau_2\}) \\
&1\{T_{1,2}\circ\theta_\epsilon\circ\theta_{t\wedge\bar\tau_2}\le(t\wedge\bar\tau_1-\bar\tau_2-\epsilon)_+\}\,.
\end{aligned}
$$

Using this decomposition it follows that the absolute value of the expression (1.17) is bounded by:

$$
(1.18)\quad
\begin{aligned}
&E_{\mu\otimes\mu}[(N1\{T_{1,2}\circ\theta_\epsilon\circ\theta_{t\wedge\bar\tau_2}\le(t\wedge\bar\tau_1-\bar\tau_2-\epsilon)_+\} \\
&- c_d\int_0^{(t\wedge\bar\tau_1-\bar\tau_2-\epsilon)_+} p_{s+\epsilon}(X^1_{\bar\tau_2+s+\epsilon}-X^2_{\bar\tau_2})ds)h(X^1)]| \\
&+ \|h\|_\infty(2E_{\mu\otimes\mu}[N1\{\bar\tau_2\wedge t\le T_{1,2}\le\bar\tau_2\wedge t+\epsilon\}] \\
&+ E_{\mu\otimes\mu}[N1\{T_{1,2}\circ\theta_\epsilon\circ\theta_{t\wedge\bar\tau_2}=0\}] \\
&E_{\mu\otimes\mu}[N1\{T_{1,2}\circ t\wedge\bar\tau_2\}1\{T_{1,2}\circ\theta_{t\wedge\bar\tau_2+\epsilon}\le t\}])
\end{aligned}
$$

It is now fairly standard to see that the last two terms converge to zero (uniformly over $X^5_., \dots, X^N_.)$). As for the second term integrating out the initial condition X^1_0, it is bounded by

$$2\|h\|_\infty E_{\mu\otimes\mu}[N \int u_0(dy)\ 1\{y \in W_{N,\epsilon}(X^2_{t\wedge\bar{\tau}_2} + \tilde{B}^2_. - \tilde{B}^1_. - B^1_{t\wedge\bar{\tau}_2})\}] \,,$$

where $\tilde{B}^i_. = X^i_{t\wedge\bar{\tau}_2+.} - X^i_{t\wedge\bar{\tau}_2}$, $i = 1, 2$, are standard independent Brownian motions. This last quantity is smaller than $2\|h\|_\infty \|V\|_\infty E_{\mu\otimes\mu}\ [N|W_{N,\epsilon}(B^2_. - B^1_.)|]$ which converges to $2\|h\|_\infty \|v\|_\infty c_d \epsilon$, as N converges to infinity.

Let us now look at the first term of (1.18). Conditioning over $X^1_.$, we have

$$\begin{aligned}
&E^2_\mu[N 1\{T_{1,2} \circ \theta_\epsilon \circ \theta_{t\wedge\bar{\tau}_2} \le (t \wedge \bar{\tau}_1 - \bar{\tau}_2 - \epsilon)_+\}] \\
&= E^2_\mu[N \tilde{E}^2_{X^2_{t\wedge\bar{\tau}_2}} [1\{T_{1,2} \circ \theta_\epsilon(w^1_{t\wedge\bar{\tau}_2+.}, \tilde{w}^2) \le (t \wedge \bar{\tau}_1 - \bar{\tau}_2 - \epsilon)_+\}]] \\
&= E^2_\mu[C^N_{(t\wedge\bar{\tau}_1-\bar{\tau}_2-\epsilon)_+}(X^1_{t\wedge\bar{\tau}_2+\epsilon+.}, p_\epsilon(y - X^2_{t\wedge\bar{\tau}_2})dy)] \,.
\end{aligned}$$

After this surgery the first term of (1.18) is now

$$\begin{aligned}
&|E_{\mu\otimes\mu}[(C^N_{t\wedge\bar{\tau}_1-\bar{\tau}_2-\epsilon)_+}(X^1_{t\wedge\bar{\tau}_2+\epsilon+.}, p_\epsilon(y - X^2_{t\wedge\bar{\tau}_2})dy) \\
&- c_d \int_0^{(t\wedge\bar{\tau}_1-\bar{\tau}_2-\epsilon)_+} p_{s+\epsilon}(X^1_{t\wedge\bar{\tau}_2+s+\epsilon} - X^2_{t\wedge\bar{\tau}_2})ds)\ h(X^1)]| \\
&\le \|h\|_\infty E_{\mu\otimes\mu}[\sup_{v\le t,x} |C^N_v(X^1_{t\wedge\bar{\tau}_2+\epsilon+.}, p_\epsilon(y - x)dy) - c_d \int_0^v p_{s+\epsilon}(X^1_{t\wedge\bar{\tau}_2+s+\epsilon} - x)ds|] \\
&\le \|h\|_\infty \sup_{u\in M(\mathbb{R}^d)} E^1_u[\sup_{v\le t,x} |C^N_v(X^1_., p_\epsilon(y - x)dy) - c_d \int_0^v p_{s+\epsilon}(X^1_s - x)ds|]
\end{aligned}$$

which is independent of $(X^5_., \dots, X^N_.)$ and converges to zero as N goes to infinity thanks to Lemma 1.6.

So we have proved that

$$E_{\overline{P}_\infty}[|G(m)|] \le \text{const.}\ \|V\|_\infty \epsilon + 2\|h\|_\infty \|V\|_\infty \epsilon \,.$$

Letting ϵ go to zero, we obtain our claim.

□

As we mentioned previously, $\tilde{E}_m[\int_0^{(t\wedge\tau-\tilde{\tau})_+} p_s(X_{\tilde{\tau}+s} - \tilde{X}_{\tilde{\tau}})ds]$ $= \int_0^t 1(s < \tau)\tilde{E}_m[1(\tilde{\tau} < s)\ p_{s-\tilde{\tau}}(X_s - \tilde{X}_{\tilde{\tau}})]ds$. So in view of the characterization of the law P given by Lemma 1.2, the fact that $\overline{P}_\infty = \delta_P$, will follow from

Proposition 1.6. *For $\overline{P}_\infty$-a.e. m, for all $s \ge 0$, and $f \in bB(\mathbb{R}^d)$:*

$$\int f(x)\ E_m[p_{s-\tau}(x - X_\tau)\ 1(\tau < s)]\ dx = E_m[f(X_s)\ 1(s > \tau)] \,. \tag{1.19}$$

Proof: By the usual arguments we simply will check (1.19), for $s \in (0,\infty) \setminus D$, and $f \in C_b(R^d)$. We recall that $D = \{t : I(\overline{P}_\infty)[\{t = \tau\}] \neq 0\}$. Setting $H(m) = \langle m, (P_{(s-\tau)_+} f(X_\tau) - f(X_s)\, 1(\tau < s)\rangle$, we simply want to check that $H(m) = 0$ $\overline{P}_\infty$-a.s. Now

$$E_{\overline{P}_\infty}[H(m)^2] =$$
$$\lim_k E_{N_k}[(P_{(s-\tau_1)_+} f(X^1_{\tau_1}) - f(X^1_s))(P_{(s-\tau_2)_+} f(X^2_{\tau_2}) - f(X^2_s))\, 1(\tau_1 < s)\, 1(\tau_2 < s)]\,.$$

However on $(A^{1,2}_{N,s} \cup A^{2,1}_{N,s})^c$, $\tau_1 \wedge s = \tau_1^2 \wedge s$, and $\tau_2 \wedge s = \tau_2^1 \wedge s$. It follows that

$$E_{\overline{P}_\infty}[H(m)^2] = \lim_k E_{N_k}[E^1_\mu[(P_{(s-\tau_1^2)_+} f(X^1_{\tau_1^2}) - f(X^1_s))1(\tau_1^2 < s)]^2] = 0\,,$$

by an application of the strong Markov property at time τ_1^2, under the law E^1_μ.

□

This now concludes the proof of Theorem 1.1. In the next section, we are going to give a different line of explanation on why Theorem 1.1 holds.

2. Limit picture for chain reactions in the constant mean free travel time regime.

The result which was presented in the last section is certainly satisfactory from a purely propagation of chaos frame of reference, however it misses the nice limit algebraic picture, corresponding to the "constant mean free travel time" regime, which underlies the result. We will somehow try to motivate and describe this limit picture.

The notion of chain reactions, leading to a specific particle, say particle one, corresponding to definition (1.5) played an important role in the derivation of Theorem 1.1. It simply involves the independent Brownian trajectories and completely forgets about destruction of particles. The idea we will follow here is that one should investigate the limit structure of the chain reactions leading to particle 1, in the constant mean free travel time regime. One then should construct the interaction (annihilation) on the limit object and somehow recapture the result of Theorem 1.1 in this scheme.

As a step in the study of the limit aspect of this somewhat loose notion of chain reaction between independent particles leading to particle 1, one may look at the limiting aspect of the first collision. For each $i \in [1, N]$, we set

$$T_i = \inf_{j \neq i} T_{i,j}\,, \quad \text{when} \quad \inf_{j \neq i} T_{i,j} \leq 1\,, \quad +\infty \text{ otherwise,}$$

and $Y^i_\cdot$ "the first colliding trajectory", is defined on the set $0 < T_i \le 1$, as X^j where j is the only index such that $T_i = T_{i,j}$, and otherwise, it is set to the constant trajectory 0.

Let us quote the following propagation of chaos result from [45].

Theorem 2.1. *The* $(X^i_\cdot, T_i, Y^i_\cdot)_{1\le i\le N}$ *are Q-chaotic where Q is the law on* $C \times ([0,1] \cup \{\infty\}) \times C$ *such that under Q*

- *X has the distribution of a Brownian motion with initial law* u_0.
- $Q[T > t|X] = \exp\{-c_d \int_0^{t\wedge 1} V(s, X_s)ds\}$, $t \ge 0$.
- *Conditionally on* X, T, $0 < T \le 1$, $Y_\cdot$ *is distributed as a Brownian motion with initial law* u_0 *conditioned to be equal to* X_T *at time T, and otherwise it is the constant trajectory* 0.

This result motivates that the natural limit "chain reaction tree" should be constructed in the following way: One starts with one Brownian particle X_s, the ancestor, running until time t, with initial density u_0 (having the role of X^1). One then constructs the trajectories having a collision with the ancestor: conditionally on X_s, one picks a Poisson distribution of points on $[0,t]$ with intensity $c_d V(s, X_s)ds$. Now conditionally on the times $0 < t_1 < \ldots < t_n < t$, one considers independent Brownian bridges: $W^1, \ldots, W^n$ where W_ℓ for $\ell \le n$ has the law of Brownian motion in time $[0, t_\ell]$, with initial distribution u_0 conditioned to be equal to X_{t_ℓ} at time t_ℓ. These $W^1, \ldots, W^n$ constitute the first generation in the chain reaction leading to $X_\cdot$. Note that we disregard the trajectory of $W^\ell_\cdot$, after time t_ℓ, since Theorem 1.1 indicates that there should not be any recollision in the limit picture.

Now one performs the same thing on each of the first generation trajectories W^ℓ, $1 \le \ell \le n$, as just described on $X_\cdot$. One obtains in this way the second generation of trajectories, and so on.

We are now going to give a precise description of this limit tree of chain reactions. As the explanation above indicates we will build a marked tree. Each "individual" being marked by a trajectory. We use Neveu's notion of tree [33], namely a tree is a subset of the set of finite sequences $U = \cup_{k\ge 0}(\mathbb{N}^*)^k$, (the vertices) containing the sequence $\emptyset$ (only element of $(\mathbb{N}^*)^0$), such that $u \in \pi$ when $uj \in \pi$, $j \in \mathbb{N}^*$, and for $u \in \pi$ there is a $\nu_u \in \mathbb{N}$, with $uj \in \pi$ if and only if $1 \le j \le \nu_u$. We will be interested in marked trees that is $\omega = (\pi, (\phi_u, u \in \pi))$, where $\phi_u \in D$, D the set of marks which is for us $\cup_{t>0} C([0,t], \mathbb{R}^d) \sim (0, +\infty) \times C([0,1], \mathbb{R}^d)$.

trajectorial picture: marked tree description:

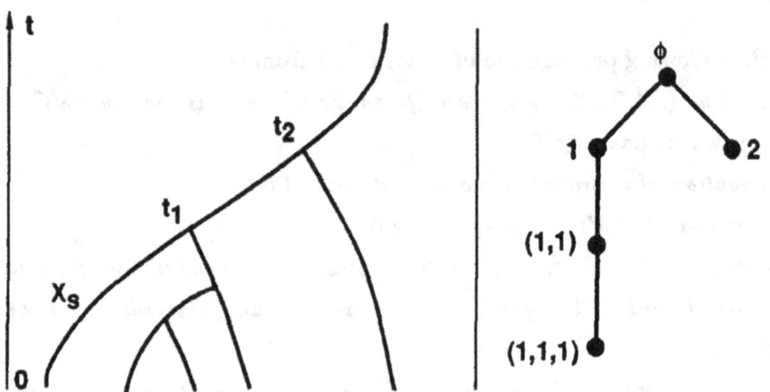

Following Neveu's notations, for the marked tree ω and $u \in \pi$, $T_u\omega$ is the tree translated at u, G_n is the nth generation of the tree: $\pi \cap (\mathbb{N}^*)^n$, and F_n the σ-field generated by G_k, $0 \le k \le n$, and ϕ^u, $u \in G_k$, $0 \le k \le n$. For a trajectory $\psi_. \in D$, we construct the probability R_ψ on the set of marked trees Ω which satisfies

- $R_\psi[\phi^\emptyset = \psi] = 1$ (the ancestor's mark is ψ).
- For f_u, $u \in U$ a collection of nonnegative measurable functions on Ω:

$$E_\psi[\prod_{u \in G_n} f_u \circ T_u \ / \ F_n] = \prod_{u \in G_n} E_{\phi^u}[f_u]$$

(Branching property).

The last requirement which describes the reproduction law is given in a somewhat more synthetic form than what was explained before:

- The random point measure on D: $\sum_{1 \le \ell \le \nu_\emptyset} \delta_{\phi^\ell}$ is a Poisson point process with intensity measure

(2.1) $$\int_0^t ds \; V(s, \psi_s) \; P^{s,\psi_s}(d\phi) \,,$$

and $0 < \tau(\phi^1) < \cdots < \tau(\phi^\emptyset) < t = \tau(\psi)$, a.s. Here $\tau(\phi)$ denotes the time duration of a particle ($(0,\infty)$ valued component on D), and $P^{s,x}$ is the law of Brownian motion with initial u_0 on time $[0,s]$, conditioned to be equal to x at time s. We have normalized here c_d as being equal to 1.

Lemma 2.1.

- R_ψ-a.s. *the tree* ω *is finite.*
- $M(\omega) = \sum_{1 \le \ell \le \nu_\emptyset} \delta_{T_\ell(\omega)}$ *(random point measure on* Ω*)*

is a Poisson point measure with intensity

$$Q_\psi = \int_0^t ds\ V(s,\psi_s)\ R^{s,\psi_s}\ ds\ , \tag{2.2}$$

where $R^{t,x} = \int R_\psi\ P^{t,x}(d\psi)$.

Proof: Let us check the first point: for $0 \le p \le n$, set

$$v_p = E_\psi[\sum_{u\in G_{n-p}} \frac{1}{p!}(c\ \tau(\phi^u))^p]\ , \quad \text{with} \quad c = \|V\|_\infty\ .$$

so that $v_0 = E_\psi[\#G_n]$ and $v_n = \frac{1}{n!}(ct)^n$. Then for $0 \le p < n$:

$$\begin{aligned} v_p &= E_\psi[E_\psi[\sum_{\substack{u\in G_{n-p-1}\\ uj\in G_{n-p}}} \frac{1}{p!}(c\ \tau(\phi^{uj}))^p/F_{n-p-1}]] \\ &= E_\psi[\sum_{u\in G_{n-p-1}} E_{\phi^u}[\sum_{1\le j\le \nu_\phi} \frac{1}{p!}(c\ \tau(\phi^j))^p]] \\ &= E_\psi[\sum_{u\in G_{n-p-1}} \int_0^{\tau(\phi^u)} \frac{1}{p!}(ct)^p\ V(t,\phi_t^u)dt] \le v_{p+1}\ . \end{aligned}$$

From this $E_\psi[\#G_n] \le \frac{1}{n!}(ct)^n$, and summing over n, we find the result. As for the second statement, if F is a positive function on Ω:

$$\begin{aligned} E_\psi[\exp\{-\langle M,F\rangle\}] &= E_\psi[\prod_{1\le\ell\le\nu_\phi} e^{-F(T_\ell(\omega))}] \\ &= E_\psi[\prod_{1\le\ell\le\nu_\phi} R_{\phi^\ell}[e^{-F}]] \quad \text{by the branching property)} \\ &= \exp\{\int_0^t V(s,\psi_s)ds \int dP^{s,\psi_s}(\phi)\ (R_\phi[e^{-F}]-1)) \quad \text{by (2.1)}, \\ &= \exp\{\int dQ(e^{-F}-1)\} \end{aligned}$$

which proves our claim.

□

Our tree ω is R_ψ a.s. finite. Now we can build the interaction on the limit tree corresponding to annihilation by adding a mark Z^u, $u \in \pi$, equal to zero or 1. $Z^u = 0$, will mean that the particle with trajectory ϕ^u is already destroyed by the time $\tau(\phi^u)$ at which it meets its direct ancestor, whereas $Z^u = 1$, will mean that it has not yet been destroyed.

So it is quite natural to impose the following recursive rule, to determine Z^u, $u \in \omega$: If $\nu_u = 0$ (no descendents), $Z^u = 1$. If $Z^{uj} = 1$, for some $1 \le j \le \nu_u$ (one of the direct descendents is alive), then $Z^u = 0$.

We now define $R^t = \int R^{t,x}\ V(t,x)\ dx$, for which the ancestor trajectory is distributed as a Brownian motion in time $[0,t]$, with initial condition u_0. Then we set

$$u(t,x) = V(t,x)\ R^{t,x}[1(Z^\emptyset = 1)]\ , \quad t > 0\ . \tag{2.3}$$

In fact u is the density of presence of ψ_t, under R^t, when the ancestor trajectory $\psi_.$ is not already destroyed at time t.

Lemma 2.2. *For* $f \in bB(R^d)$:

$$\int u(t,x)\ f(x)\ dx = E_{R^t}[f(\psi_t)\ 1(Z^\emptyset = 1)]\ .$$

Proof:

$$\begin{aligned}\int u(t,x)\ f(x) dx &= \int V(t,x)\ R^{t,x}[Z^\emptyset = 1]\ f(x)\ dx \\ &= \int V(t,x) dx \int P^{t,x}(d\psi)\ R_\psi[Z = 1]\ f(\psi_t) \\ &= E_{R^t}[f(\psi_t)\ 1(Z^\emptyset = 1)]\ .\end{aligned}$$

□

The result we show now tells us that one obtains the nonlinear equation (in integral form) (1.2), by constructing the interaction directly on the "limit chain reaction tree".

Theorem 2.3. $u(t,x)$, $0 \le t \le T$, *is the solution in* $L^\infty([0,T] \times R^d, ds\ dm)$ *of the integral equation:*

$$w(t,x) = V(t,x) - \int_0^t \int w^2(s,y)\ p_{t-s}(y,x)\ ds\ dy\ . \tag{2.4}$$

Proof: The uniqueness is standard, see [44] for details. Let us check that u is a solution of (2.4). Observe that

$$\begin{aligned}R_\psi[Z^\emptyset = 1] &= \exp\{-\int_0^t V(s,\psi_s)\ R^{s,\psi_s}[Z^\emptyset = 1] ds\} \\ &= \exp\{-\int_0^t u(s,\psi_s) ds\}\ .\end{aligned} \tag{2.5}$$

Then

$$u(t,x) = V(t,x)\, R^{t,x}[Z=1]$$
$$= V(t,x)\Big(1 - \int_0^t ds\; P^{t,x}(d\psi)\; u(s,\psi_s)\; \exp\{-\int_0^s u(r,\psi_r)dr\}\Big)$$
$$= V(t,x) - \int_0^t ds \int dP^t(\psi) p_{t-s}(\psi_s, x)\; u(s,\psi_s)\; \exp\{-\int_0^s u(r,\psi_r)dr\}$$
$$= V(t,x) - \int_0^t ds \int dy\; V(s,y)\; p_{t-s}(y,x)\; u(s,y) \int P^{s,y}(d\psi) \exp\{-\int_0^s u(r,\psi_r)dr\}$$
$$= V(t,x) - \int_0^t ds \int dy\; p_{t-s}(y,x)\; u^2(s,y)\;,$$

which shows that u satisfies (2.4).

□

3) Some comments

Let us finally give some comments on the results presented in sections 1 and 2.

- The results presented in section 2, strongly suggest that in fact when one studies collisions (without any destructions) between independent particles $(X^1_., \dots, X^N_.)$, a better result than Theorem 2.1 should hold. Somehow one should have a propagation of chaos result at the level of the "trees of chain reactions" leading to each particle X^i, $1 \le i \le N$. The natural "limit" should in fact be precisely the noninteracting marked tree presented in section 2.
- The proof of Theorem 2.3 is very algebraic. The fact that equation (2.4) arises when one constructs the interaction directly on the marked tree remains true when one applies the same construction to a "general Markov process", (see [44]). One then obtains the integral form of the nonlinear equation:

$$\partial_t u = L^* u - u^2 \,, \tag{3.1}$$

where L is the formal generator of the Markov process. In fact by a slight variation on the construction of the marked tree, and of the interaction, one obtains the integral equation corresponding to

$$\partial_t u = L^* u - u^{k+1} \,, \quad (k \ge 1, \text{ integer}). \tag{3.2}$$

- Not only does the construction work for a "general Markov process" but for Brownian motion in one dimension as well! Consider for instance in this case, the random times $0 < t_1 < \cdots < t_n < t$, picked with a Poisson distribution of intensity $V(s, X_s)ds$, on $[0,t]$, when X_s, $s \in [0,t]$ is the ancestor trajectory. They are not at all the first "collision times" of $X_.$ with the first generation trajectories $W^1_., \dots, W^n_.$. Indeed the $W^\ell_.$ are distributed as independent bridges, being conditioned to be equal to X_{t_ℓ} at time t_ℓ. Of course in dimension $d \ge 2$, the trajectory $W^\ell_.$ meets the ancestor trajectory $X_.$ only at time t_ℓ, but in dimension $d = 1$, this need not be the case.

  This point should be viewed as the fact that the limit structure constructed in section 2, exists very generally whether or not it comes as a limit picture from an approximating constant mean free travel time regime.
- One may wonder what the appropriate collision regime should be if one tries to handle $(k+1)$-particle collisions with possible limit survival equation (3.2). As mentioned previously in Remark 1.2 5), the right guess should not be dictated by

a notion of distance of interaction. Much more naturally it should be picked as a suitable level set in the $(k+1)$ particle configuration space of some potential generated by a measure sitting on the "diagonal" (which is polar) (see [45]). A plausible guess would be to look at the set $h(x_1,\dots,x_{k+1}) \geq N^k$, where

$$h(x_1,\dots,x_{k+1}) = \int_0^\infty e^{-s} ds \int_{R^d} \prod_1^{k+1} p_s(x_i,z)\, dz = g_1^{(kd)}(D)\,.$$

where $g_1^{(kd)}(|x-x'|)$ is the 1-Green's function of Brownian motion in R^{kd} and D is the distance of $(x_1,\dots,x_{k+1})$ to the diagonal $\{(z,\dots,z)\}$ in $(R^d)^{k+1}$. If one wants to benefit from Brownian scaling, in dimension d, with $kd>2$, one can use instead of h,

$$f(x_1,\dots,x_{k+1}) = \int_0^\infty ds \int_{R^d} \prod_1^{k+1} p_s(x_i,z)\, dz = \frac{2}{c_{kd}} D^{2-kd}\,,$$

$(c_{kd} = (kd-2)vol(S_{kd}))$, which is continuous with values in $(0,\infty]$, finite except on the diagonal $x_1 = \dots = x_{k+1}$, and homogeneous of degree $2-kd$. The level set $\{f \geq N^k\}$ is then the homothetic of ratio $N^{-1/(d-2/k)}$ of the set $\{f \geq 1\}$.

IV. Uniqueness for the Boltzmann process

In this chapter, we are going to explain how ideas somewhat similar to those used in the "tree construction" of Chapter III, section 2, can be put at work to produce a uniqueness result of the nonlinear process associated to the spatially homogeneous Boltzmann equation for hard spheres.

The spatially homogeneous Boltzmann equation for hard spheres describes the time evolution of the density $u(t,v)$ of particles with velocity v, in a dilute gas, under an assumption of spectral homogeneity and hard spheres collisions as:

$$\partial_t u = \int_{R^n \times S_n} (u(t,\tilde{v})u(t,\tilde{v}') - u(t,v)\, u(t,v'))\, |(v'-v)\cdot n|\; dv'\; dn$$

$$\text{with } \tilde{v} = v + (v'-v)\cdot nn$$

$$\tilde{v}' = v' + (v - v')\cdot nn$$

If f is some "nice test function", disregarding integrability problems, a change of variable yields:

$$\partial_t\langle u,\ f\rangle = \langle u\otimes u\ ,\ (f(v+(v'-v)\cdot nn) - f(v))\ |(v'-v)\cdot n|\rangle\ .$$

On this latter form, the equation naturally appears as the forward equation of a "non-linear jump process", with Levy system:

$$\int M_t(v,d\tilde{v})\ h(\tilde{v}) = \int_{R^n\times S_n} h((v'-v)\cdot nn)\ |(v'-v)\cdot n|\ u(t,v')\ dv'\ dn$$

In section 2 we will introduce the nonlinear process as the solution to a certain (non-linear) martingale problem. The unboundedness of the factor $|(v-v')\cdot n|$ will create a serious source of difficulty in seeing that this martingale problem is well posed.

1) Wild's formula

The theme of this section will be a formula of Wild [56], for the solution of the spatially homogeneous Boltzmann equation, which for instance covers the case of "cut-off hard spheres" (that is a collision intensity $|(v'-v)\cdot n| \wedge C$). We will also give some probabilistic interpretations of the formula for closely related to the formula.

The setting is the following: we suppose that we have a Markovian kernel Q_1 : $R^n \times R^n \times R^n$, and for μ_1, μ_2 two probabilities (or two bounded measures) on R^n, we define the probability (a bounded measure) $\mu_1 \circ \mu_2$ as:

$$\langle \mu_1 \circ \mu_2\rangle = \langle \mu_1 \otimes \mu_2, Q_1 f\rangle\ . \tag{1.1}$$

It is clear that for the variation norm we have the estimate $\|u_1 \circ u_2\| \leq \|u_1\| \, \|u_2\|$. Wild's formula will give a series expansion for the solution of

$$\partial_t u = u \circ u - u \tag{1.2}$$
$$u_{t=0} = u_0 \in M(R^n) \, .$$

The spatially homogeneous Boltzmann equation for hard spheres with cutoff collision kernel $|(v'-v)\cdot n| \wedge 1$, for instance, will correspond to the choice of kernel:

$$Q_1 f(v,v') = \int_0^1 d\alpha \int_{S_n} dn[f(v+(v'-v)\cdot nn)\, 1(\alpha \leq |(v'-v)\cdot n| \wedge 1)$$
$$+ f(v)\, 1(\alpha > |(v'-v)\cdot n| \wedge 1)] \, .$$

Proposition 1.1. *For any $u_0 \in M(R^n)$, here is a unique strongly continuous solution of*

$$u_t - u_0 = \int_0^t (u_s \circ u_s - u_s)\, ds \, , \tag{1.3}$$

given by Wild's sum:

$$u_t = e^{-t} \sum_{k \geq 1} (1-e^{-t})^{k-1}\, u^k \, , \tag{1.4}$$

where $u^1 = u_0$ and $u^{n+1} = \frac{1}{n} \sum u^k \circ u^{n+1-k}$.

Proof: The uniqueness part follows from a classical O.D.E. result. Let us show that (1.4) does provide a solution to (1.3).

Define $u_t^1 \equiv u_0$, and for $n \geq 1$,

$$u_t^{n+1} = e^{-t} u_0 + \int_0^t e^{-(t-s)}\, u_s^n \circ u_s^n \, ds \, , \tag{1.5}$$

then by induction one sees that for $n \geq 1$:

$$u_t^n \geq e^{-t} \sum_1^n (1-e^{-t})^{k-1}\, u^k \, .$$

From this one easily concludes that u_t^n converges uniformly on bounded time intervals in variation norm to u_t given by (1.4). Because of (1.5), u_t is a solution of (1.3).

□

Wild's formula expresses u_t as a barycenter of the sequence u^n, with weights $e^{-t}(1-e^{-t})^{n-1}$. It can in several instances be used to derive asymptotic properties of u_t from those of the sequence u^n, see Ferland-Giroux [10], McKean [25], Tanaka [49], Murata-Tanaka [30].

Wild's formula has a nice interpretation in terms of continuous time binary branching trees, see also Ueno [55], [56]. Indeed $[e^{-t}(1-e^{-t})^{n-1}]_{n\geq 1}$, is the distribution of the total number of particles of such a branching tree at time t, provided each particle branches with unit intensity. For each n, u^n is a convex combination of the various ways of inserting parentheses in a monomial $u_0 \circ \cdots \circ u_0$ of degree n. The ways of inserting parentheses in such a monomial of degree n are in natural correspondence with the tree subsets (in the sense given in section 2 of Chapter III) of the set of vertices $V = \cup_{k\geq 0}\{1,2\}^k$, which possesses exactly n bottom vertices (with no descendants). For instance $u \circ (u \circ u)$ is associated to

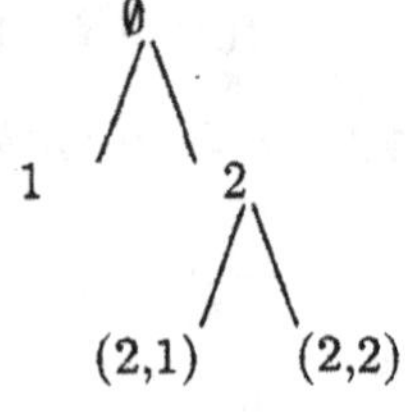

$(u \circ u) \circ (u \circ u)$ is associated to

(1.6)

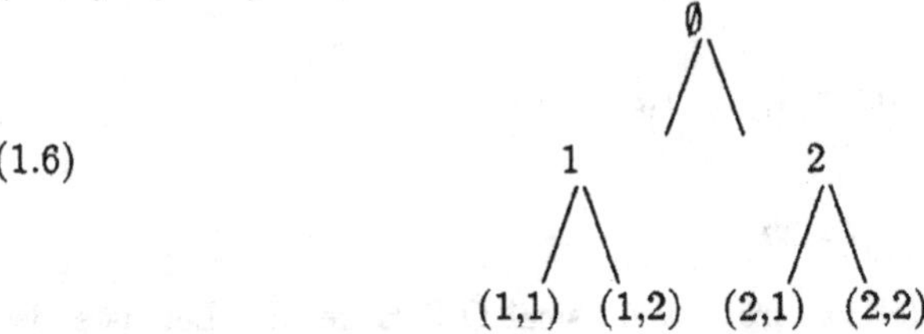

The coefficients appearing in the convex combinations expressing u^n in terms of the various ways of inserting parentheses in a monomial of degree n of u yield precisely the conditional time binary tree, given that it has n bottom vertices at time t.

On the other hand one can also keep track of the time order at which branching occurs. For instance in (1.6) one distinguishes between two ordered trees depending on whether 1 or 2 branched first. Looking at the skeleton of successive jumps, it is easy to see that conditionally on the fact that there are n individuals at time t, there are $(n-1)!$ equally likely such ordered trees.

If one uses a perturbation expansion of

$$u_t = u_0 e^{-t} + \int_0^t e^{-(t-s)}\, u_s \circ u_s \, ds \;,$$

one finds

(1.7)

$$u_t = u_0 e^{-t} + e^{-t}(1-e^{-t}) u_0 \circ u_0 + \cdots + \frac{e^{-t}}{(n-2)!}(1-e^{-t})^{n-2}\, u_0^{(n-1)} + e^{-t} \int_{0,s_{n-1}<\ldots<s_1<t} ds_{n-1} \ldots ds_1\, u_{s_{n-1}}^{(n)} \,.$$

Here for $v \in M(R^n)$, $v^{(n)}$ is defined as:

$$v^{(n)} = \sum_{\sigma \in S_{n-1}} v \underset{\sigma(1)}{\circ} v \underset{\sigma(2)}{\circ} \cdots \underset{\sigma(n-1)}{\circ} v$$

and the permutation σ dictates the order in which the operation $\circ$ is performed. Observe for instance that $(u \circ u) \circ (u \circ u)$ corresponds both to $u \overset{\circ}{_1} u \overset{\circ}{_3} u \overset{\circ}{_2} u$ and to $u \underset{2}{\circ} u \underset{3}{\circ} u \underset{1}{\circ} u$.

Of course by letting n go to infinity in formula (1.7), one finds (since the last term converges to zero):

(1.8)
$$u_t = \sum_{k \geq 1} \frac{e^{-t}}{(k-1)!}(1-e^{-t})^{k-1}\, u_0^{(n-1)} \,,$$

which only differs from Wild's formula because one uses the ordered trees as operation schemes. In Wild's formla one precisely forgets the ordering and simply keeps track of the skeleton subset of $\cup_{k \geq 0}\{1,2\}^k$ induced by the continuous time binary tree.

So we have somewhat informally presented the interpretation of Wild's formula in terms of a continuous time binary branching tree.

We are now going to explain a slightly modified point of view, which will be parallel to the construction of section 2 of Chapter III, and also close in spirit to the uniqueness proof for the Boltzmann process which will be provided in the next section.

First we suppose that we have a measurable map, $\psi(v, v', y)$, where y belongs to an auxiliary Polish space E, and a probability $\nu(dy)$, such that

(1.9)
$$Q_1 f(v, v') = \int_E f(\psi(v, v', y))\, d\nu(y) \,.$$

Here the variable y plays the role of a collision parameter.

We basically keep the notations of section 2 Chapter III, and as a first step we will construct a "noninteracting tree". As before we consider Ω the set of marked trees $\omega = (\pi, (\phi^u,\ u \in T))$, where the marks ϕ now take their values in $D = \{(\tau, v, y) \in (0,\infty) \times R^n \times E\}$. Here τ represents a trajectory duration, v an initial velocity and y a collision parameter.

Now for any $(\tau, v, y) \in D$, we consider the probability $R_{(\tau,v,y)}$ on Ω, such that:

- $R_{(\tau,v,y)}[\phi^{\emptyset} = (\tau, v, y)] = 1$ (the ancestor's mark is (τ, v, y)).
- For f_u, $u \in U$ a collection of nonnegative measurable functions on Ω,

$$E_{(\tau,v,y)}[\prod_{u \in G_n} f_u \circ T_u / F_n] = \prod_{u \in G_n} E_{\phi^u}[f_u]$$

(Branching property).
- The reproduction law is given by the fact that the point measure on $D : \sum_{1 \leq \ell \leq \nu_\emptyset} \delta_{\theta^\ell}$ is a Poisson point process with intensity measure on D:

$$1(s < \tau)\; ds \otimes u_0(dv) \otimes \nu(dy) \;,$$

and $R_{(\tau,v,y)}$- a.s., $0 < \tau^1 < \ldots < \tau^{\alpha(1)} < \tau$.
Now very similarly to section 1 of Chapter III, we have:
Lemma 1.2.
- $R_{(\tau,v,y)}$-a.s., ω is a finite tree.
- Under $R_{(\tau,v,y)}$, the random point measure on Ω : $M(\omega) = \sum_{1 \leq \ell \leq \nu_\phi} \delta_{T_\ell(\omega)}$, is a Poisson point measure with intensity

$$Q_{(\tau,v,y)} = \int_{[0,\tau] \times R^n \times E} d\tau'\, du_0(v')\, d\nu(y')\, R_{\tau',v',y'} \;.$$

We can now construct the interaction on the tree $\omega = (\pi, (\tau^u, v^u, y^u),\; u \in \pi)$ as a supplementary mark x^u, which will represent the velocity of the particle $u \in \pi$, at time τ^u. The mark v^u corresponds to the initial velocity of this particle and the collision parameter y^u, will be used in calculating the effect of paricle u, on its direct ancestor.

More precisely, using the fact that the tree is a.s. finite, we set: $z^u = v^u$ if u has no descendants, $z^u = z_{\nu_u}$, otherwise, where the sequence z_j, $0 \leq j \leq \nu_u$ is defined by:

$$\begin{aligned} z_j &= \psi(z_j,\; z^{uj}, v^{uj}) \;, \quad 1 \leq j \leq \nu_u \\ z_0 &= v^u \;. \end{aligned} \tag{1.10}$$

We now denote by R^t for $t > 0$, the measure $\int du_0(v)\; d\nu(y)\; R_{t,v,y}$. With this notation, Lemma 1.2 says that $M(\omega)$ under R^t is a Poisson point process with intensity measure $\int_0^t ds\; R^s$.

The corresponding result to Theorem 2.3 of section 2, now tells us that under R^t, the law of the supplementary mark $z^\emptyset$, solves the equation (1.3).

Proposition 1.3. *The law u_t of $z^\emptyset$ under R^t, satisfies*

$$u_t - u_0 = \int_0^t (u_s \circ u_s - u_s)\; ds \;.$$

Proof: With notation (1.10), working under R^t:

$$z^\theta = v^\theta\ 1(\nu_\theta = 0) + \psi(z_{\nu_\theta-1}, z^{\nu_\theta}, y^{\nu_\theta})\ 1(\nu_\phi \geq 1)\ .$$

Now, with the help of Lemma 1.2, conditionally on $\nu_\theta \geq 1$, and $\tau^{\nu_\theta} = s$, $z_{\nu_\theta-1}, z^{\nu_\theta}, y^{\nu_\theta}$ are independent, with, $z_{\nu_\theta-1}$, z^{ν_θ}, u_s distributed. So we find that $u_t = u_0 e^{-t} + \int_0^t e^{-(t-s)} u_s \circ u_s\ ds$.

Our claim follows from this.

□

Proposition 1.3 provides a representation formula for the solution of spatially homogeneous Boltzmann equations with cutoff, which is also giving an intuition for the proof off uniqueness of the hard sphere Boltzmann process given in the next section.

2) Uniqueness for the Boltzmann process.

In this section we will look at the Boltzmann process (spatially homogeneous for hard spheres), as the solution of a certain martingale problem. The unboundedness of the collision intensity $|(v'-v)\cdot n|$, is a source of difficulty for the uniqueness of the solution, especially if one does not want to assume too many moment integrability conditions on the solution.

The existence of the solution, in the theorem we are now going to state, comes from a tightness result on the solutions corresponding to the cutoff problems (with collision function $|(v'-v)\cdot n| \wedge N$). For details on the existence part, we refer the reader to [39].

Theorem 2.1. *Let $u_0 \in M(R^n)$, be such that $\int |v|^3\ u_0(dv) < \infty$. There is a unique probability P on $D(R_+, R^n)$, such that*

i) *for $T < \infty$, $\sup_{t \leq T} \int |X_t|^3\ dP < \infty$*

ii) $X_0 \circ P = u_0$

iii) *for $f \in bB(R^n)$,*

$$\begin{aligned} f(X_t) - f(X_0) - \int_0^t \int_{D\times S_n} &[f(X_s + (X_s(\omega') - X_s)\cdot nn) \\ &- f(X_s)]\ |X_s(\omega') - X_s)\cdot n|\ dP(\omega')\ dn\ ds \end{aligned}$$

is a P-martingale.

We are now going to explain in a number of steps the proof of uniqueness.

Let us first recall that for a solution P of such a martingale problem, one can give a trajectorial representation as follows. Denote by $M_p(R_+ \times R_+ \times S_n \times D)$, the set of simple pure point measure on $R_+ \times R_+ \times S_n \times D$, finite on any compact restriction of the first two coordinates (with at mot one atom on each $\{t\} \times R_+ \times S_n \times D$, and none if t = 0).

Now on the product space $R^n \times M_p$ one can put the product of the probabilitiy u_0 and of Poisson measure with intensity $dt \otimes d\theta \otimes dn \otimes dP(\omega')$. P is now the law of the unique solution Z_s of the equation

$$
(2.1)\quad \begin{aligned}
Z_t = Z_0 + \int_0^t \int_{R_+\times S_n\times D} & (X_s(\omega') - Z_{s-})\cdot n\, n\, 1\{\theta < |(Z_{s-} - X_s(\omega'))\cdot n|\} \\
& N(ds\, d\theta\, dn\, d\omega') \\
\text{with } \int_0^t \int_{R_+\times S_n\times D} & 1\{\theta < |(Z_{s-} - X_s(\omega'))\cdot n|\}\, N(ds\, d\theta\, dn\, d\omega') < \infty\,, \\
\text{for all } t < \infty\,, &
\end{aligned}
$$

Here Z_0 is the R^n-valued coordinate on the space $R^n \times M_p$, and N is the canonical Poisson measure induced by the second coordinate.

Consider then P_1 and P_2 two solutions of the martingale problem. We are first going to construct a coupling measure P_0 on $D(R_+,R^n)^2$, of P_1 and P_2, whch will also satisfy a nonlinear martingale problem.

Lemma 2.2. *There exists a probability P_0 on $D(R_+,R^n)^2$ such that*

1) $X^1_\cdot \circ P_0 = P_1$, $X^2_\cdot \circ P_0 = P_2$.

2) *for* $f \in bB(R^n \times R^n)$:

$$
\begin{aligned}
f(X^1_t,X^2_t) - f(X^1_0,X^2_0) - \int_0^t \int_{D^2\times S_n} & \\
[f(\tilde X^1_s,\tilde X^2_s) \mid (X^1_s - X^1_s(\omega'))\cdot n| & \wedge |(X^2_s - X^2_s(\omega'))\cdot n| \\
+ f(\tilde X^1_s, X^2_s)\, (|(X^1_s - X^1_s(\omega'))\cdot n| & - |(X^2_s - X^2_s(\omega'))\cdot n|)_+ \\
+ f(X^1_s,\tilde X^2_s)\, (|(X^2_s - X^2_s(\omega'))\cdot n| & - |(X^1_s - X^1_s(\omega'))\cdot n|)_+ \\
- f(X^1_s,X^2_s)\, (|(X^2_s - X^1_s(\omega'))\cdot n| & \vee |(X^2_s - X^2_s(\omega'))\cdot n|)]\, dn\, dP_0(\omega')\, ds
\end{aligned}
$$

is a P_0-martingale, with the notation for $i = 1,2$:

$$
\tilde X^i_s = X^i_s + (X^i_s(\omega') - X^i_s(\omega))\cdot n\, n
$$

3) $(X^2_0,X^2_0)\circ P_0 = \text{diag } u_0$,*the diagonal image of u_0 on $R^n \times R^n$.*

Proof: The set $\mathcal{C}$ of probabilities on $D(R_+,R^n)^2$ with respective projections on the first and second coordinates givenby P_1 and P_2, is a weakly compact and convex set.

Denote by F the map which associates to $Q \in \mathcal{C}$, the law on $D(R_+,R^n)^2$ of $(Z^1_\cdot, Z^2_\cdot)$ solutions on $R^n \times R^n \times M_p(R_+ \times R_+ \times S_n \times D(R_+,R^n)^2)$ with measure diag $u_0\otimes$ Poisson $(dt \otimes d\theta \otimes dn \otimes dQ)$ of the equations for $i = 1, 2$.

$$Z^i_t = Z^i_0 + \int_0^t \int_{R_+\times S_n \times D(R_+,R^n)^2} (X^i_s(\omega') - Z^i_{s-}) \cdot nn \; 1\{\theta < |(Z^i_s - X^i_s(\omega')) \cdot n|\} \\ N(ds\; d\theta\; dn\; d\omega') ,$$

with

$$\int_0^t \int 1\{\theta < |(Z^i_{s-} - X^i_s(\omega')) \cdot n|\} \; N(ds\; d\theta\; dn\; d\omega') < \infty .$$

In view of representation (2.1) F maps $\mathcal{C}$ into $\mathcal{C}$. Moreover $F(Q)$ is characterized by 1), 2), 3), with the replacement of $dP_0(\omega')$ integration by $dQ(\omega')$ integration. From this one sees that F is weakly continuous, and the existence of a fixed point to the map F now follows from Tychonov's theorem. This yields our claim.

□

Let us now introduce some notations convenient for what follows. We set for $x = (x_1, x_2)$, $x' = (x'_1, x'_2)$:

$$k(x,x') = (1 + |x_i| + |x'_i|) \vee (1 + |x_2| + |x'_2|) ,$$
$$\phi(x) = 9(1 + |x_1|^2 + |x_2|^2) ,$$
$$Qh(x,x') = \int h(x + \Phi(x,x',y),\; x' + \Phi(x',x,y))\; d\nu(y)$$

where $y = (\alpha, n) \in [0,1] \times S_n = E$, $d\nu = d\alpha \otimes dn$, and Φ is the map from $R^{2n} \times R^{2n} \times E$ into R^{2n}, given by:

$$\begin{aligned} \Phi(x,x',y) &= ((x'_1 - x_1)\cdot nn,\; (x'_2 - x_2)\cdot nn), \;\text{ if } \alpha \le (a_1 \wedge a_2)/k, \\ &= ((x'_1 - x_1)\cdot nn) \text{ if } a_1/k \ge \alpha > (a_1 \wedge a_2)/k, \\ &= (0,\; (x'_2 - x_2)\cdot nn) \text{ if } a_2/k \ge \alpha > (a_1 \wedge a_2)/k, \\ &= (0,0) \quad \text{ if } \alpha \ge (a_1 \vee a_2)/k, \end{aligned}$$

with $a_i = |(x_i - x'_i)\cdot n|$, $k = k(x,x')$. We will also write for $f \in bB(R^n)$: $Q_1 f = Q(f \otimes 1)$.

If we put the probability diag $u_0\otimes$ Poisson$(ds \otimes d\theta \otimes d\nu \otimes dP_0(\omega'))$, on $R^n \times R^n \times M_p(R_+ \times R_+ \times E \times D)$ (with $D = D(R_+,R^n)^2$), P_0 can also be represented as the law of the solution of

$$Z_t(\omega) = Z_0(\omega) + \int_0^t \int_{R_+\times E \times D} \Phi(Z_{s-}(\omega),\omega'(s),y)\; 1(\theta < k(Z_{s-}(\omega),\omega'(s))) \\ N(\omega, ds\; d\theta\; dy\; d\omega') , \tag{2.3}$$

with

$$\int_0^t \int_{R_+\times E\times D} 1(\theta < k(Z_{s^-}(\omega), \omega'(s)))\ N(\omega, ds\ d\theta\ dy\ d\omega')\ .$$

Moreover we have the estimates:

$$(2.4)\qquad \begin{aligned} &k(x,x') \le \phi(x)^{1/2}\ \phi^{1/2}(x')\,,\\ &Q(\phi\oplus\phi) = \phi\oplus\phi \quad (\text{where } \phi\oplus\phi(x,x') = \phi(x)+\phi(x'))\,. \end{aligned}$$

Another nice feature of our coupling measure P_0, as seen from (2.2), is that if x and x' are "diagonal", that is $x_1 = x_2$ and $x_1' = x_2'$, then $\Phi(x,x',y)$ is diagonal as well.

The representation formula (2.3) for the coupling measure P_0, should be viewed as a way to keep track thanks to the marks "$\omega_\cdot'$" of the first generation of trajectories, contributing to the determination of the bi-particle trajectory $Z.$ on $[0,t]$. The scheme is now to construct the successive generations, using first a similar formula to (2.3) for each mark ω' (this yields the second generation), and iterating the procedure. On this "projective object", the problem will now be to see that the ancestor trajectory is in fact the result of a calculation on a finite tree whicch preserves the diagonal property of the initial conditions. This will prove that P_0 is in fact supported by the diagonal, and will yield uniqueness.

The first step (compare with Lemma 1.2, and Lemma 2.1 of Chapter III).

Lemma 2.3. *There is an auxiliary space $(\tilde\Omega, P)$, endowed with a Poisson point measure $N(\tilde\omega, ds\ d\theta\ dy\ d\tilde\omega')$ of intensity $ds\otimes d\theta\otimes d\nu(y)\otimes dP(\tilde\omega')$, and with a P_0-distributed process $(Z.(\omega))$ such that Z_0 is independent of the point measure N and* diag u_0 *distributed and*

$$\text{for } t>0,\ \int_0^t \int_{R_+\times E\times\tilde\Omega} 1\{\theta < k(Z_{s^-}(\tilde\omega),\ Z_s(\tilde\omega'))\}\ N(\tilde\omega, ds\ d\theta\ dy\ d\tilde\omega') < \infty$$

$$\begin{aligned} Z_t(\tilde\omega) = Z_0(\tilde\omega) + &\int_0^t \int_{R_+\times E\times\tilde\Omega} \Phi(Z_{s^-}(\tilde\omega),\ Z_s(\tilde\omega'), y)\\ &1\{\theta < k(Z_{s^-}(\tilde\omega), Z_s(\tilde\omega'))\}\ N(\omega, ds\ d\theta\ dy\ d\omega')\ , \end{aligned}$$

for the proof of this we refer the reader to [39]. The idea of the construction as alluded to before is that (2.3) gives a natural map from $R^{2n}\times M_p(R_+\times R_+\times E\times D)$ into D. Iterating this one has a natural map from $R^{2n}\times M_p(R_+\times R_+\times E\times R^{2n}\times M_p(R_+\times R_+\times E\times D))$ in the space $R^{2n}\times M_p(R_+\times R_+\times E\times D)$. The space $\tilde\Omega$ essentially arises as a projective limit of this scheme.

Now on our extended space $\tilde{\Omega}$, we can define the number $K_t(\tilde{\omega})$ of generations which contribute to the state of the "ancestor trajectory" $Z_.(\tilde{\omega})$ on time $[0,t]$, by stating:

$$\{K \geq 0\} = R_+ \times \tilde{\Omega}$$

$$\{K \geq 1\} = \{(t,\tilde{\omega})/\exists s \in (0,t],\ N(\tilde{\omega},\{s\} \times R_+ \times E \times \tilde{\Omega}) = 1,$$
$$\text{and } {}^s\theta < k(Z_{s-}(\tilde{\omega}), Z_s({}^s\tilde{\omega}'))\}$$

here ${}^s\theta$ and ${}^s\tilde{\omega}'$ are the marks of $N(\tilde{\omega},\cdot)$ at time s.

$$\{K \geq n+1\} = \{(t,\omega)/\exists s \in (0,t],\ N(\tilde{\omega},\{s\} \times R \times E \times \tilde{\Omega}\} = 1,$$
$${}^s\theta < k(Z_{s-}(\tilde{\omega}), Z_s({}^s\tilde{\omega}')) \text{ and } K_s({}^2\tilde{\omega}') \geq n\}\ ,$$

and

$$\{K \geq \infty\} = \bigcap_n \{K \geq n\}\ .$$

Our next step is that only a finite number of generations play a role in the determination of $Z_.(\tilde{\omega})$.

Lemma 2.4. *For* $t > 0$, $K_t(\omega) < \infty$, P-a.s.

Proof: Set $\bar{Z}_t(\tilde{\omega}) = Z_t(\tilde{\omega})$ when $K_t < \infty$, δ otherwise. So $\bar{Z}_.$ is $R^n \cup \{\delta\}$ valued. Now for $f \in bB(R^n)$ (equal to zero on δ):

$$f(\overline{Z}_t) - f(\bar{Z}_0) = \int_0^t \int_{R_+\times E\times\tilde{\Omega}} [f(Z_s(\tilde{\omega}))\ 1\{K_{s-}(\tilde{\omega}) \vee K_s(\tilde{\omega}') < \infty\}$$
$$- f(\bar{Z}_{s-}(\tilde{\omega}))]\ 1\{\theta < k(Z_{s-}(\tilde{\omega}), Z_s(\tilde{\omega}'))\}\ N(\omega,\ ds\, d\theta\, d\nu(y)\, dP(\tilde{\omega}'))$$

From this after integration, we find:

$$E[f(\bar{Z}_t)] - E[f(Z_0)] = \int_0^t \int_{\Omega\times\Omega} Q_1 f(\bar{Z}_s(\tilde{\omega}), \bar{Z}_s(\tilde{\omega}'))\ k(\bar{Z}_s(\tilde{\omega}), \bar{Z}_s(\tilde{\omega}'))$$
$$- f(\bar{Z}_s(\tilde{\omega}))\ k(\bar{Z}_s(\tilde{\omega}), Z_s(\tilde{\omega}'))\ dP(\tilde{\omega})\ dP(\tilde{\omega}')\ ds \tag{2.5}$$

In view of our integrability assumptions on P_1, and P_2, we can apply (2.5) with ϕ instead of f. And by a very similar argument using $Q(\phi \oplus \phi) = \phi \oplus \phi$ as stated in (2.2), we get:

$$E[\phi(Z_t)] = E[\phi(Z_0)]$$

It then follows that

$$E[\phi(Z_t) - \phi(\bar{Z}_t)] = \int_0^t \int_{\Omega\times\Omega} -\frac{1}{2}[Q(\phi \oplus \phi) - \phi \oplus \phi) - \phi \oplus \phi] \times k(\bar{Z}_s(\tilde{\omega}), \bar{Z}_s(\tilde{\omega}'))$$
$$+ \phi(\bar{Z}_s(\tilde{\omega}))[k(\bar{Z}_s(\tilde{\omega}), Z_s(\tilde{\omega}')) - k(\bar{Z}_s(\tilde{\omega}), \bar{Z}_s(\tilde{\omega}'))]$$
$$dP(\tilde{\omega})\ dP(\tilde{\omega}')\ ds$$

Now the first term in the integral is zero, as for the second, using $k(x,x') \leq \phi^{1/2}(n)\ \phi^{1/2}(x')$, we find:

$$E[\phi(Z_t)\ 1\{K_t = \infty\}] \leq \int_0^t E[\phi^{3/2}(Z_s)]\ E[\phi^{1/2}(Z_s)\ 1\{K_s = \infty\}]\ ds$$

From Gronwall's lemma, we now find $E[\phi(Z_t)\ 1\{K_t = \infty\}] = 0$, from which our claim follows.

□

We now have the required ingredients to see that our coupling probability P_0 is diagonally supported.

Indeed when $K_t(\tilde{\omega}) = 0$, then $Z_t^1(\tilde{\omega}) = Z_0^1(\tilde{\omega}) = Z_0^2(\tilde{\omega}) = Z_t^2(\tilde{\omega})$, since no jump occurs.

Suppose now that we know that $K_T(\tilde{\omega}) \leq n$ implies $Z_t^1(\tilde{\omega}) = Z_t^2(\tilde{\omega})$, for $t \leq T$. Take now $\tilde{\omega} \in \tilde{\Omega}$ with $K_T(\tilde{\omega}) = n+1$, then for $t \leq T$:

$$Z_t(\tilde{\omega}) = Z_0(\tilde{\omega}) + \int_0^t \int_{R^n \times E \times \tilde{\Omega}} \Phi(Z_{s^-}(\tilde{\omega}), Z_s(\tilde{\omega}'), y)\ 1\{\theta < k(Z_{s^-}(\tilde{\omega}), Z_s(\tilde{\omega}'))\}$$
$$1\{K_s(\tilde{\omega}') \leq n\}\ N(\tilde{\omega}, ds\ d\theta\ dy\ d\tilde{\omega}')\ ,$$

since only mark ${}^s\tilde{\omega}'$ for which $K_s({}^s\tilde{\omega}') \leq n$ come in the determination of $Z_t(\tilde{\omega})$, $t \leq T$. Now as observed already if x and x' are diagonal so is $\Phi(x, x', y)$. It then follows that $Z_t(\tilde{\omega})$, $t \leq T$ is diagonal. Now our claim follows by induction, thanks to Lemma 2.4.

So we obtain that the coupling probability P_0 is diagonally supported, and this proves that $P_1 = P_2$.

□

References

1. Baxter, J. R. - Chacon, R. V. - Jain, N. C.: Weak limits of stopped diffusions, Trans. Amer. Math. Soc., 293, 2, 767-792, (1986).
2. Billingsley, P.: Convergence of Probability Measures, Wiley, New York (1968).
3. Calderoni, P. - Pulvirenti, M.: Propagation chaos for Burger's equation, Ann. Inst. H. Poincaré, série A, N.S. 39, 85-97 (1983).
4. Cercignani, C.: The grad limit for a system of soft spheres, Comm. Pure Appl. Math 26, 4 (1983).
5. Dawson, D. A.: Critical dynamics and fluctuations for a mean field model of cooperative behavior, J. Stat. Phys. 31, 29-85 (1978).
6. Dawson, D. A. - Gärtner, J.: Large deviations from the McKean-Vlasov limit for weakly interacting diffusions, Stochastics 20, 247-308 (1987).
7. De Masi, A. - Ianiro, N. - Pellegrinotti, A. - Pressutti, E.: A survey of the hydrodynamical behavior of many particle systems, in Nonequilibrium phenomena II, Ed.: J. L. Lebowitz, E. W. Montroll, Elsevier (1984).
8. Dobrushin, R. L.: Prescribing a system of random variables by conditional distributions, Th. Probab. and its Applic. 3, 469 (1970).
9. ———— Vlasov equations, Funct. Anal. and Appl. 13, 115 (1979).
10. Ferland, R., Giroux, G.: Cutoff Boltzmann equation: convergence of the solution, Adv. Appl. Math. 8, 98-107 (1987).
11. Funaki, T.: The diffusion approximation of the spatially homogeneous Boltzmann equation, Duke Math. J. 52, 1-23, (1985).
12. Goodman, J., Convergence of the random value method, IMA, vol. 9, G. Papanciolaou ed., Hydrodynamic Behavior and Interacting Particles, 99-106, Springer, Berlin (1987).
13. Guo, M. - Papanicolaou, G. C.: Self diffusion of interacting Brownian particles, Taniguchi Symp., Katata 1985, 113-151.
14. Gutkin, E.: Propagation of chaos and the Hopf-Cole transformation, Adv. Appl. Math. 6, 413-421, 1985.
15. Kac, M.: Foundation of kinetic theory, Proc. Third Berkeley Symp. on Math. Stat. and Probab. 3, 171-197, Univ. of Calif. Press (1956).
16. ———— Some probabilistic aspects of the Boltzmann equation, Acta Physica Austraiaca, suppl. X, Springer, 379-400 (1979).
17. Karandikar, R. L.,Horowitz, J.: Martingale problems associated with the Boltzmann equation, preprint, 1989.
18. Kipnis, C. - Olla, S. - Varadhan, S.R.S.: Hydrodynamics and large deviation for simple exclusion processes Comm. Pure Appl. Math., 42, 115-137, (1989).
19. Kurtz, T.: Approximation of population processes CMBS-NSF Reg. Conf. Sci. Appl. Math. Vol. 36, Society for Industrial and Applied Mathematics, Philadelphia (1981).
20. Kusuoka, S. - Tamura, Y.: Gibbs measures with mean field potentials, J. Fac. Sci. Tokyo Univ., sect. 1A, 31, 1, 223-245 (1984).
21. Kotani, S. - Osada, H.: Propagation of chaos for Burgers' equation, J. Math. Soc. Japan, 37, 275-294 (1985).
22. Lanford, O. E., Time evolution of large classical systems, Lecture Notes in Physics 38, 1-111, Springer, Berlin, (1975).
23. Lang, R. - Nguyen, X.X.: Smoluchowski's theory of coagulation in colloids holds rigorously in the Boltzmann-Grad limit, Z. Wahrscheinlichkeitstheor. Verw. Gebiete 54, 227-280 (1980).

24. Liggett, T. M.: Interacting particle systems, Springer, Berlin (1985).
25. McKean, H. P.: Speed of approach to equilibrium for Kac's caricature of a Maxwellian gas, Arch. Rational Mech. Anal. 21, 347-367 (1966).
26. ———— A class of Markov processes associated with nonlinear parabolic equations, Proc. Nat. Acad. Sci. 56, 1907-1911 (1966).
27. ———— Propagation of chaos for a class of nonlinear parabolic equations, Lecture series in differential equations 7, 41-57, Catholic University, Washington, D. C. (1967).
28. ———— Fluctuations in the kinetic theory of gases, Comm. Pure Appl. Math. 28, 435-455 (1975).
29. Marchioro, C. – Pulvirenti, M.: Hydrodynamics in two dimensions and vortex theory, Comm. Math. Phys. 84, 483-504 (19820.
30. Murata, H. - Tanaka, H.: An inequality for certain functionals of multidimensional probability distributions, Hiroshima Math. J., 4, 75-81 (1974).
31. Nagasawa, M. – Tanaka, H.: Diffusion with interaction and collisions between colored particles and the propagation of chaos, Probab. Th. Rel. Fields 74, 161-198 (1987).
32. ———— On the propagation of chaos for diffusion processes with coefficients not of average form, Tokyo Jour. Math. 10 (2), 403-418 (1987).
33. Neveu,J.: Arbres et processus de Galton-Watson, Ann. Inst. Henri Poincaré Nouv. Ser. B, 22, 2,199-208 (1986).
34. Oelschläger K.: A law of large numbers for moderately interacting diffusion processes, Z. Wahrscheinlichkeitstheor. Verw. Gebeite 69, 279-322 (1985).
35. Osada, H.: Limit points of empirical distributions of vortices with small viscosity, IMA, vol. 9, G. Papanicolaou ed., Hydrodynamic behavior and interacting particles, 117-126, Springer, Berlin (1987).
36. ———— Propagation of chaos for the two dimensional Navier-Stokes equation.
37. Scheutzow, M.: Periodic behavior of the stochastic Brusselator in the mean field limit, Prob. Th. Re. Fields 72, 425-462, (1986).
38. Shiga, T. – Tanaka, H.: Central limit theorem for a system of Markovian particles with mean field interactions, Z. Wahrscheinlichkeitstheor. Verw. Gebiete 69, 439-445 (1985).
39. Sznitman, A. S.: Equations de type Boltzmann spatialement homogènes, Z. Wahrscheinlichkeitstheor. Verw. Gebiete 66, 559-592 (1984).
40. ———— Nonlinear reflecting diffusion process and the propagation of chaos and fluctuations associated, J. Funct. Anal. 56 (3), 311-336 (1984).
41. ———— A fluctuation result for nonlinear diffusions, infinite dimensional analysis, S. Albeverio, ed., 145-160, Pitman, Boston (1985).
42. ———— A propagation of chaos result for Burgers' equation, Z. Wahrscheinlichkeitstheor. Verw. Gebiete 71, 581-613 (1986).
43. ———— Propagation of chaos for a system of annihilating Brownian spheres, Comm. Pure Appl. Math. 60, 663-690 (1987).
44. ———— A trajectorial representation for certain nonlinear equations, Astérisque, 157-158, 363-370 (1988).
45. ———— A limiting result for the structure of collisions between many independent diffusions, Probab. Th. Rel Fields 81, 353-381 (1989).
46. Sznitman, A. S. – Varadhan, S.R.S.: A multidimensional process involving local time, Z. Wahrscheinlichkeitstheor. Verw. Gebiete 71, 553-579 (1986).

47. Spohn, H.: The dynamics of systems with many particles, statistical mechanics of local equilibrium states (preprint).
48. Tamura, Y.: On asymptotic behaviors of the solution of a nonlinear diffusion equation, J. Fac. Sci. Tokyo Univ., sect. IA, 31, 1, 195-221 (1984).
49. Tanaka, H.: Probabilistic treatment of the Boltzmann equation of Maxwellian molecules, Z. Wahrscheinlichkeitstheor. Verw. Gebiete 46, 67-105 (1978).
50. ———— Some probabilistic problems in the spatially homogeneous Boltzmann equation, Proc. IFIP–ISI conf. on appl. of random fields, Bangalore, Jan. 82.
51. ———— Limit theorems for certain diffusion processes with interaction, Taniguchi Symp. S. A. Katata, 469-488 (1982).
52. Tanaka, H. – Hitsuda, M.: Centtal limit theorem for a simple model of interacting particles, Hiroshima Math. J. 11, 415-423 (1981).
53. Uchiyama, K.: On the Boltzmann Grad limit for the Broadwell model of the Boltzmann equation, J. Stat. Phys. 52, 331-355 (1988).
54. ———— Derivation of the Botlzmann equation from particle dynamics, Hiroshima Math. J. 18, 2 (1988).
55. Ueno, T.: A class of Markov processes with bounded nonlinear generators, Japanese J. Math. 38, 19-38 (1968).
56. ———— A path space and the propagation of chaos for Boltzmann's gas model, Proc. Japan Acad. 6 (47) 529-533 (1971).
57. Wild, E.: On the Boltzmann equation in the kinetic theory of gases, Proc. Cambridge Phil. Soc., 47, 602-609 (1951).
58. Zvonkin, A. K.: A transformation of the phase space of a diffusion process that removes the drift, Math. USSR Sbornik, 22, 1, 129-149, (1974).

TEN LECTURES

ON PARTICLE SYSTEMS

Rick DURRETT

Originally published in: *Ecole d'Eté de Probabilités de Saint-Flour XXIII – 1993*, Lecture Notes in Mathematics, Vol. **1608**, 97–201, DOI: 10.1007/BFb0095747, © Springer-Verlag Berlin Heidelberg 1995, Reprint by Springer-Verlag Berlin Heidelberg 2012

Preface. These lectures were written for the 1993 St. Flour Probability Summer School. Their aim is to introduce the reader to the mathematical techniques involved in proving results about interacting particle systems. Readers who are interested instead in using these models for biological applications should instead consult Durrett and Levin (1993).

In order that our survey is both broad and has some coherence, we have chosen to concentrate on the problem of proving the existence of nontrivial stationary distributions for interacting particle systems. This choice is dictated at least in part by the fact that we want to make propaganda for a general method of solving this problem invented in joint work with Maury Bramson (1988): comparison with oriented percolation. Personal motives aside, however, the question of the existence of nontrivial stationary distributions is the first that must be answered in the discussion of any model.

Our survey begins with an overview that describes most of the models we will consider and states the main results we will prove, so that the reader can get a sense of the forest before we start investigating the individual trees in detail. In Section 2 we lay the foundations for the work that follows by proving an existence theorem for particle systems with translation invariant finite range interactions and introducing some of the basic properties of the resulting processes. In Section 3 we give a second construction that applies to a special class of "additive" models, that makes connections with percolation processes and that allows us to define dual processes for these models.

The general method mentioned above makes its appearance in Section 4 (with its proofs hidden away in the appendix) and allows us to prove a very general result about the existence of stationary distributions for attractive systems with state space $\{0,1\}^S$. The comparison results in Section 4 are the key to our treatment of the threshold contact and voter models in Section 5, the cyclic systems in Section 6, the long range contact process in Section 7, and the predator prey system in 9.

In Section 7 we explore the first of two methods for simplifying interacting particle systems: assuming that the range of interaction is large. In Section 8 we meet the second: superimposing particle motion at a fast rate. The second simplification leads to a connection with reaction diffusion equations which we exploit in Section 9 to prove the existence of phase transitions for predator prey systems.

The quick sketch of the contents of these lectures in the last three paragraphs will be developed more fully in the overview. Turning to other formalities, I would like to thank the organizers of the summer school for this opportunity to speak and write about my favorite subject. Many of the results presented here were developed with the support of the National Science Foundation and the Army Research Office through the Mathematical Science Institute at Cornell University. During the Spring semester of 1993, I gave 10 one and a half hour lectures to practice for the summer school and to force myself to get the writing done on time. You should be grateful to the eight people who attended this dress rehersal: Hassan Allouba, Scott Arouh, Itai Benjamini, Carol Bezuidenhout, Elena Bobrovnikova, Sungchul Lee, Gang Ma, and Yuan-Chung Sheu, since their suffering has lessened yours.

Although it is not yet the end of the movie, I would like to thank the supporting cast now: Tom Liggett, who introduced me to this subject; Maury Bramson, the co-discoverer of the comparison method and long range limits, to whom I turn when my problems get too hard; David Griffeath, my electronic colleague who introduced me (and the rest of the world) to the beautiful world of the Greenberg Hastings and cyclic cellular automata; Claudia Neuhauser, my former student who constantly teachs me how to write; and Ted Cox, with whom I have written some of my best papers. The field of interacting particle systems has grown considerably since Liggett's 488 page book was published in 1985, so it is inevitable that more is left out than is covered in these notes. The most overlooked researcher in this treatment is Roberto Schonmann whose many results on the contact process, bootstrap percolation, and metastability in the Ising model did not fit into our plot.

1. Overview

In an interacting particle system, there is a countable set of spatial locations S called *sites*. In almost all of our applications $S = \mathbf{Z}^d$, the set of points in d dimensional space with integer coordinates. Each site can be in one of a finite set of *states* F, so the state of the system at time t is $\xi_t : S \to F$ with $\xi_t(x)$ giving the state of x. To describe the evolution of these models, we specify an *interaction neighborhood*

$$\mathcal{N} = \{z_0, z_1, \dots z_k\} \subset \mathbf{Z}^d$$

with $z_0 = 0$ and define *flip rates*

$$c_i(x,\xi) = g_i(\xi(x+z_0), \xi(x+z_1), \dots, \xi(x+z_k))$$

In words, the state of x flips to i at rate $c_i(x,\xi)$ when the state of the process is ξ. In symbols, if $\xi_t(x) \neq i$ then

$$\frac{P(\xi_{t+s}(x) = i | \xi_t = \xi)}{s} \to c_i(x,\xi) \quad \text{as } s \to 0$$

The formula for c_i indicates that our interaction is *finite range*, i.e., the flip rates depend only on the state of x and of a finite number of neighbors; and *translation invariant*, i.e., the rules applied at x are just a translation of those applied at 0.

To explain what we have in mind when making these definitions, we now describe two famous concrete examples. In this section and throughout these lectures (with the exception of Sections 2 and 3) we will suppose that

$$\mathcal{N} = \{x : \|x\|_p \le r\}$$

Here $r \ge 1$ is the *range* of the interaction and $\|x\|_p$ is the usual L^p norm on $\mathbf{R}^d$. That is, $\|x\|_p = (x_1^p + \dots + x_d^p)^{1/p}$ when $1 \le p < \infty$ and $\|x\|_\infty = \sup_i |x_i|$. In most of our models the flip rates are based on the number of neighbors in state i, so we introduce the notation:

$$n_i(x,\xi) = |\{z \in \mathcal{N} : \xi(x+z) = i\}|$$

where $|A|$ is the number of points in A.

Example 1.1. The basic contact process. To model the spread of a plant species we think of each site x as representing a square area in space with $\xi_t(x) = 0$ if that area is vacant and $\xi_t(x) = 1$ if there is a plant there, and we formulate the dynamics as follows:

$$\begin{aligned} c_0(x,\xi) &= \delta \qquad \text{if } \xi(x) = 1 \\ c_1(x,\xi) &= \lambda n_1(x,\xi) \qquad \text{if } \xi(x) = 0 \end{aligned}$$

In words, plants die at rate δ independent of the state of their neighbors, while births at vacant sites occur at a rate proportional to the number of occupied neighbors. Note that flipping to i has no effect when $\xi(x) = i$ so the value of $c_i(x, \xi)$ on $\{\xi(x) = i\}$ is irrelevant and we could delete the qualifying phrases "if $\xi(x) = 1$" and "if $\xi(x) = 0$" if we wanted to.

Example 1.2. The basic voter model. This time we think of the sites in $\mathbf{Z}^d$ as representing an array of houses each of which is occupied by one individual who can be in favor of ($\xi_t(x) = 1$) or against ($\xi_t(x) = 0$) a particular issue or candidate. Our simple minded voters change their opinion to i at a rate that is equal to the number of neighbors with that opinion. That is,

$$c_i(x, \xi) = n_i(x, \xi)$$

The first question to be addressed for these models is:

Do the rates specify a unique Markov process?

There is something to be proved since there are infinitely many sites and hence no first jump, but for our finite range translation invariant models, a result of Harris (1972) allows us to easily show that the answer is Yes. (See Section 2.) The main question we will be interested in is:

When do interacting particle systems have a nontrivial stationary distributions?

To make this question precise we need a few definitions. The state space of our Markov process is F^S, the set of all functions $\xi : S \to F$. We let $\mathcal{F}$ = all subsets of F and equip F^S with the usual product σ-field $\mathcal{F}^S$, which is generated by the *finite dimensional sets*

$$\{\xi(y_1) = i_1, \ldots, \xi(y_k) = i_k\}$$

So any measure π on $\mathcal{F}^S$ can be described by giving its *finite dimensional distributions*

$$\pi(\xi(y_1) = i_1, \ldots, \xi(y_k) = i_k)$$

As in the theory of Markov chains, π is said to be a *stationary distribution* for the process if when we start from an initial state ξ_0 with distribution π (i.e., $\pi(A) = P(\xi_0 \in A)$ for $A \in \mathcal{F}^S$) then ξ_t has distribution π for all $t > 0$. Since our dynamics are translation invariant, we will have a special interest in stationary distributions that are *translation invariant*, i.e., ones in which the probabilities $\pi(\xi(x + y_1) = i_1, \ldots, \xi(x + y_k) = i_k)$ do not depend upon x.

To explain the term "nontrivial" we note that in Example 1.1 the "all 0" state ($\xi(x) \equiv 0$) and in Example 1.2 for any i the all i state are *absorbing states*. That is, once the process enters these states it cannot leave them. If S were finite this fact (and enough irreducibility, which is present in Examples 1.1 and 1.2) would imply that all stationary distributions were *trivial*, i.e., concentrated on absorbing states. However, when S is infinite this argument fails and indeed, as the next few results show it is possible to have a nontrivial stationary distributions.

Theorem 1. Consider the basic contact process with $\mathcal{N} = \{x : \|x\|_p \le r\}$ with $r \ge 1$. If $\lambda|\mathcal{N}| \le \delta$ then there is only the trivial stationary distribution. If $\delta/\lambda < \delta_0$ then there is a nontrivial translation invariant stationary distribution.

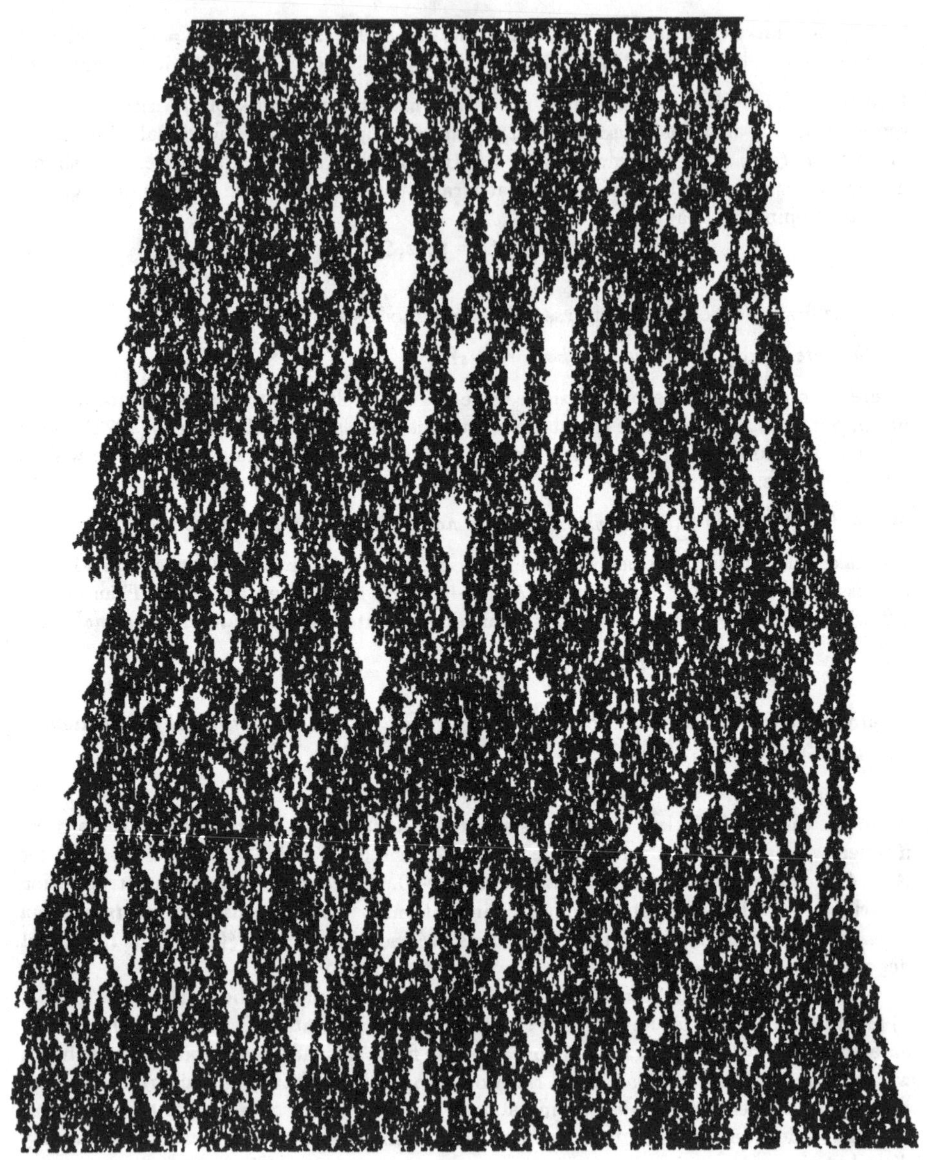

Figure 1.1. Nearest neighbor contact process in $d = 1$ with $\lambda = 2$.

The first result is easy to see. If the contact process has k particles then the number drops to $k-1$ at rate δk and increases to $k+1$ at rate $\leq \lambda|\mathcal{N}|$ with the upper bound achieved when all particles are isolated (i.e., no two particles are neighbors). The reader should attempt to prove the converse before we hit it with our sledgehammer in Section 4. By a simple comparison that you will learn about in Section 2, it is enough to prove the result when $\mathcal{N} = \{x : \|x\|_1 = 1\}$ and $d = 1$. A simulation of this case with $\lambda = 2$ is given in Figure 1.1. A result of Holley and Liggett (1978) implies that in this situation there is a nontrivial stationary distribution. In our simulation we have started with the interval $[180, 540]$ occupied at time 0 at the top of the page. As time runs down the page from 0 to 720, it is clear that the region occupied by particles is growing linearly, as predicted by a result of Durrett (1980).

Turning to the voter model, the classic paper of Holley and Liggett (1975) tells us that

Theorem 2A. *Clustering* occurs in $d \leq 2$. That is, for any ξ_0 and $x, y \in \mathbf{Z}^d$ we have

$$P(\xi_t(x) \neq \xi_t(y)) \to 0 \text{ as } t \to \infty$$

Theorem 2B. Let ξ_t^θ denote the process starting from an initial state in which the events $\{\xi_0^\theta(x) = 1\}$ are independent and have probability θ. In $d \geq 3$ as $t \to \infty$, $\xi_t^\theta \Rightarrow \xi_\infty^\theta$, a translation invariant stationary distribution in which $P(\xi_\infty^\theta(x) = 1) = \theta$.

Here $\Rightarrow$ denotes *weak convergence*, which in this setting is just convergence of finite dimensional distributions. That is, for any $x_1, \ldots x_m \in \mathbf{Z}^d$ and $i_1, \ldots, i_m \in \{1, 2 \ldots, \kappa\}$ we have

$$P(\xi_t^\theta(x_1) = i_1, \ldots \xi_t^\theta(x_m) = i_m) \to P(\xi_\infty^\theta(x_1) = i_1, \ldots \xi_\infty^\theta(x_m) = i_m)$$

We will say that *coexistence* occurs if there is a translation invariant stationary distribution in which each of the possible states in F has positive density. Theorems 2A and 2B say that in the voter model coexistence is possible in $d \geq 3$ but not in $d \leq 2$. We will see in Section 3 that this is a consequence of the fact that if we take two independent random walks with jumps uniformly distributed on $\mathcal{N}$ then they will hit with probability 1 in $d \leq 2$ but with probability < 1 in $d \geq 3$.

Figure 1.2 gives a simulation of a voter model with five opinions on $\{0, 1, \ldots, 119\}^2$. Here and in the next six simulations in this section, we use periodic boundary conditions. That is, sites on the top row are neighbors of those on the bottom row, and those on the left edge are neighbors of those on the right edge. We started at time 0 by assigning a randomly chosen opinion to each site. Figure 1.2 shows the state at time 500 suggesting that the clustering asserted in Theorem 2A occurs very slowly. Results of Cox (1988) imply that the expected time for our system to reach consensus is about

$$4 \ln(5/4) \cdot \frac{2}{\pi} (120)^2 \ln 120 = 39,173$$

The conclusions just derived for the voter depend on the fact that the flip rates are linear. Nonlinear flip rates can produce quite different behavior:

Figure 1.2. Five opinion two dimensional voter model at time 500

Example 1.3. The threshold voter model. Cox and Durrett (1991) introduced a modification of the voter model in which

$$c_i(x,\xi) = \begin{cases} 1 & \text{if } n_i(x,\xi) \geq \theta \\ 0 & \text{if } n_i(x,\xi) < \theta \end{cases}$$

In words, these voters change their opinion at rate 1 if at least θ neighbors disagree with them. This change in the rules changes the behavior of the model drastically.

We start with the case $\theta = 1$:

Theorem 3A. If $d = 1$ and $\mathcal{N} = \{-1, 1\}$ then clustering occurs.

Theorem 3B. In all other cases (recall we supposed that $\mathcal{N} = \{z : \|z\|_p \leq r\}$ with $r \geq 1$) we have coexistence. That is, there is a nontrivial translation invariant stationary distribution μ_{12} in which 1's and 2's each have density 1/2.

Here as in many other cases, the one dimensional nearest neighbor case is an exception. Cox and Durrett (1991) proved Theorem 3A and that coexistence occurs in some cases (e.g., $d = 1$ and $r \geq 7$) but the sharp Theorem 3B is due to Liggett (1992). Note that in the threshold voter model coexistence occurs in all but one case, while in the basic voter model coexistence occurs only in $d \geq 3$. A second difference is that when coexistence occurs the basic voter model has a one parameter family of nontrivial stationary distributions constructed in Theorem 2B but we believe

Conjecture 3C. When coexistence occurs in the threshold one voter model there is a unique spatially ergodic translation invariant stationary distribution in which 1's and 2's have positive density.

Here, we say that π on F^S is *spatially ergodic* if under π the family of random variables $\{\xi(x) : x \in \mathbf{Z}^d\}$ is an ergodic stationary sequence, i.e., the σ-field of events invariant under all spatial shifts is trivial. We need the assumption of spatial ergodicity to rule out nontrivial convex combinations

$$a\mu_1 + b\mu_2 + (1 - a - b)\mu_{12}$$

where μ_i is the point mass on the all i state, and μ_{12} is the measure constructed in Theorem 3B. In general, the set of translation invariant stationary distributions for an interacting particle system is a convex set and in most examples, the extreme points of the set are the stationary distributions that are spatially ergodic. However, there is no general result that shows this is true. See Problem 7 on page 178 of Liggett (1985).

While the threshold 1 case is fairly well understood, there are many open problems concerning higher thresholds. To illustrate these we observe that computer simulations suggest

Conjecture 3D. For the Moore neighborhood $\mathcal{N} = \{z : \|z\|_\infty = 1\}$ in $d = 2$ the threshold voter model has the following behaviors

Figure 1.3. Threshold 2 voter model, Moore neighborhood

Figure 1.4. Threshold 3 voter model, Moore neighborhood

coexistence	$\theta = 1, 2$
clustering	$\theta = 3, 4$
fixation	$\theta \geq 5$

Here, *fixation* means that each sites flips only a finite number of times. To see that the last line is a reasonable guess note that an octagon of 1's cannot flip to 0 since each 1 has at most 4 neighbors that are 0

$$\begin{matrix} 0 & 0 & 0 & 0 & 0 & 0 \\ 0 & 0 & 1 & 1 & 0 & 0 \\ 0 & 1 & 1 & 1 & 1 & 0 \\ 0 & 1 & 1 & 1 & 1 & 0 \\ 0 & 0 & 1 & 1 & 0 & 0 \\ 0 & 0 & 0 & 0 & 0 & 0 \end{matrix}$$

We will prove the result about fixation for $\theta \geq 5$ and coexistence for $\theta = 1$ in Section 5. The other conclusions are open problems. In support of our conjectures we introduce Figures 1.3 and 1.4 which give simulations at time 50 of the case $\theta = 2$ on $\{0, 1, \ldots, 89\}^2$ and $\theta = 3$ on $\{0, 1, \ldots, 179\}^2$ starting from product measure with density 1/2.

Our next two systems model the competition of biological species. We begin with

Example 1.4. The multitype contact process. The set of states is $F = \{0, 1, \ldots, \kappa\}$, where 0 indicates a vacant site and $i > 0$ indicates a site occupied by one plant of type i. The flip rates are linear

$$c_0(x, \xi) = \delta_{\xi(x)}$$
$$c_i(x, \xi) = \lambda_i n_i(x, \xi) \quad \text{if } \xi(x) = 0$$

Here and in what follows the rates we do not mention are 0. Suppose for simplicity that $\kappa = 2$. Neuhauser (1992) has shown

Theorem 4A. Suppose $\delta_1 = \delta_2$ and $\lambda_1 > \lambda_2$. If ξ_0 is translation invariant and has a positive density of 1's then $P(\xi_t(x) = 2) \to 0$.

In words, the species with the higher birth rate wins out ("survival of the fittest"). The following stronger result should be true but Neuhauser's proof relies heavily on the assumption that $\delta_1 = \delta_2$.

Conjecture 4B. Suppose $\lambda_1/\delta_1 > \lambda_2/\delta_2$. If ξ_0 contains infinitely many 1's then $P(\xi_t(x) = 2) \to 0$.

When $\lambda_1 = \lambda_2$ and $\delta_1 = \delta_2$, Neuhauser showed that the multitype contact process behaves like the voter model.

Theorem 4C. *Clustering* occurs for translation invariant initial states in $d \leq 2$. That is, if ξ_0 is translation invariant, then for any $x, y \in \mathbf{Z}^d$, and $1 \leq i < j \leq \kappa$ we have

$$P(\xi_t(x) = i, \xi_t(y) = j) \to 0 \text{ as } t \to \infty$$

Theorem 4D. Let ξ_t^θ denote the process starting from an initial state in which the events $\{\xi_0^\theta(x) = i\}$ are independent and have probability θ_i with $\theta_2 = 1 - \theta_1$. In $d \geq 3$, as $t \to \infty$, $\xi_t^\theta \Rightarrow \xi_\infty^\theta$, a translation invariant stationary distribution in which

$$P(\xi_t(x) = i) = \begin{cases} 1-\rho & \text{when } i = 0 \\ \rho\theta_i & \text{when } i > 0 \end{cases}$$

where ρ is the equilibrium density of occupied sites in the one type contact process.

The last result is a little disturbing for biological applications. It says that if species compete on an equal footing then coexistence is not possible in $d = 2$ even if the birth and death rates are exactly the same. (This situation may sound unlikely to occur in nature but it occurs, for example, if we look at the competition of *genets* genetically identical individuals of the same species.) Somewhat surprisingly, if species 2 dominates species 1, we get coexistence for an open set of parameter values.

Example 1.5. Successional dynamics. We suppose that the set of states at each site are 0 = grass, 1 = a bush, 2 = a tree and we formulate the dynamics as

$$\begin{aligned} c_0(x,\xi) &= \delta_{\xi(x,\xi)} \\ c_1(x,\xi) &= \lambda_1 n_1(x,\xi) \quad \text{if } \xi(x) = 0 \\ c_2(x,\xi) &= \lambda_2 n_2(x,\xi) \quad \text{if } \xi(x) \leq 1 \end{aligned}$$

The title of this example and its formulation are based on the observation that if an area of land is cleared by a fire, then regowth will occur in three stages: first grass appears then small bushes and finally trees, with each species growing up through and replacing the previous one. With this in mind, we allow each type to give birth onto sites occupied by lower numbered types. As in the threshold voter model, the one dimensional nearest neighbor case is an exception.

Theorem 5A. Coexistence is not possible in the one dimensional nearest neighbor case, i.e., $d = 1$, $\mathcal{N} = \{-1, 1\}$.

Conjecture 5B. In all other cases (recall we supposed that $\mathcal{N} = \{z : \|z\|_p \leq r\}$ with $r \geq 1$) we have coexistence for an open set of values $(\delta_1, \lambda_1, \delta_2, \lambda_2)$.

Figure 1.5 shows a simulation of the nearest neighbor model on $\{0, 1, \ldots, 89\}^2$ with parameters $\lambda_1 = 5/4$, $\delta_1 = 1$, $\lambda_2 = 1.9/4$, and $\delta_2 = 1$ run until time 100, which presumably represents the equilibrium state. Sites in state 1 are gray; those in state 2 are black.

Proving that coexistence occurs in the two dimensional nearest neighbor case of this model seems to be a difficult problem, since computer simulations indicate that the open set referred to in Conjecture 5B is rather small. However, if we assume that the range of interaction is large, we can get very accurate results about the coexistence region. Let $\beta_i = \lambda_i |\mathcal{N}|$.

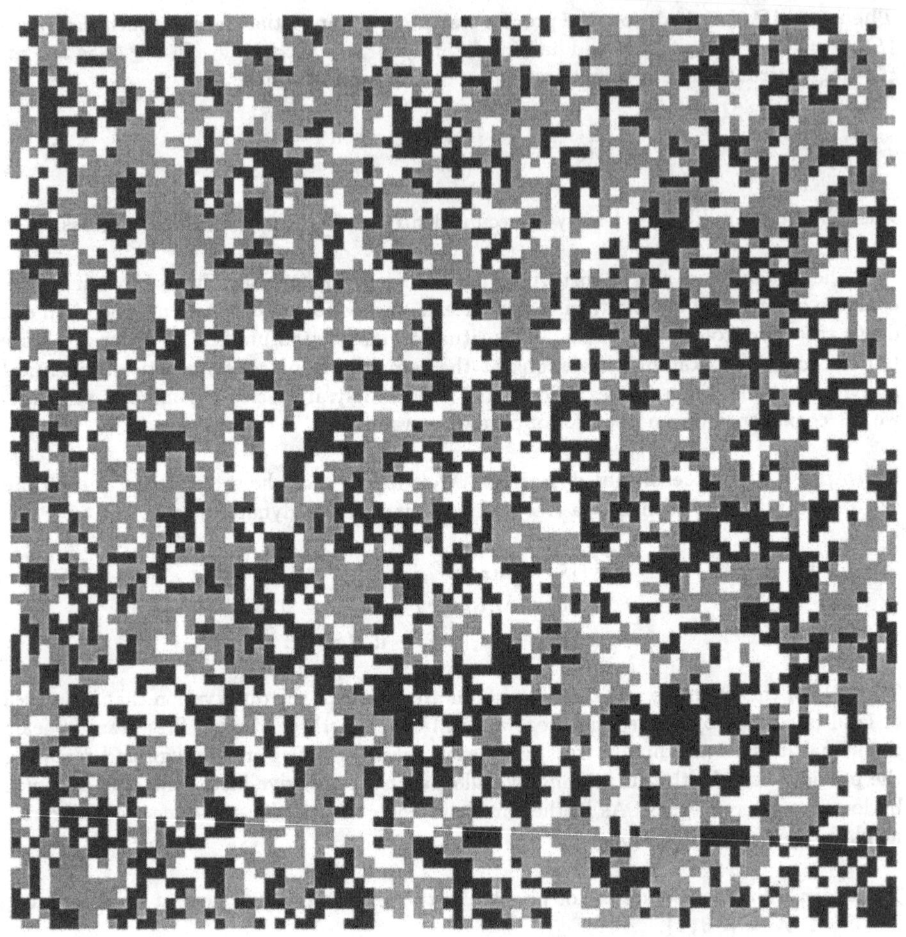

Figure 1.5. Two dimensional nearest neighbor succesional dyanmics, $\beta_1 = 5$, $\beta_2 = 1.9$

Theorem 5C. Suppose that

$$(\star) \qquad \beta_1 \cdot \frac{\delta_2}{\beta_2} > \delta_1 + \beta_2 \cdot \frac{\beta_2 - \delta_2}{\beta_2}$$

If r is large then coexistence occurs.

Theorem 5D. Suppose that

$$\beta_1 \cdot \frac{\delta_2}{\beta_2} < \delta_1 + \beta_2 \cdot \frac{\beta_2 - \delta_2}{\beta_2}$$

If r is large then coexistence is impossible.

Theorems 5A and 5C are due to Durrett and Swindle (1991), while the converse in 5D is due to Durrett and Schinazi (1993). To explain the condition in Theorems 5C and 5D, we begin by observing that if we assumed that $u(t) = P(\xi_t(x) = 2)$ does not depend on x and the states of neighboring sites were independent, then writing $y \sim x$ to denote "y is a neighbor of x"

$$(1.1) \qquad \begin{aligned} \frac{du}{dt} &= -\delta_2 P(\xi_t(x) = 2) + \sum_{y\sim x} \lambda_2 P(\xi_t(x) < 2, \xi_t(x) = 2) \\ &= -\delta_2 u + \beta_2 u(1-u) \end{aligned}$$

where the first equality is true in general and the second follows from our assumptions and the fact that $\beta_2 = |\mathcal{N}|\lambda_2$. Dropping the $-\beta_2 u^2$ term

$$\frac{du}{dt} \leq (\beta_2 - \delta_2)u$$

so if $\delta_2 > \beta_2$ all solutions tend to 0 exponentially fast. If $\delta_2 < \beta_2$ and we let $u^* = (\beta_2 - \delta_2)/\beta_2$ then

$$-\delta_2 u + \beta_2 u(1-u) \begin{cases} > 0 & \text{for } 0 < u < u^* \\ < 0 & \text{for } u > u^* \end{cases}$$

so if $u(0) > 0$, $u(t) \to u^*$ as $t \to \infty$.

Applying the reasoning that led to (1.1) to $v(t) = P(\xi_t(x) = 1)$ we see that

$$(1.2) \qquad \begin{aligned} \frac{dv}{dt} &= -\delta_1 P(\xi_t(x) = 1) - \sum_{y\sim x} \lambda_2 P(\xi_t(x) = 1, \xi_t(y) = 2) \\ &\quad + \sum_{y\sim x} \lambda_1 P(\xi_t(x) = 0, \xi_t(y) = 1) \\ &= -\delta_1 v - \beta_2 vu + \beta_1(1-u-v)v \end{aligned}$$

where again the first equality is true in general and the second follows from our assumptions and the fact that $\beta_i = |\mathcal{N}|\lambda_i$. To analyze (1.2), we note that if the 2's are in equilibrium and the density of 1's is very small, then

$$u = (\beta_2 - \delta_2)/\beta_2 \qquad (1-u-v) \approx \delta_2/\beta_2$$

$$\text{1's are born at rate } \approx \beta_1 \cdot \frac{\delta_2}{\beta_2} \cdot v$$

$$\text{1's die at rate } \approx \left(\delta_1 + \beta_2 \cdot \frac{\beta_2 - \delta_2}{\beta_2}\right) v$$

So if ($\star$) holds a small density of 1's will grow in time, while if we reverse the inequality in ($\star$) and use $(1-u-v) \le (\beta_2 - \delta_2)/\beta_2$ then the birth rate always exceeds the death rate and $v(t) \to 0$.

The practice of calculating how densities evolve when we suppose that adjacent sites are independent is called *mean field theory*. Theorems 5C and 5D are one instance of the general principle that when the range of interaction is large mean field calculations are almost correct. A second method of making mean field calculations correct, which leads to connections with nonlinear partial differential equations, is to introduce particle motion at a fast rate.

Example 1.6. Predator prey systems. In this model we think of 0 = vacant, 1 = occupied by a fish, and 2 = occupied by a shark and we have the following flip rates

$$c_1(x,\xi) = \beta_1 n_1(x,\xi)/2d \quad \text{if } \xi(x) = 0$$
$$c_2(x,\xi) = \beta_2 n_2(x,\xi)/2d \quad \text{if } \xi(x) = 1$$
$$c_0(x,\xi) = \begin{cases} \delta_1 & \text{if } \xi(x) = 1 \\ \delta_2 + (\gamma n_2(x,\xi)/2d) & \text{if } \xi(x) = 2 \end{cases}$$

In words, fish die at rate δ_1 and are born at vacant sites at a rate proportional to the number of fish at neighboring sites. So in the absence of sharks, the fish are a contact process.

Sharks die of natural causes at rate δ_2 and kill a neighboring shark at rate $\gamma/2d$. The birth rate for sharks may look a little strange at first: fish turn into sharks at rate proportional to the number of shark neighbors. This is not what happens in the ocean but it does capture an essential feature of the interaction: when the density of fish is too low then the sharks die faster than they give birth. A second justification of this mechanism is that, as we will see in Section 9, in a suitable limit we get standard predator-prey equations.

Here $n_i(x,\xi) = |\{z \in \mathcal{N} : \xi(x+z) = i\}|$ as usual, but for reasons that will become clear in a moment we take $S = \epsilon\mathbf{Z}^d$ and $\mathcal{N} = \{z : |z| = \epsilon\}$ the nearest neighbors. We use a small lattice so that we can introduce *stirring* at a fast rate, i.e., for each $x, y \in \epsilon\mathbf{Z}^d$ with $|x-y| = \epsilon$ we exchange the values at x and y at rate ϵ^{-2}. That is, we change the configuration from ξ to $\xi^{x,y}$ defined by

$$\xi^{x,y}(x) = \xi(y), \quad \xi^{x,y}(y) = \xi(x), \quad \xi^{x,y}(z) = \xi(z) \text{ if } z \ne x, y$$

The combination of the space scale of ϵ and the time scale of ϵ^{-2} means that the individual values will perform Brownian motions in the limit $\epsilon \to 0$. The fast stirring keeps the states of neighboring sites independent, so using mean field reasoning leads to the following result due to DeMasi, Ferrari and Lebowitz (1986).

Theorem 6A. Suppose $\xi_0^\epsilon(x)$, $x \in \epsilon\mathbf{Z}^d$, are independent and let $u_i^\epsilon(t,x) = P(\xi_t(x) = i)$. If $u_i^\epsilon(0,x) = g_i(x)$ is continuous then as $\epsilon \to 0$, $u_i^\epsilon(t,x)$ converges to $u_i(t,x)$ the bounded solution of

$$\begin{aligned}\frac{\partial u_1}{\partial t} &= \Delta u_1 + \beta_1 u_1(1-u_1-u_2) - \beta_2 u_1 u_2 - \delta_1 u_1 \\ \frac{\partial u_2}{\partial t} &= \Delta u_2 + \beta_2 u_1 u_2 - \delta_2 u_2 - \gamma u_2^2\end{aligned} \tag{1.3}$$

with $u_i(0,x) = g_i(x)$.

Here the Δu_i terms reflect the fact that in the limit the individual values are performing Brownian motions run at rate 2. The other terms can be seen by using the reasoning that led to (1.1) and (1.2).

If we suppose that the initial functions $g_i(x)$ are constant then this is true at later times $u_i(t,x) = v_i(t)$ and the v_i satisfy

$$\begin{aligned}\frac{\partial v_1}{\partial t} &= v_1((\beta_1 - \delta_1) - \beta_1 v_1 - (\beta_1 + \beta_2)v_2) \\ \frac{\partial v_2}{\partial t} &= v_2(-\delta_2 + \beta_2 v_1 - \gamma v_2)\end{aligned} \tag{1.4}$$

Here we have rearranged the right hand side to show that it is the standard predator-prey equations with limited growth. (See for example Hirsch and Smale (1974) p. 263.) To determine the conditions for coexistence, we start by finding the fixed points of the dynamical systems, i.e., points (ρ_1, ρ_2) so that $v_i(t) \equiv \rho_i$ is a solution of (1.4). There are three

(i) $\rho_1 = \rho_2 = 0$. No sharks or fish, the trivial equilibrium.

(ii) We have a solution with $\rho_2 = 0$ and $\rho_1 = (\beta_1 - \delta_1)/\beta_1$ if $\beta_1 > \beta_2$. This forumla is the same as the one in the last example because in the absence of sharks, fish are a contact process.

(iii) There is a fixed point with $\rho_i = \sigma_i > 0$ if and only if

$$\frac{\beta_1 - \delta_1}{\beta_1} > \frac{\delta_2}{\beta_2} \tag{1.5}$$

(which implies $\beta_1 > \delta_1$). We do not have an intuitive explanation for the last condition. It is simply what results when we solve the two equations in two unknowns.

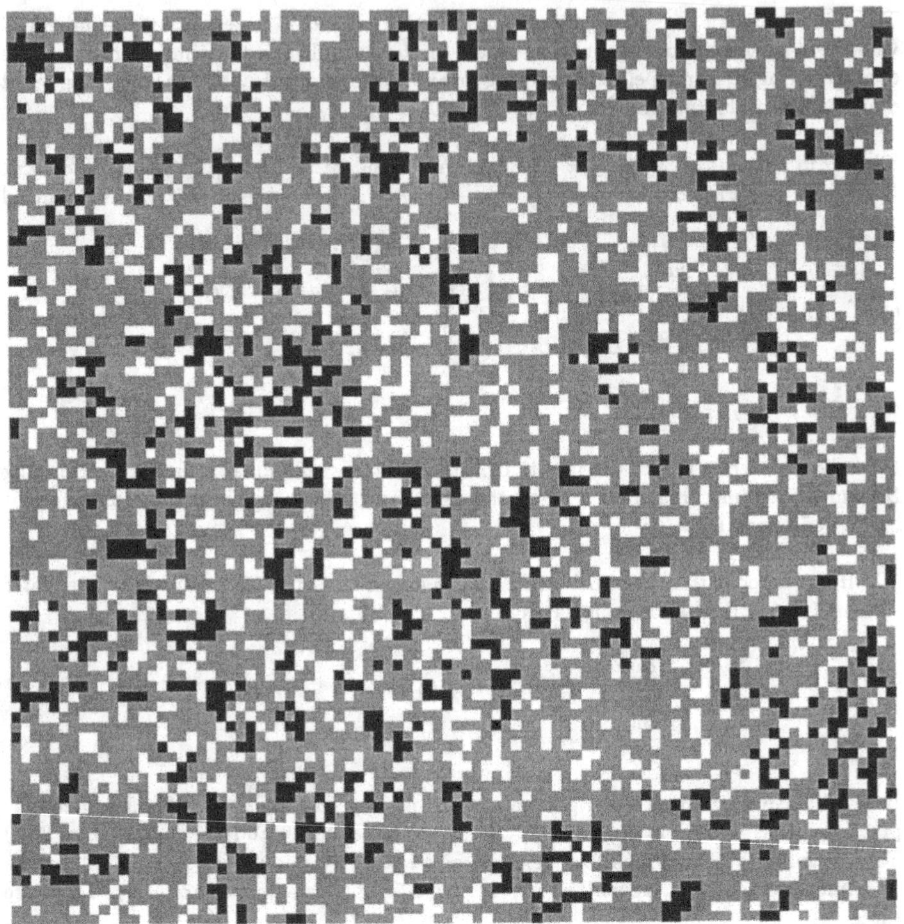

Figure 1.6. Predator prey model $\beta_1 = \beta_2 = 3$, $\delta_1 = \delta_2 = 1$, $\gamma = 1$

By exploiting the connection between the particle system and the partial differential equation given in Theorem 6A, we can prove

Theorem 6B. If (1.5) holds then for small ϵ coexistence occurs.

It would be nice to prove coexistence results without fast stirring. Figure 1.6 shows a simulation of the system on $\{0, 1, \ldots, 79\}^2$ at time 50 with $\beta_1 = \beta_2 = 3$, $\delta_1 = \delta_2 = 1$, $\gamma = 1$ and no stirring. Again sites in state 1 are gray; those in state 2 are black.

Example 1.7. Epidemic model. In this example, we think of $\mathbf{Z}^2$ as representing an array of houses each of which is occupied by one individual who can be (0) susceptible = healthy but capable of getting the disease, (1) infected with the disease, or (2) immune to further infection. The flip rates are

$$\begin{aligned} c_1(x, \xi) &= \lambda n_1(x, \xi) \quad \text{if } \xi(x) = 0 \\ c_2(x, \xi) &= \delta \quad \text{if } \xi(x) = 1 \\ c_0(x, \xi) &= \alpha \quad \text{if } \xi(x) = 2 \end{aligned}$$

As usual, the rates we did not mention are 0. In words, a susceptible individual gets infected at a rate proportional to the number of infected neighbors. Infected individuals become removed at rate δ. Here $1/\delta$ is the mean duration of the disease and to obtain the Markov property we have assumed that the duration of the disease has an exponential distribution. If we want to model the short term behavior of a measles or flu epiemic then we set $\alpha = 0$ since recovered individuals are immune to the disease. If we want to examine longer time properties then immune individuals will die (or move out of town) and new susceptibles will be born (or move into town) so to keep a fixed population size of one individual per site, we combine the two transitions into one.

To describe the conditions for coexistence we begin with case $\alpha = 0$ and consider the behavior of the model starting from one infected individual at 0 in the midst of an otherwise susceptible population. Let $\eta_t = \{x : \xi_t(x) = 1\}$ be the set of the infected individuals at time t and let $\tau = \inf\{t : \eta_t = \emptyset\}$. We will have $\eta_t = \emptyset$ for all $t > \tau$ so we say the infection *dies out* at time τ. Let $\delta_c = \inf\{\delta : P(\tau = \infty) = 0\}$. The faster people recover the harder it is for the epidemic to propagate so we have $P(\tau = \infty) = 0$ for all $\delta > \delta_c$.

If we restrict our attention to the nearest neighbor case, then results of Cox and Durrett (1988) describe the asymptotic behavior of the epidemic when $\delta < \delta_c$ and $\tau = \infty$. Building on those results Durrett and Neuhauser (1991) have shown

Theorem 7. Suppose $d = 2$ and $\mathcal{N} = \{x : |x| = 1\}$. If $\delta < \delta_c$ and $\alpha > 0$ then coexistence occurs.

Zhang has generalized the results of Cox and Durrett (1988) to finite range interactions. Presumably one can also prove the result of Durrett and Neuhauser (1991) in that level of generality but no one has had the courage to try to write out all the details.

Closely related to the epidemic model is

Example 1.8. Greenberg Hastings Model. In this model, we think of having a neuron at each $x \in \mathbf{Z}^d$ that is connected to each of its neighbors. The states of each neuron are $F = \{0, 1, \ldots, \kappa - 1\}$ where 1 is excited, $2, \ldots, \kappa - 1$ are a sequence of recovery states, and 0 indicates a fully rested neuron that is capable of being excited. These interpretations motivate the following flip rates

$$\begin{aligned} c_1(x,\xi) &= 1 \quad \text{if } \xi(x) = 0 \text{ and } n_i(x,\xi) \geq \theta \\ c_i(x,\xi) &= 1 \quad \text{if } i \neq 1 \text{ and } \xi(x) = i - 1 \end{aligned}$$

Here arithmetic is done modulo κ so $0 - 1 = \kappa - 1$. The second rule says that once excited, the neuron progresses through the recovery states at rate 1 until it is fully rested; the first that a rested neuron becomes excited at rate 1 if the number of its neighbors that are excited is at least the threshold θ. The next result, due to Durrett (1992), gives a regime in which this model has (somewhat boring) stationary distributions.

Theorem 8A. Let $\epsilon > 0$ and suppose $\theta \leq (1 - \epsilon)|\mathcal{N}|/2\kappa$. If r is large then there is a stationary measure close to the uniform product measure.

Here the *uniform product measure* is the one in which the coordinates $\xi(x)$ are independent and $P(\xi(x) = i) = 1/\kappa$. Based on the analogy with the epidemic model where if $\delta < \delta_c$ there is a coexistence for any $\alpha > 0$, we expect that

Conjecture 8B. There is a constant $a > 0$ so that if $\theta \leq a|\mathcal{N}|$ then coexistence occurs for any κ.

Computer simulations indicate that in this regime the excitation sustains itself by producing moving fronts. See Figure 1.7 for a simulation of the system with $\mathcal{N} = \{x : \|x\|_\infty \leq 2\}$, threshold $\theta = 3$, and $\kappa = 8$. Excited states are black, rested sites are white, recovering sites are appropriate shades of gray.

The analogue of Conjecture 8B has been proved by Durrett and Griffeath (1993) for the Greenberg Hastings cellular automaton in which $\xi_{n+1}(x) = \xi_n(x) + 1$ if $\xi_n(x) > 0$ or $\xi_n(x) = 0$ and $n_i(x, \xi_n) \geq \theta$; $\xi_{n+1}(x) = \xi_n(x)$ otherwise. See Figure 1.8 for a simulation of the cellular automaton with the same color scheme and parameters: $\mathcal{N} = \{x : \|x\|_\infty \leq 2\}$, threshold $\theta = 3$, and $\kappa = 8$ run until it has become periodic with period 8. For more on the cellular automaton consult Fisch, Gravner and Griffeath (1991), (1992), (1993), and Gravner and Griffeath (1993).

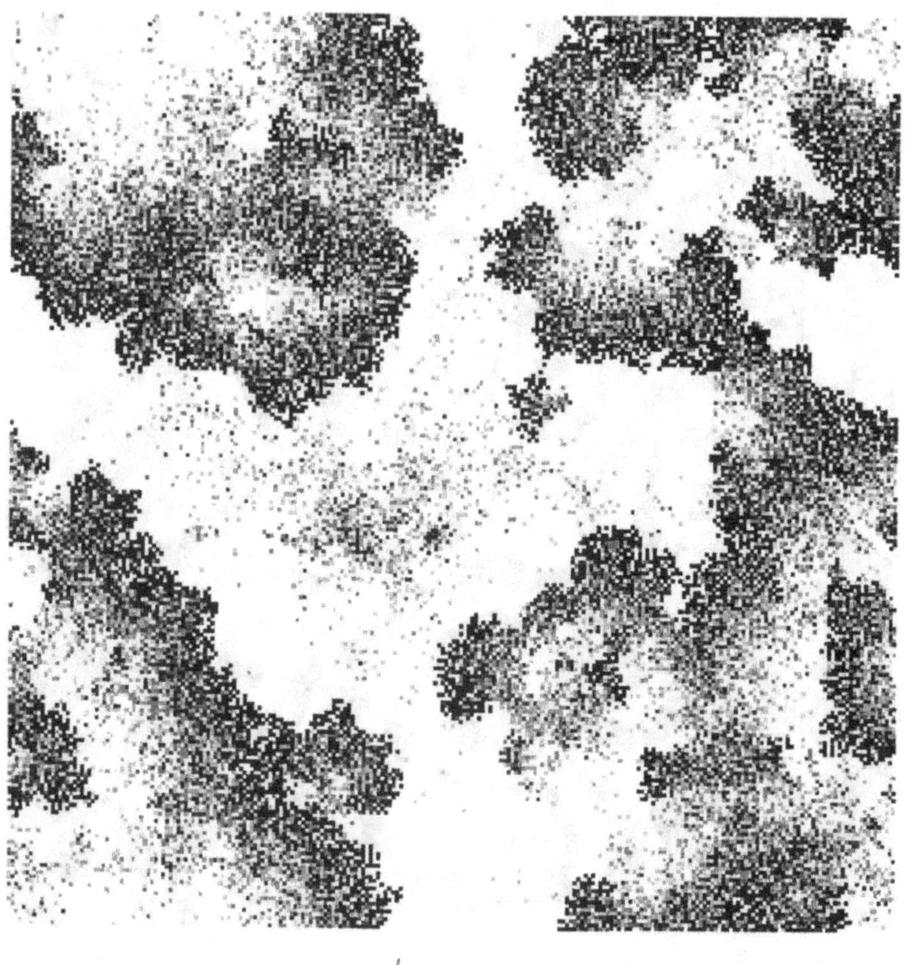

Figure 1.7. Greenberg Hastings model. $\mathcal{N} = \{x : \|x\|_\infty \leq 2\}$, $\theta = 3$, $\kappa = 8$

Figure 1.8. Greenberg Hastings cellular automaton. $\mathcal{N} = \{x : \|x\|_\infty \leq 2\}$, $\theta = 3$, $\kappa = 8$

2. Construction, Basic Properties

To construct an interacting particle system from given translation invariant finite flip rates

$$c_i(x,\xi) = g_i(\xi(x+z_0), \xi(x+z_1), \dots, \xi(x+z_k))$$

based on a neighborhood set $\mathcal{N} = \{z_0, z_1, \dots, z_n\}$ we can, by changing the time scale, assume that $c_i(x,\xi) \le 1$. For each $x \in \mathbf{Z}^d$ and $i \in F$, let $\{T_n^{x,i} : n \ge 1\}$ be the arrival times of independent rate 1 Poisson processes (i.e., if we set $T_0^{x,i} = 0$ then the increments $T_n^{x,i} - T_{n-1}^{x,i}$ are independent and have an exponential distribution with mean 1) and let $U_n^{x,i}$ be independent and uniform on (0,1). At time $t = T_n^{x,i}$ site x will flip to state i if $U_n^{x,i} < c_i(x, \xi_{t-})$ and stay unchanged otherwise. To see that this recipe produces the desired flip rates recall the *thinnning property* of the Poisson process: if we keep the points from $\{T_n^{x,i} : n \ge 1\}$ that have $U_n^{x,i} < p$ then the result is a Poisson process with rate p.

Since there are infinitely many Poisson processes, and hence no first arrival, we have to show that we can use our recipe to compute the time evolution. To do this, we use an argument of Harris (1972). Let t_0 be a small positive number to be chosen later. We draw an unoriented arc between x and y if $y - x \in \mathcal{N}$ and for some i, $T_1^{x,i} < t_0$. The presence of an arc between x and y indicates that a Poisson arrival has caused x to look at y to see if it should flip or caused y to look at x. Conversely, if there is no arc between x and y then neither site has looked at the other. The last observation implies that the sites in two different components of the resulting random graph have not influenced each other by time t_0 and hence their evolutions can be computed separately. To finish the construction then it suffices to show

(2.1) Theorem. If t_0 is small enough then with probability one, all the connected components of our random graph are finite.

For then in each component there is a first flip and we can compute the effects of the changes sequentially. This allows us to construct the process up to time t_0 but t_0 is independent of the initial configuration, so iterating we can construct the process for all time.

PROOF OF (2.1): Let $\mathcal{N}^* = \{z_1, \dots, z_k, -z_1, \dots, -z_k\}$ be the set of possible displacements along edges of the graph. (In this section alone, we will allow $\mathcal{N}$ to be a general finite set not just $\{x : \|x\|_p \le r\}$.) We say that $y_0, y_1, \dots y_n$ is a *path* of length n if $y_m - y_{m-1} \in \mathcal{N}^*$ when $0 < m \le n$. We call a path *self-avoiding* if $y_i \ne y_j$ when $0 \le i < j \le n$. Let $R = \max\{|z_i| : z_i \in \mathcal{N}\}$. (Here $|z| = \|z\|_2$.) We claim that

(a) If 0 is connected to some point with $|z| > M$ then there is a self-avoiding path of length $\ge M/R$ starting at 0.

To see this note that if there is a path from 0 to z, then by removing loops we can make it self-avoiding. Since each step along the path moves us a distance $\le R$, there must be at least $|z|/R$ such steps. The next ingredient in the proof is

(b) If x, y, z, w are distinct, the presence of edges from x to y and from z to w are independent events.

To see this note that the presence of an edge from x to y is determined by the Poisson processes $T_n^{x,i}$ and $T_n^{y,i}$ with $i \in F$. From (b) it is easy to see

(c) Let $N = |\mathcal{N}^*|$ and $\kappa = |F|$. The probability of a self avoiding path of length $2n-1$ starting at a given point x is at most

$$N^{2n-1}(1-e^{-2\kappa t_0})^n$$

The first factor is the number of paths of length $2n-1$ and hence an upper bound on the number of self-avoiding paths. To see the second factor note that the presence of the edges $(z_0, z_1), (z_2, z_3), \ldots (z_{2n-2}, z_{2n-1})$ are independent events that have probability $1-e^{-2\kappa t_0}$ since the probability of no arrival by time t_0 in one of the 2κ Poisson processes $T_n^{x,i}$ and $T_n^{y,i}$ is $e^{-2\kappa t_0}$.

If we pick t_0 small enough then $N^2(1-e^{-2\kappa t_0}) \leq 1/2$, so the probability of a self-avoiding path of length $2n-1$ decreases to 0 exponentially fast, and it follows from (a) that with probability 1 the cluster containing any given point x is finite. □

An immediate consequence of the construction is

(2.2) **Corollary.** If ξ_0 is translation invariant then ξ_t is.

PROOF: The family of Poisson processes is translation invariant, so if the initial state is, then so is the result of our computation. □

It should also be clear from the construction that ξ_t is a Markov process, i.e., if we know the state at time s, information about ξ_r for $r < s$ is irrelevant for computing the evolution for $t > s$. Being a Markov process there is an associated family of operators defined by

$$T_t f(\xi) = E_\xi f(\xi_t)$$

where E_ξ denotes the expected value starting from $\xi_0 = \xi$. The Markov property of ξ_t implies that the T_t form a *semigroup*. That is, $T_s T_t = T_{s+t}$. If you are not familiar with semi-groups don't worry. We will only use the most basic results that can be found in Chapter 1 of Dynkin (1965) or in Chapter ? of Revuz and Yor (1991), and we will only use those facts in this section. The first thing we want to prove is

(2.3) **Corollary.** T_t is a *Feller semigroup*, i.e., if f is continuous with respect to the product topology on F^S then $T_t f$ is continuous.

PROOF: Note that our construction defines on the same probability space the process starting from any initial configuration. If $t \leq t_0$ then proof of (2.1) shows that up to time t_0, $\mathbf{Z}^d$ breaks up into a collection of finite non-interacting islands. From the last fact it follows easily that if $\xi_0^n \to \xi_0$, (which means that for each fixed x, $\xi_0^n(x) \to \xi_0(x)$) then $\xi_t^n \to \xi_t$ almost surely. If f is continuous it follows that $f(\xi_t^n) \to f(\xi_t)$ almost surely. Since F^S is compact in the product topology, any continuous function is necessarily bounded, and it follows from the bounded convergence theorem that $Ef(\xi_t^n) \to Ef(\xi_t)$. This proves the result for $t \le t_0$. Using the semigroup property $T_{t+s} = T_s T_t$, it follows that the result holds for $t \le 2t_0$, $t \le 3t_0$, and hence for all t. □

Our next step is to compute the generator of the semigroup. Let $\xi^{x,i}$ denote the configuration ξ flipped to i at x. That is,

$$\xi^{x,i}(x) = i \qquad \xi^{x,i}(y) = \xi(y) \quad \text{otherwise}$$

Suppose $f(\xi)$ only depends on the values of finitely many coordinates and let

$$Lf = \sum_{x \in \mathbf{Z}^d, i \in F} c_i(x,\xi)\left(f(\xi^{x,i}) - f(\xi)\right)$$

The sum converges since only finitely many terms are nonzero. Our next result says that L is the generator of T_t.

$$\text{(2.4)} \qquad \frac{d}{dt} T_t f(\xi)\bigg|_{t=0} = Lf(\xi)$$

If you have seen the generator of a Markov process with a discrete state space the formula should not be surprising. The proof of (2.4) is much like the proof for that case so we will only give a quick sketch.

PROOF: Suppose f only depends on the values of ξ in $[-L, L]^d$ and recall we have defined $R = \max\{|z_i| : z_i \in \mathcal{N}\}$. If t is small then with high probability there is at most one site $x \in [-L - 2R, L + 2R]^d$ and one value of $i \in F$ with $T^{x,i} < t$. By considering the various possible values of x and i and noting that the probability that $\xi_0 = \xi$ changes to $\xi^{x,i}$ is $\sim tc_i(x,\xi)$, the result follows easily. □

For the rest of this section, we will restrict our attention to the case $F = \{0,1\}$, in which case we think of $1 =$ occupied by a particle and $0 =$ vacant. Since we think of 1's are particles we call $c_1(x,\xi)$ the *birth rates* and call $c_0(x,\xi)$ the *death rates* . We say that the birth rates $c_1(x,\xi)$ are *increasing* if

$$\xi(y) \le \zeta(y) \text{ for all } y \ne x \text{ and } \xi(x) = \zeta(x) = 0 \text{ implies } c_1(x,\xi) \le c_1(x,\zeta)$$

We say that *death rates* $c_0(x,\xi)$ are *decreasing* if

$$\xi(y) \le \zeta(y) \text{ for all } y \ne x \text{ and } \xi(x) = \zeta(x) = 1 \text{ implies } c_1(x,\xi) \ge c_1(x,\zeta)$$

A process with increasing birth rates and decreasing death rates is said to be *attractive.* The last term comes from analogies with the Ising model in statistical mechanics. This assumption is not very attractive for biological systems since there the death rate usually increases due to crowding, but the attractive property is what we need to prove the following useful result.

(2.5) **Theorem.** For an attractive process, if we are given initial configurations with $\xi_0(x) \le \zeta_0(x)$ for all x then the processes defined by our construction have $\xi_t(x) \le \zeta_t(x)$ for all x and t.

PROOF: Intuitively this is true since each flip preserves the inequality. To check this suppose that $\xi_{s-}(y) = 0$ and a birth event $T_n^{y,1}$ occurs at time s. If $\zeta_{s-}(y) = 1$ then $\zeta_s(y) = 1$ and the inequality will certainly hold after the flip. If $\zeta_{s-}(y) = 0$ and the inequality holds before the flip, then since our birth rates are increasing $c(y, \xi_{s-}) \le c(y, \zeta_{s-})$. By considering the possible values of $U_n^{y,i}$ se see that in all cases the inequality holds after the flip.

value of $U_n^{y,i}$	change in ξ	change in ζ
$[0, c(y, \xi_{s-}))$	flips to 1	flips to 1
$[c(y, \xi_{s-}), c(y, \zeta_{s-}))$	stays 0	flips to 1
$[c(y, \zeta_{s-}), 1)$	stays 0	stays 0

A similar argument applies if $\xi_{s-}(y) = 1$ and a death event $T_n^{y,0}$ occurs at time s.

To turn the intuitive argument in the last paragraph into a proof, suppose that the inequality fails at some point x at some time $t \le t_0$. Let C_x be the connected component containing x for the random graph defined in the proof of (3.1), and let $s > 0$ be the first time the property fails at some point $y \in C_x$. By the definition of s the inequality holds on C_x before time s. Since C_x contains all the neighbors of any site in C_x that flips by time t_0 it follow from the argument in the last paragraph that the inequality will hold until the next flip after time s. Since C_x is a finite set, the next flip will occur at a time $> s$, contradicting the defintion of s and showing that the inequality must hold up to time t_0. Iterating the last conclusion we see that the result holds for all time. □

To explain our interest in (2.2), (2.4), and (2.5) we will now prove that

(2.6) **Theorem.** If $\lambda|\mathcal{N}| < \delta$ then the contact process has no nontrivial stationary distribution.

PROOF: Consider the contact process starting from all sites occupied, i.e., suppose $\xi_0^1(x) = 1$ for all x. It follows from (2.2) that $P(\xi_t^1(x) = 1)$ is independent of x, so writing $y \sim x$ for "y is a neighbor of x" and using $\frac{d}{dt} T_t f = T_t L f$ we have

$$\begin{aligned}\frac{d}{dt} P(\xi_t^1(x) = 1) &= -\delta P(\xi_t^1(x) = 1) + \sum_{y \sim x} \lambda P(\xi_t^1(x) = 0, \xi_t^1(y) = 1) \\ &\le -\delta P(\xi_t^1(x) = 1) + \lambda |\mathcal{N}| P(\xi_t^1(y) = 1)\end{aligned}$$

If $\lambda|\mathcal{N}| < \delta$ then the last inequality implies that $P(\xi_t^1(x) = 1) \to 0$ as $t \to \infty$. Now any initial configuration has $\xi_0(x) \le 1 = \xi_0^1(x)$ for all x, so by (2.5), we have $\xi_t(x) \le \xi_t^1(x)$ for all t and x and it follows that $P(\xi_t(x) = 1) \to 0$ for any initial configuration. If we pick ξ_0 to have a stationary distribution then $P(\xi_t(x) = 1)$ is independent of t, so the last conclusion implies this probability is 0 and the result follows. □

The last argument shows that if we start an attractive process with all sites occupied and find $P(\xi_t^1(x) = 1) \to 0$ then there is no nontrivial stationary distribution. Our next result proves the converse. Recall that $\Rightarrow$ denotes weak convergence, which in this setting is just convergence of finite dimensional distribution.

(2.7) **Theorem.** As $t \to \infty$, $\xi_t^1 \Rightarrow \xi_\infty^1$. The limit is a stationary distribution which is stochastically larger than any other stationary distribution and called the *upper invariant measure.*

PROOF: The key to the proof is the following observation:

(2.8) **Lemma.** For any set $A \subset \mathbf{Z}^d$, $t \to P(\xi_t^1(x) = 0 \text{ for all } x \in A)$ is increasing.

PROOF: Let $\zeta_0 = \xi_s^1$. Clearly, $\xi_0^1(x) \geq \zeta_0(x)$ for all x so (2.5) implies that for all t and x, $\xi_t^1(x) \geq \zeta_t(x)$. Since ζ_t has the same distribution as ξ_{s+t}^1 it follows that

$$P(\xi_t^1(x) = 0 \text{ for all } x \in A) \leq P(\xi_{s+t}^1(x) = 0 \text{ for all } x \in A) \qquad \square$$

Let $\phi(A) = P(\xi(x) = 0 \text{ for all } x \in A)$ and $B = \{x_1, \ldots, x_m\}$ Using the inclusion exclusion formula on the events $E_i = \{\xi_t(x) = 0\}$ on $A \cup \{x_i\}$, we can express any finite dimensional distribution in terms of the $\phi(C)$.

$$\begin{aligned} 1 - P(\xi(x) = 0 \text{ for all } x \in A, \xi(x) = 1 \text{ for all } x \in B) &= P(\cup_{i=1}^m E_i) \\ = \sum_{i=1}^m \phi(A \cup \{x_i\}) - \sum_{i<j} \phi(A \cup \{x_i, x_j\}) + \ldots &+ (-1)^{m+1}\phi(A \cup B) \end{aligned}$$

So (2.8) implies convergence of all finite dimensional distributions. $\square$

The fact that ξ_∞^1 is a stationary distribution follows from a general result.

(2.9) **Lemma.** Suppose the Markov process X has a Feller semigroup and $X_t \Rightarrow X_\infty$ then (the distribution of) X_∞ is a stationary distribution.

PROOF: Recall that if X_0 has distribution μ then the probability measure μT_t defined by

$$\int (\mu T_t)(dx) f(x) = \int \mu(dx) T_t f(x) = \int \mu(dx) E_x f(X_t)$$

for all bounded continuous functions f gives the distribution of X_t when X_0 has distribution μ. The key to the proof of (2.9) is the following general fact:

(2.10) If T_t is a Feller semigroup and $\mu_s \Rightarrow \mu$ then $\mu_s T_t \Rightarrow \mu T_t$.

To prove (2.10) we note that $T_t f$ is bounded and continuous

$$\begin{aligned} \lim_{s \to \infty} \int (\mu_s T_t)(dx) f(x) &= \lim_{s \to \infty} \int \mu_s(dx) T_t f(x) \\ &= \int \mu(dx) T_t f(x) = \int (\mu T_t)(dx) f(x) \end{aligned}$$

where the second inequality follows from the fact that $T_t f$ is continuous and $\mu_s \Rightarrow \mu$. To prove (2.9) now, let μ_s be the distribution of X_s and note that the Markov property implies $\mu_s T_t = \mu_{s+t}$. The right hand side converges to μ, and by (2.10) the left hand side converges to μT_t, so $\mu T_t = \mu$, i.e., μ is a stationary distribution. □

Finally we have to explain and show the claim "ξ^1_∞ is stochastically larger than any other stationary distribution π." By *stochastically larger* we mean that if f is any increasing function which depends on only finitely many cooordinates then

$$Ef(\xi^1_\infty) \geq \int f(\xi)d\pi(\xi) \tag{2.11}$$

Here f is *increasing* means that if $\xi(x) \leq \zeta(x)$ for all x then $f(\xi(x)) \leq f(\zeta(x))$. To prove the claim let ζ_0 have distribution π. Clearly, $\xi^1_0(x) \geq \zeta_0(x)$ for all x so (3.5) implies that $\xi_t(x) \geq \zeta_t(x)$ for all t and x. Now if f is increasing

$$Ef(\xi_t) \geq Ef(\zeta_t) = \int f(\xi)d\pi(\xi)$$

since π is a stationary distribution. If f depends on only finitely many coordinates then it is continuous and

$$Ef(\xi^1_t) \to Ef(\xi^1_\infty)$$

Combining the last two conclusions, proves our claim and completes the proof of (3.7). □

(2.12) **Remark.** A result of Holley implies that since ξ^1_∞ is stochastically larger than π, we can define random variables ξ and ζ with these distributions on the same probability space so that $\xi(x) \geq \zeta(x)$.

Later we will need a variation of (2.9). The next result and (3.15) are not needed until Section 5, so I suggest that you wait until later to read the rest of this section.

(2.13) **Theorem.** Suppose the Markov process X has a compact state space Λ and a Feller semigroup T_t. Let μ_t be the distribution of X_t and ν_t the Cesaro average defined by

$$\nu_t(A) = \frac{1}{t}\int_0^t \mu_s(A)$$

If $t_k \to \infty$ and $\nu_{t_k} \Rightarrow \nu$ then ν is a stationary distribution.

(2.14) **Corollary.** Since the set of probability measures on Λ is compact in the weak topology, this implies in particular that stationary distributions exist.

PROOF: Since $\mu_s T_r = \mu_{s+r}$ we have

$$\begin{aligned}\nu_{t_k} T_r &= \frac{1}{t_k}\int_0^{t_k} \mu_s T_r ds = \frac{1}{t_k}\int_r^{r+t_k} \mu_s ds \\ &= \nu_{t_k} + \frac{1}{t_k}\int_{t_k}^{r+t_k} \mu_s ds - \frac{1}{t_k}\int_0^r \mu_s ds\end{aligned}$$

The two error terms on the right hand side have each total mass r/t_k and hence converge weakly to 0. Since $\nu_{t_k} \Rightarrow \nu$ it follows that $\nu_{t_k}T_r \Rightarrow \nu$. On the other hand it follows from (3.10) that $\nu_{t_k}T_r \Rightarrow \nu T_r$ so we have $\nu T_r = \nu$ as desired. □

In Section 5, we will also need the following result:

(2.15) **Theorem.** The upper invariant measure ξ^1_∞ is spatially ergodic.

PROOF: We begin with the observation that

(2.16) for each t, ξ^1_t is spatially ergodic.

To prove (2.16) we let $V^x = (\{T^{x,i}_n, n \geq 1\}, \{U^{x,i}_n, n \geq 1\}, i = 0, \dots, \kappa - 1)$. $\{V^x, x \in \mathbf{Z}^d\}$ are i.i.d. and $\xi_t(x)$ is a function of the V_x so the result follows from a generalization of (1.3) in Chapter 6 of Durrett (1992). In words, functions of ergodic sequences are ergodic.

To let $t \to \infty$, we note that the proof of (2.8) shows ξ^1_t is stochastically larger than ξ^1_∞ so (2.12) implies that we can construct the two processes on the same space so that $\xi^1_t(x) \geq \xi^1_\infty(x)$ for all x. Let f be an increasing function that depends on only finitely many coordinates. The ergodic theorem implies that as $L \to \infty$

$$\frac{1}{(2L+1)^d} \sum_{x:\|x\|_\infty \leq L} \xi^1_t(x) \to Ef(\xi_t)$$

$$\frac{1}{(2L+1)^d} \sum_{x:\|x\|_\infty \leq L} \xi^1_\infty(x) \to E(f(\xi_\infty)|\mathcal{I})$$

The last result and our comparison imply that $E(f(\xi_\infty)|\mathcal{I}) \leq Ef(\xi_t)$ where $\mathcal{I}$ is the σ-field of shift invariant events. Letting $t \to \infty$ we have $E(f(\xi_\infty)|\mathcal{I}) \leq Ef(\xi_\infty)$ and since the left hand side has expected value $Ef(\xi_\infty)$, it follows that

$$E(f(\xi_\infty)|\mathcal{I}) = Ef(\xi_\infty) \quad \text{a.s.} \tag{2.17}$$

At this point we have shown that (2.17) holds for increasing functions that depends on only finitely many coordinates. Now every function on $\{0,1\}^k$ is a difference of two increasing functions so (2.17) holds for any function of finitely many coordinates. Taking limits and using the inequality

$$E|E(X-Y|\mathcal{I})| \leq E(|X-Y||\mathcal{I}) = E|X-Y|$$

shows that (2.17) holds for all bounded f so $\mathcal{I}$ is trivial. □

3. Percolation Substructures, Duality

In this section we introduce a variation of the construction used in Section 2, due to Harris (1976) and Griffeath (1979), which applies to a special class of models with state space $\{0,1\}^S$ and leads to a "duality relationship." For these purposes it is convenient to write our systems as set valued processes in which the state at time t is the set of sites occupied by 1's. We begin with

Example 3.1. The basic contact process. We let $\mathcal{N}$ be a finite set of neighbors of 0, say that y is a neighbor of x if $y - x \in \mathcal{N}$, and formulate the dynamics as follows:

(i) Particles die at rate 1.

(ii) A particle is born at a vacant site x at rate λ times the number of occupied neighbors.

To construct the process we introduce independent Poisson processes $\{U_n^x, n \geq 1\}$ with rate 1 and $\{T_n^{x,y}, n \geq 1\}$ with rate λ for each $x, y \in \mathbf{Z}^d$ with $y - x \in \mathcal{N}$. At the space time points (x, U_n^x) we write a δ to indicate that a death will occur if x is occupied, and we draw an arrow from $(y, T_n^{x,y})$ to $(x, T_n^{x,y})$ to indicate that if y is occupied then there will be a birth from y to x.

Given the Poisson processes and forgetting about the special marks, we could construct the process using the algorithm described in the last section. We introduce the special marks to make contact with percolation: we imagine fluid entering the bottom of the picture at the points in ξ_0 and flowing up the structure. The δ's are dams, the arrows are pipes that allow the fluid to flow in the direction of the arrow, and ξ_t is the set of sites that are wet at time t.

An example of the *percolation substructure* and the corresponding realization of ξ_t starting from $\xi_0 = \{0,1\}$ is given in Figure 3.1. The thick lines indicate the sites that are occupied. To be able to define the dual process, we need an explicit recipe for constructing ξ_t from the picture. We say that there is a *path from* $(x,0)$ *to* (y,t) if there is a sequence of times $s_0 = 0 < s_1 < s_2 < s_n < s_{n+1} = t$ and spatial locations $x_0 = x, x_1, \ldots, x_n = y$ so that

(i) for $i = 1, 2, \ldots, n$ there is an arrow from x_{i-1} to x_i at time s_i

(ii) the vertical segments $\{x_i\} \times (s_i, s_{i+1})$, $i = 0, 1, \ldots n$ do not contain any δ's.

(Exercise: Find a path from $(2,0)$ to $(3,t)$ in Figure 3.1.) Intutitively the arrows are births that will occur if there are no δ's in the intervals in (ii), so to define the process starting from $\xi_0^A = A$ we let

$$\xi_t^A = \{y : \text{ for some } x \in A \text{ there is a path from } (x,0) \text{ to } (y,t)\} \tag{3.1}$$

It should be clear from the definitions that ξ_t^A is the contact process with one small modification: because of the open intervals in (ii) and the strict inequality in $s_n < s_{n+1} = t$, the process we have constructed is left continuous. For example, if there is a death at x at time t, the particle will not be dead at time t but it will be dead at time $t + \epsilon$ when ϵ

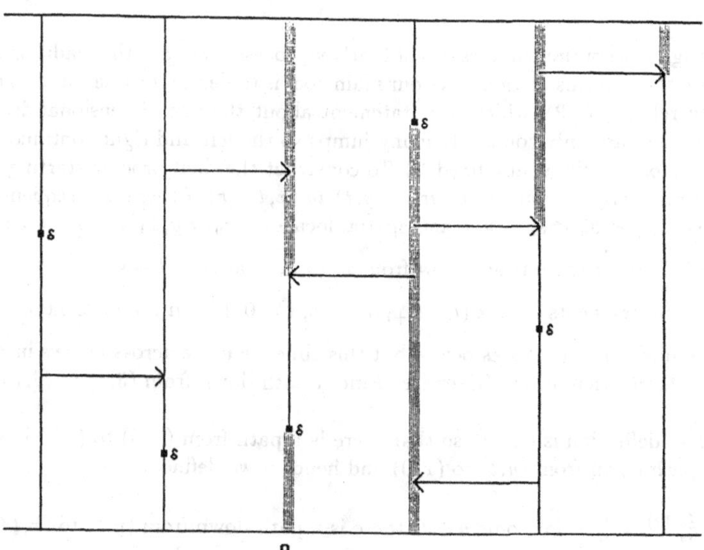

Figure 3.1. Contact process

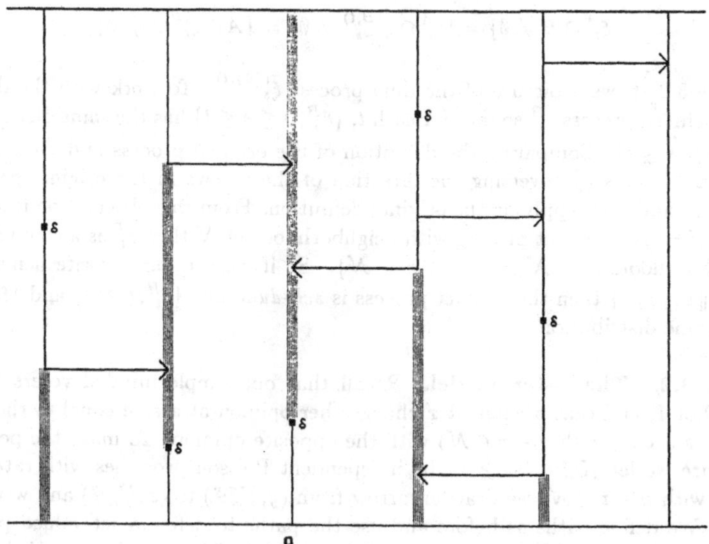

Figure 3.2. Dual of the contact process

is small.

Although left continuous versions of Markov processes are not the traditional ones, we will tolerate them in this section since our main goal is to define the dual process and derive the duality relation (3.2), which is a statement about the one dimensional distributions. (Note that there are only countably many jumps so the left and right continuous versions are equal almost surely at any fixed t.) To construct the dual process starting from time t, we say that there is a *path down from* (y,t) *to* $(x,t-r)$ if there is a sequence of times $s_0 = 0 < s_1 < s_2 < s_n < s_{n+1} = r$ and spatial locations $x_0 = y, x_1, \dots, x_n = x$ so that

(i) for $i = 1, 2, \dots, n$ there is an arrow from x_i to x_{i-1} at time $t - s_i$

(ii) the vertical segments $\{x_i\} \times (t - s_{i+1}, t - s_i)$, $i = 0, 1, \dots n$ do not contain any δ's.

That is, we have to avoid δ's as before but this time we move across arrows in a direction opposite to their orientation. (Exercise: Find a path down from $(3,t)$ to $(2,0)$ in Figure 3.1.)

The last definition is chosen so that there is a path from $(x,0)$ to (y,t) if and only if there is a path down from (y,t) to $(x,0)$ and hence if we define

$$(3.2) \qquad \hat{\xi}_s^{(B,t)} = \{x : \text{ for some } y \in B \text{ there is a path down from } (y,t) \text{ to } (x, t-s)\}$$

then $\{\xi_t^A \cap B \neq \emptyset\} = \{A \cap \hat{\xi}_t^{(B,t)} \neq \emptyset\}$. With a little more thought one sees that for any $0 \le s \le t$

$$(3.3) \qquad \{\xi_t^A \cap B \neq \emptyset\} = \{\xi_s^A \cap \hat{\xi}_{t-s}^{(B,t)} \neq \emptyset\} = \{A \cap \hat{\xi}_t^{(B,t)} \neq \emptyset\}$$

Figure 3.2 shows a picture of the dual process $\hat{\xi}_s^{(\{0\},t)}$. To work with the dual, it is useful to define a process $\hat{\xi}_s^B$ so that for each t, $\{\hat{\xi}_s^B; 0 \le s \le t\}$ has the same distribution as $\{\hat{\xi}_s^{(B,t)} : 0 \le s \le t\}$. Comparing the definition of the original process and the dual shows that we can do this by reversing the direction of the arrows in the original percolation substructure and then applying the original definition. From this observation it should be clear that if ξ_t^A is a contact process with neighborhood set $\mathcal{N}$ then $\hat{\xi}_t^B$ is a contact process with neighborhood set $-\mathcal{N} = \{-x : x \in \mathcal{N}\}$. So if we use our favorite neighborhood $\mathcal{N} = \{x : \|x\|_p \le r\}$ then the contact process is *self-dual*, i.e., $\{\hat{\xi}_t^B, t \ge 0\}$ and $\{\xi_t^B, t \ge 0\}$ have the same distribution.

Example 3.2. The voter model. Recall that our simple minded voters have two opinions 0 or 1, and that a voter at x changes her opinion at a rate equal to the number of neighbors (i.e., y with $y - x \in \mathcal{N}$) with the opposite opinion. To make the percolation substructure we let $\{U_n^{x,y} : n \ge 1\}$ be independent Poisson processes with rate 1 when $x, y \in \mathbf{Z}^d$ with $y - x \in \mathcal{N}$, we draw an arrow from $(y, U_n^{x,y})$ to $(x, U_n^{x,y})$ and write a δ at $(x, U_n^{x,y})$. We define paths as before and use the paths to define a set valued process in which the state at time t is the set of sites with opinion 1. Writing 1 for occupied and 0 for vacant and thinking about the defintion it is easy to see that the effect of an "arrow-delta" from y to x is as follows:

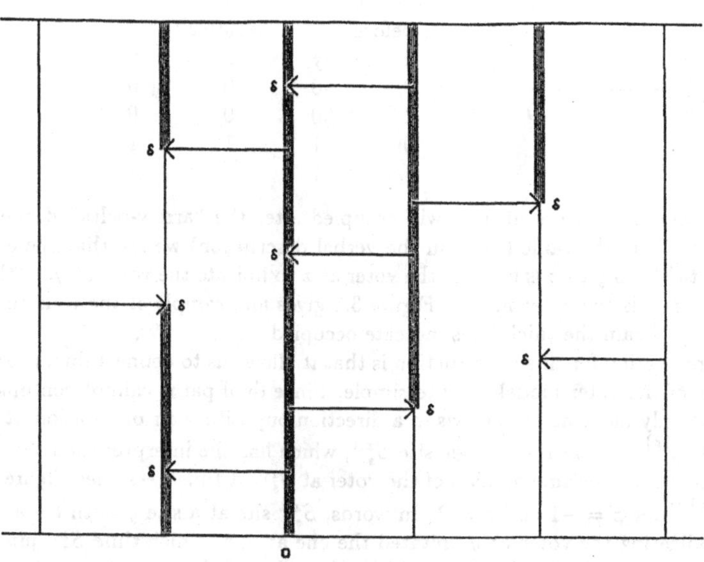

Figure 3.3. Voter model

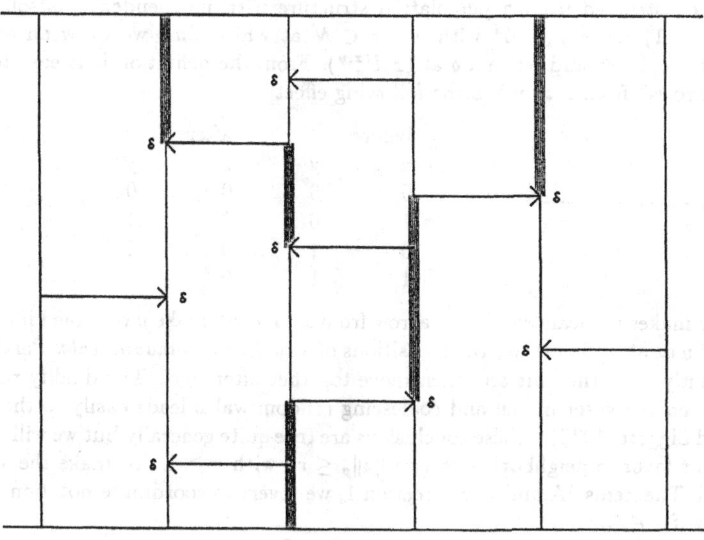

Figure 3.4. Dual of the voter model

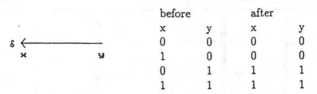

δ ← x y

before		after	
x	y	x	y
0	0	0	0
1	0	0	0
0	1	1	1
1	1	1	1

In words, because of the δ at x, x will occupied after the "arrow-delta" if and only if y is occupied. From the table (or from the verbal description) we see that the effect of an "arrow-delta" from y to x is to force the voter at x to imitate the voter at y, so the process defined by (3.1) is the voter model. Figure 3.3 gives an example of the construction with $\xi_0 = \{-1, 0\}$. Again the thick lines indicate occupied sites.

The motivation for this construction is that it allows us to define a dual process which in the case of the voter model is quite simple. Since dual paths cannot continue through δ's and can only move across arrows in a direction opposite their orientation, it is easy to check that $\hat{\xi}_s^{(\{x\},t)}$ is always a single site $S_s^{x,t}$, which has the interpretation that the voter at x at time t has the same opinion of the voter at $S_s^{x,t}$ at time $t-s$. See Figure 3.4 which shows $\hat{\xi}_s^{(\{x\},t)}$ for $x=-1$ and $x=2$. In words, $S_s^{x,t}$ sits at a site y until $t-s = U_n^{y,z}$ for some z, indicating the voter at y imitated the one at z, at which time $S_s^{x,t}$ jumps from y to z. From the last description it should be clear that $S_s^{x,t}$ is a continuous time random walk that for each $w \in \mathcal{N}$ jumps from y to $y+w$ at rate 1.

To determine the behavior of the dual starting from more than one point, we note that it is constructed from a percolation structure with independent Poisson processes $\{U_n^{x,y} : n \geq 1\}$ for $x, y \in \mathbf{Z}^d$ with $y - x \in \mathcal{N}$ at which time we draw an arrow from $(x, U_n^{x,y})$ to $(y, U_n^{x,y})$ and write a δ at $(x, U_n^{x,y})$. From the definition it is easy to see that a "delta-arrows" from x to y has the following effect

δ → x y

before		after	
x	y	x	y
0	0	0	0
1	0	0	1
0	1	0	1
1	1	0	1

The δ at x makes it vacant while the arrow from x to y will make y occupied if there was a particle at y or at x. These are the transitions of a *coalescing random walk.* Particles move independently until they hit and then move together after that. The duality relationship (3.3) between the voter model and coalescing random walks leads easily to the results of Holley and Liggett (1975). These conclusions are true quite generally but we will state them only for our favorite neighborhoods $\{z : \|z\|_p \leq r\}$ with $r \geq 1$. To make the statements here match Theorems 2A and 2B in Section 1, we revert to coordinate notation: $\xi_t(x) = 1$ if and only if $x \in \xi_t$.

Theorem 3.1. *Clustering* occurs in $d \leq 2$. That is, for any ξ_0 and $x, y \in \mathbf{Z}^d$ we have

$$P(\xi_t(x) \neq \xi_t(y)) \to 0 \text{ as } t \to \infty$$

Theorem 3.2. Let ξ_t^θ denote the process starting from an initial state in which the events $\{\xi_0^\theta(x) = 1\}$ are independent and have probability θ. In $d \geq 3$ as $t \to \infty$, $\xi_t^\theta \Rightarrow \xi_\infty^\theta$, a translation invariant stationary distribution in which $P(\xi_\infty(x) = 1) = \theta$.

PROOF OF THEOREM 3.1. From our discussion of the dual it should be clear that

$$P(\xi_t(x) \neq \xi_t(y)) \leq P(S_t^{(x,t)} \neq S_t^{(y,t)})$$

since if the two sites x and y trace their opinions back to the same site at time 0 then they will certainly be equal at time t. Now the difference $S_s^{(x,t)} - S_s^{(y,t)}$ is a random walk stopped when it hits 0, and the random walk has jumps that have mean 0 and finite variance. Such random walks are *recurrent*, and since ours is also an irreducible Markov chain, it will eventually hit 0. Since 0 is an absorbing state for $S_s^{(x,t)} - S_s^{(y,t)}$ it follows that $P(S_t^{(x,t)} \neq S_t^{(y,t)}) \to 0$ and the proof is complete. □

Remark. The reader should not misinterpret Theorem 3.1 as saying that the voter model is boring in $d \leq 2$. Cox and Griffeath (1986) have proved a number of interesting results about the clustering in $d = 2$, which is rather exotic since two dimensional random walk is just barely recurrent.

PROOF OF THEOREM 3.2. From the proof of (2.8) we see that it is enough to prove the convergence of $P(\xi_t \cap B = \emptyset)$ for each B. To treat these probabilities we observe that

$$P(\xi_t \cap B = \emptyset) = E\{(1-\theta)^{|\hat\xi_t^{(B,t)}|}\}$$

since by duality there are no particles in B at time t if and only if none of the sites in $\hat\xi_t^{(B,t)}$ is occupied at time 0, an event with probability $(1-\theta)^{|\hat\xi_t^{(B,t)}|}$. To analyze the right hand side we note that $\hat\xi_t^{(B,t)}$ has the same distribution as $\hat\xi_t^B$ constructed from the percolation substructure that has the directions of all the arrows reversed. Since $\hat\xi_t^B$ is a coalescing random walk, $|\hat\xi_t^B|$ is a decreasing function of t and has a limit. Since $0 \leq (1-\theta)^{|\hat\xi_t^B|} \leq 1$ it follows from the bounded convergence theorem that

$$\lim_{t\to\infty} E\{(1-\theta)^{|\hat\xi_t^{(B,t)}|}\} \text{ exists}$$

and the proof is complete. □

Since the ξ_t^θ are translation invariant (by (2.2)), it follows that the limits ξ_∞^θ are.

$$P(x \in \xi_t^\theta) = P(S_t^{x,t} \in \xi_0^\theta) = \theta$$

for all t so $P(x \in \xi_\infty^\theta) = \theta$. Holley and Liggett (1975) showed that the ξ_∞^θ are spatially ergodic and give all the stationary distributions for the voter model. That is, all stationary distributions are a convex combination of the (distributions of the) ξ_∞^θ. For proofs of this result see the original paper by Holley and Liggett (1975) or Chapter V of Liggett (1985).

Using duality we can prove a convergence theorem due to Harris (1976) for a general class of processes that contains the contact process as a special case. We begin by introducing the models we will consider.

Additive processes. For each finite $A \subset \mathbf{Z}^d$ and $x \in \mathbf{Z}^d$ we introduce independent Poisson processes $\{T_n^{x,A}, n \geq 1\}$ and $\{U_n^{x,A}, n \geq 1\}$ with rates $\lambda(A)$ and $\delta(A)$. (To have a finite range interaction, we only allow finitely many of the rates to be nonzero.) At times $T_n^{x,A}$ we draw arrows from $x+z$ to x for all $z \in A$ and there will be a birth if some site in $x+A$ is occupied. At times $U_n^{x,A}$ we write a δ at x, draw arrows from $x+z$ to x for all $z \in A$, and there will be a death at x unless some point in $x+A$ is occupied. The process is then obtained from the percolation substructure by using (3.1). In the new notation our two examples may be written as (the rates we do not mention are 0):

The contact process. $\lambda(A) = \lambda$ if $A = \{x\}$ with $x \in \mathcal{N}$; $\delta(\emptyset) = 1$.

The voter model. $\delta(A) = 1$ if $A = \{x\}$ with $x \in \mathcal{N}$.

It should be clear that for any additive process the birth rates are increasing and the death rates are decreasing so these systems are attractive. To see that additive processes are a fairly small subclass of the attractive models, we will now consider

Example 3.3. Nonlinear Contact Processes. In these systems the flip rates are

$$c_0(x,\xi) = 1$$
$$c_1(x,\xi) = b(|\{y \in \mathcal{N} : \xi(x+y) = 1\}|)$$

where $b(0) = 0$. To get the desired death rates we set $\delta(\emptyset) = 1$ and $\delta(A) = 0$ otherwise. To see what birth rates we can create we begin with the special case

(i) $d = 1$, $\mathcal{N} = \{-1, 1\}$. In this situation we must have

$$\lambda(\{1\}) = \lambda(\{-1\}) = a_1 \qquad \lambda(\{1,-1\}) = a_2$$

and the other $\lambda(A) = 0$, so $b(1) = a_1 + a_2$ and $b(2) = 2a_1 + a_2$ which is possible with $a_1, a_2 \geq 0$ if and only if

$$b(1) \leq b(2) \leq 2b(1)$$

The extreme case $b(2) = 2b(1)$ is the basic contact process, the other extreme $b(2) = b(1) = b$ is called the *threshold contact process* because the birth rate is b if there is at least one occupied neighbor. An example of a system not covered by this construction is the *sexual reproduction model* which has $b(1) = 0$ and $b(2) = \lambda$.

(ii) Suppose $|\mathcal{N}| = 4$ and think about $\mathcal{N} = \{-2,-1,1,2\}$ in $d = 1$ or $\mathcal{N} = \{z : \|z\|_1 = 1\}$ in $d = 2$. (The geometry of the set $\mathcal{N}$ does not enter into the decision as to whether or not a system is additive.) In this case $\lambda(A) = a_i$ if $A \subset \mathcal{N}$ with $|A| = i$ (and 0 otherwise) so

$$\begin{aligned} b(1) &= a_1 + 3a_2 + 3a_3 + a_4 \\ b(2) &= 2a_1 + 5a_2 + 4a_3 + a_4 \\ b(3) &= 3a_1 + 6a_2 + 4a_3 + a_4 \\ b(4) &= 4a_1 + 6a_2 + 4a_3 + a_4 \end{aligned}$$

To see the equation of $b(2)$ say, note that any two element subset of $\mathcal{N}$ touches 2 of the singleton subsets of $\mathcal{N}$, all but one of the 6 two element subsets, all 4 of the three element subsets, and the four element subset. Subtracting the equations gives

$$\begin{aligned} b(4) - b(3) &= a_1 \\ b(3) - b(2) &= a_1 + a_2 \\ b(2) - b(1) &= a_1 + 2a_2 + a_3 \\ b(1) - b(0) &= a_1 + 3a_2 + 3a_3 + a_4 \end{aligned}$$

and taking differences again

$$\begin{aligned} a_1 &= b(4) - b(3) \\ a_2 &= (b(3) - b(2)) - (b(4) - b(3)) \\ a_3 &= (b(2) - b(1)) - 2(b(3) - b(2)) + (b(4) - b(3)) \\ a_4 &= ((b(1) - b(0)) - 3(b(2) - b(1)) + 3(b(3) - b(2)) - (b(4) - b(3)) \end{aligned}$$

The process is additive if and only if these quantities are nonnegative. These conditions are monotonicity and convexity properties of the sequence of birth rates $b(i)$. A result for general neighborhoods can be found in Harris (1976), see (6.4) on page 184. The conclusions we would like the reader to draw from this computation are that (i) the additive processes are a small subset of the attractive processes but (ii) when we consider nonlinear contact processes with $|\mathcal{N}| = 4$ additive processes are a four dimensional subset of the four dimensional set of models.

Harris' convergence theorem for additive processes. Before getting started we need to introduce a technical condition. Let ξ_t^0 denote the process starting from a single particle at the origin. We say ξ_t is *irreducible* if for any x and $t > 0$ $P(x \in \xi_t) > 0$. Recall that in Section 2, we let ξ_t^1 denote the process starting from $\xi_0^1 = \mathbf{Z}^d$ and showed that for any attractive process $\xi_t^1 \Rightarrow \xi_\infty^1$, a translation invariant stationary distribution.

Theorem 3.3. Suppose ξ_t is an irreducible additive process with $\delta(\emptyset) > 0$. If ξ_0 is translation invariant and assigns 0 probability to the empty configuration then $\xi_t \Rightarrow \xi_\infty^1$ as $t \to \infty$.

Corollary. ξ_∞^1 is the only translation invariant stationary distribution that assigns 0 probability to the empty configuration.

Remarks. The condition $\delta(\emptyset) = 0$ eliminates the voter model for which the conclusion of Theorem 3.3 is always false. Our result is only for translation invariant initial distributions. With a lot more work one can prove a *complete convergence theorem*:

Theorem 3.4 Suppose ξ_t is an irreducible additive process with $\delta(\emptyset) > 0$. Then for any A,

$$\xi_t^A \Rightarrow P(\tau^A < \infty)\delta_\emptyset + P(\tau^A = \infty)\xi_\infty^1$$

where $\delta_\emptyset$ denotes the pointmass on the emptyset and we are using ξ^1_∞ to denote its distribution.

In words, if the process does not die out, then at large times it looks like the process starting from all 1's. This implies that all stationary distributions have the form $\theta\delta_\emptyset + (1-\theta)\xi^1_\infty$. For the contact process, this result is due to Bezuidenhout and Grimmett (1990). To prove this in the general case you will need to consult Bezuidenhout and Gray (1993).

PROOF OF THEOREM 3.3. To begin we note that the duality equation (3.3) implies

$$\begin{aligned} P(\xi^1_t \cap B \neq \emptyset) &= P(\hat{\xi}^{(B,t)}_t \cap \xi^1_0 \neq \emptyset) \\ &= P(\hat{\xi}^B_t \neq \emptyset) \to P(\hat{\tau}^B = \infty) \end{aligned}$$

as $t \to \infty$. As in the proof of Theorem 3.2, the argument in (2.8) shows that it is enough to prove $P(\xi_t \cap B \neq \emptyset) \to P(\hat{\tau}^B = \infty)$. Half of this is very easy. By duality and the fact that $\xi_0 \subset \mathbf{Z}^d$

$$P(\xi_t \cap B \neq \emptyset) = P(\xi_0 \cap \hat{\xi}^{(B,t)}_t \neq \emptyset) \leq P(\hat{\tau}^B > t)$$

so

$$\limsup_{t\to\infty} P(\xi_t \cap B \neq \emptyset) \leq P(\hat{\tau}^B = \infty)$$

To prove the other direction, we let t_0 be the constant in (2.1) and observe that (3.3) implies

$$P(\xi_{t+t_0} \cap B \neq \emptyset) = P(\xi_{t_0} \cap \hat{\xi}^{(B,t+t_0)}_t \neq \emptyset)$$

To get the right hand side to converge to $P(\hat{\tau}^B = \infty)$ we need to show that when $\hat{\xi}^{(B,t+t_0)}_t \neq \emptyset$ then it will intersect ξ_{t_0} with high probability. The first step in doing this is to show that when $\hat{\xi}^{(B,t+t_0)}_t \neq \emptyset$, it will contain a large number of points with high probability. To do this, let

$$\Lambda = \sum_A |A|(\lambda(A) + \delta(A))$$

be the rate at which an isolated particle gives birth to a new particle and let $\alpha = (1 - e^{-\delta(\emptyset)})e^{-\Lambda}$ be a lower bound on the probability that in one unit of time an isolated particle is killed and does not give birth. Now for any K

$$P(t < \hat{\tau}^B \leq t+1) \geq \alpha^K P(0 < |\hat{\xi}^{(B,t+t_0)}_t| \leq K)$$

To see this note that the events that each particle is killed by a δ are independent, and write the statement that no particle gives birth in terms of Poisson processes in the percolation substructure. Since $P(t < \hat{\tau}^B \leq t+1) \to 0$ as $t \to \infty$, and α^K is a positive constant, it follows that

$$P(0 < |\hat{\xi}^{(B,t+t_0)}_t| \leq K) \to 0 \tag{3.4}$$

To complete the proof now it suffices to show

(3.5) **Lemma.** If $\epsilon > 0$ then we can pick K large enough so that if $|A| \geq K$ then $P(\xi_{t_0} \cap A = \emptyset) \leq 3\epsilon$.

For then it follows that from (3.5) and (3.4) that

$$\liminf_{t\to\infty} P(\xi_{t_0} \cap \hat{\xi}_t^{(B,t+t_0)} \neq \emptyset) \geq (1-3\epsilon) \liminf_{t\to\infty} P(|\hat{\xi}_t^{(B,t+t_0)}| \geq K)$$
$$\geq (1-3\epsilon) P(\hat{\tau}^B > t)$$

Remark. For the conclusion in (3.5) it is important that we let the process run for a positive amount of time. The initial configuration ξ_0 that is $2\mathbf{Z}$ with probability 1/2 and $2\mathbf{Z}+1$ with probability 1/2 is translation invariant but $P(\xi_0 \cap \{2,4,\ldots,2K\}) = 1/2$ for all K.

PROOF OF (3.5): For this proof it is convenient to use the coordinate representation of the process, i.e., $\xi_t(x) = 1$ if x is occupied at time t and 0 otherwise. Let μ be the distribution of ξ_0 (i.e., the induced measure on $\{0,1\}^S$) and use P_ξ to denote the probability law for ξ_t when $\xi_0 = \xi$. Our assumption of irreducibility and attractiveness imply that $P_\xi(\xi_{t_0}(x) = 1) > 0$ unless $\xi \equiv 0$, an event that by assumption has probability 0, so

(3.6) For any $\epsilon > 0$ there is a $\rho < 1$ so that

$$\mu(\{\xi : P_\xi(\xi_{t_0}(x) = 0) > \rho\}) \leq \epsilon$$

Here we need translation invariance to conclude that the left hand side does not depend on x. The second ingredient is to note repeated use of Hölder's inequality gives

$$E(X_1 \cdots X_k) \leq (E|X_1^k|)^{1/k} \cdots (E|X_k^k|)^{1/k}$$

which in turn implies

(3.7) Let $X_1, \ldots, X_k$ be random variables so that $0 \leq X_i \leq 1$ and $P(X_i > \rho) \leq \epsilon$. Then

$$E(X_1 \cdots X_k) \leq \rho^k + \epsilon$$

Pick J so that $\rho_\epsilon^J \leq \epsilon$. Our proof of the next result explains why we chose the time t_0. The result is valid for any time t, see Holley (1972).

(3.8) Given $\epsilon > 0$ and J, we can pick L so that if $B \subset \mathbf{Z}^d$ with $|B| = J$ and $\|x-y\|_\infty > 2L$ whenever $x, y \in B$ with $x \neq y$ then

$$\left| E_\xi \left\{ \prod_{x\in B} (1 - \xi_{t_0}(x)) \right\} - \prod_{x\in B} \{E_\xi(1 - \xi_{t_0}(x))\} \right| \leq \epsilon$$

PROOF OF (3.8): First we compute the value of each $\xi_{t_0}(x)$ with $x \in B$ by using an independent copy of the percolation substructure $\mathcal{P}_x$. The second step is to combine all these independent substructures to make a new one $\mathcal{P}_{all}$ by taking $T_n^{y,A}$ and $U_n^{y,A}$ from $\mathcal{P}_x$ if and only if $y + A \subset D(x, L) = \{z : \|x - z\|_\infty \le L\}$ and then using another independent percolation substructure $\mathcal{P}^*$ to fill in the missing Poisson processes. Let R be the largest value of $\|x\|_\infty$ for a point in some set A with $\lambda(A)$ or $\delta(A) > 0$. R is the range of the interaction. If the cluster containing x in $\mathcal{P}_x$ defined in the proof of (2.1) lies inside $D(x, L - R)$ then it is identical with the cluster containing x in $\mathcal{P}_{all}$ and the values computed for ξ_{t_0} are the same. Since the states of x in the processes on $\mathcal{P}_x$ are independent, it follows from the proof of (2.1) that if L is large the random variables $1 - \xi_{t_0}(x)$ on $\mathcal{P}_{all}$ are equal with high probability to independent random variables and (3.8) follows. □

To complete the proof of (3.5) now, we observe that

(3.9) If $B \subset \mathbf{Z}^d$ with $|B| = J$ and if $\|x - y\|_\infty > 2L$ whenever $x, y \in B$ with $x \neq y$ then

$$P(\xi_{t_0}(x) = 0 \text{ for all } x \in B) = \int \mu(d\xi)\, E_\xi \prod_{x \in B} (1 - \xi_{t_0}(x))$$

$$\le \epsilon + \int \mu(d\xi) \prod_{x \in B} E_\xi(1 - \xi_{t_0}(x)) \le 2\epsilon + \rho_\epsilon^J \le 3\epsilon$$

by (3.8), (3.6), (3.7), and the choice of J. To get from the last result to the desired conclusion we let $K = (4L+1)^d J$ and observe that if $|A| \ge K$ we can find a subset B with $|B| = J$ that sastisfies the hypotheses of (3.9). □

Example 3.4. Multitype contact processes, defined in Section 1, have state space $\{0, 1, \kappa - 1\}^S$ where 0 indicates a vacant site and $i > 0$ indicates a site occupied by one plant of type i, and have flip rates that are linear:

$$c_0(x, \xi) = \delta_{\xi(x)}$$
$$c_i(x, \xi) = \lambda_i n_i(x, \xi) \qquad \text{if } \xi(x) = 0$$

When $\lambda_i = \lambda$ and $\delta_i = \delta$, this process can be studied by using a duality that is a hybrid of the one for the contact process and for the voter model. The first step is to construct the process as we did the contact process. We introduce independent Poisson processes $\{U_n^x, n \ge 1\}$ with rate δ and $\{T_n^{x,y}, n \ge 1\}$ with rate λ for each $x, y \in \mathbf{Z}^d$ with $y - x \in \mathcal{N}$. As before, we write a δ at (x, U_n^x) to indicate that a death will occur if x is occupied by a particle of either type, and we draw an arrow from $(y, T_n^{x,y})$ to $(x, T_n^{x,y})$ to indicate that if x is vacant and y is occupied then there will be a birth from y to x.

If we define the dual process as in (3.2) then reasoning as before we see that x will be occupied at time t if and only if some site in $\hat{\xi}_t^{(\{x\},t)}$ is occupied in ξ_0. The dual for the mutlitype contact process is the set $\hat{\xi}_t^{(\{x\},t)}$ plus an ordering of that set with the interpretation that the type of x is that of the first occupied site in the ordering. For example in the realization drawn in Figure 3.2, the ordering is $1 > 2 > -2$

The first site in $\hat{\xi}_s^{(\{x\},t)}$ in this ordering is called the ***distinguished particle***. Results of Neuhauser (1992) show that the movements of the distinguished particle are enough like those of a random walk to conclude that in $d \le 2$ the distinguished particles for the duals of two different sites will eventually be equal for large t. This is the key idea in proving Theorems 4C and 4D in Section 1. In the two type case, when $\delta_1 = \delta_2$ and $\lambda_1 < \lambda_2$ we can augment the construction above with Poisson processes of arrows that only allow the births of 2's and an easy argument gives Theorem 4A. However such an approach will never give us Conjecture 4B.

4. A Comparison Theorem

In this section we will introduce a comparison theorem that is very useful in proving the existence of nontrivial translation invariant stationary distributions. At this point we have to ask for the reader's patience: the result given in Theorem 4.3 is powerful but you will need to see a few applications to understand how it works.

Our general method for proving the existence of stationary distributions is to compare the process of interest with oriented percolation, so our first step is to introduce oriented percolation and state some of its basic properties, the proofs of which are hidden away in the appendix. Let

$$\mathcal{L}_0 = \{(x,n) \in \mathbf{Z}^2 : x+n \text{ is even}, n \geq 0\}$$

and make $\mathcal{L}_0$ into a graph by drawing oriented edges from (x,n) to $(x+1,n+1)$ and from (x,n) to $(x-1,n+1)$. Given random variables $\omega(x,n)$ that indicate whether the sites are open (1) or closed (0), we say that (y,n) can be reached from (x,m) and write $(x,m) \to (y,n)$ if there is a sequence of points $x = x_m, \ldots, x_n = y$ so that $|x_k - x_{k-1}| = 1$ for $m < k \leq n$ and $\omega(x_k,k) = 1$ for $m \leq k \leq n$. In the standard oriented percolation model the variables $\omega(x,n)$ are independent, but in almost all cases our comparisons will introduce dependencies between the $\omega(x,n)$, so we need a more general set-up. We say that the $\omega(x,n)$ are "*M dependent with density at least* $1-\gamma$" if whenever (x_i,n_i), $1 \leq i \leq I$ is a sequence with $\|(x_i,n_i)-(x_j,n_j)\|_\infty > M$ if $i \neq j$ then

$$P(\omega(x_i,n_i) = 0 \text{ for } 1 \leq i \leq I) \leq \gamma^I \tag{4.1}$$

Note: Classical M-dependence would require that the $\omega(x_i,n_i)$ considered above are independent. However the probability in (4.1) is the only one we need to control and hence the only thing we assume.

Given an initial condition $W_0 \subset 2\mathbf{Z} = \{x : (x,0) \in \mathcal{L}_0\}$, we can define a process by

$$W_n = \{y : (x,0) \to (y,n) \text{ for some } x \in W_0\}$$

In words, the sites W_n are those that are wet at time n. To keep the terminology straight, think of open sites as air spaces in a rock, and the sites in W_n as the ones that the fluid can reach (and hence wet) at level n. We use W_n^0 to denote the process that results when $W_0^0 = \{0\}$ and we let

$$\mathcal{C}_0 = \{(y,n) : (0,0) \to (y,n)\}$$

be the set of all points in space-time that can be reached by a path from $(0,0)$. (When $(0,0)$ is open $\mathcal{C}_0 = \cup_n (W_n^0 \times \{n\})$.) $\mathcal{C}_0$ is called the *cluster containing the origin.* Figure 4.1 shows a simulation of the independent oriented percolation process in which sites are open (indicated by black dots) with probability $p = 0.6$. Time goes up the page and lines connect the points of $\mathcal{C}_t$.

When the cluster containing the origin is infinite, i.e., $\{|\mathcal{C}_0| = \infty\}$ we say that *percolation occurs.* Our first result shows that if the density of open sites is high enough then percolation occurs. All that is important about the upper bound is that it is < 1 for small γ and converges to 0 as $\gamma \to 0$.

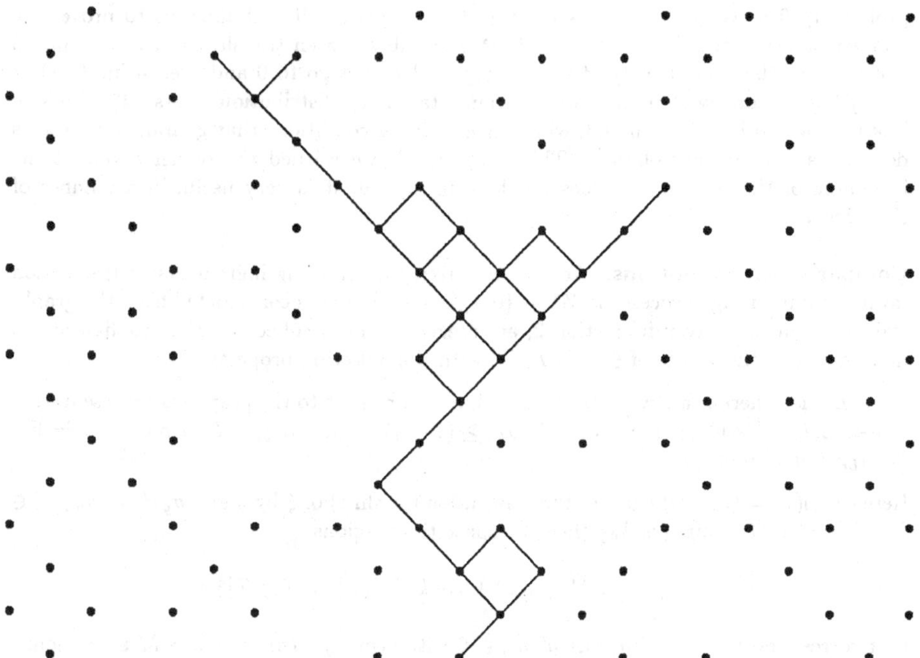

Figure 4.1

Theorem 4.1. If $\gamma \leq 6^{-4(2M+1)^2}$ then

$$P(|\mathcal{C}_0| < \infty) \leq 55\,\theta^{1/(2M+1)^2} \leq 1/20$$

In order to prove the existence of stationary distributions we need results about M dependent oriented percolation starting from the initial configuration W_0^p in which the events $\{x \in W_0^p\}$, $x \in 2\mathbf{Z}$ are independent and have probability p. We will sometimes call this a ***Bernoulli random set with density p***. Taking $p = 1$ (i.e., all sites wet initially) corresponds to computing the upper invariant measure for oriented percolation, but for some of the proofs below we will need to allow $p < 1$. Note that the estimate on the lim inf is independent of p and is 1 minus the upper bound in Theorem 4.1.

Theorem 4.2. If $p > 0$ and $\gamma \leq 6^{-4(2M+1)^2}$ then

$$\liminf_{n\to\infty} P(0 \in W_{2n}^p) \geq 1 - 55\,\gamma^{1/(2M+1)^2} \geq 19/20$$

The last result shows that if the density of open sites in oriented percolation is sufficiently high and if we start with from a Bernoulli random set with density p then the probability 0 is wet at time t does not go to 0. This result will allow us to prove in a number of situations that if we start from a suitably chosen translation invariant initial distribution, then the density of sites of type i does not go to 0 and then using (2.7) or (2.11) that a nontrivial translation invariant stationary distribution exists. The missing link is provided by Theorem 4.3, which gives general conditions that guarantee a process dominates oriented percolation. This is the result we warned the reader about at the beginning of the section – it does not look pretty but it is very useful in a number of situations.

Comparison Assumptions. We suppose given the following ingredients: a translation invariant finite range process $\xi_t : \mathbf{Z}^d \to \{0, 1, \ldots \kappa - 1\}$ that is constructed from the graphical representation given in Section 2, an integer L, and a collection H of configurations determined by the values of ξ on $[-L, L]^d$ with the following property:

if $\xi \in H$ then there is an event G_ξ measurable with respect to the graphical representation in $[-k_0L, k_0L]^d \times [0, j_0T]$ and with $P(G_\xi) \geq (1 - \gamma)$ so that if $\xi_0 = \xi$ then on G_ξ, ξ_T lies in $\sigma_{2Le_1}H$ and in $\sigma_{-2Le_1}H$.

Here $(\sigma_y\xi)(x) = \xi(x + y)$ denotes the translation (or shift) of ξ by y and $\sigma_yH = \{\sigma_y\xi : \xi \in H\}$. If we let $M = \max\{j_0, k_0\}$ then the space time regions

$$\mathcal{R}_{m,n} = (m2Le_1, nT) + \{[-k_0L, k_0L]^d \times [0, j_0T]\}$$

that correspond to points $(m, n), (m', n') \in \mathcal{L}$ with $\|(m, n) - (m', n')\|_\infty > M$ are disjoint.

For a concrete instance of the comparison assumptions consider the applications we will make to the threshold contact process in Section 5 and to the basic contact process in Section 7. In both cases $\kappa = 2$, and H is the set of configurations with at least K 1's in $[-L, L]^d$, $k_0 = 4$, and $j_0 = 1$. In words, we show that if there is a "pile" of at least K particles in $[-L, L]^d$ then with high probability there will be piles of at least K particles in $-2Le_1 + [-L, L]^d$ and in $2Le_1 + [-L, L]^d$ at time T, and the event that guarantees this is measurable with respect to the graphical representation in $[-4L, 4L]^d \times [0, T]$. Figure 4.2 below gives a picture of the event.

Using words inspired by the contact process example, our comparison assumptions say that if we have a "pile of particles" in $I_m = m2Le_1 + [-L, L]^d$ at time nT (i.e., $\xi_{nT} \in \sigma_{m2Le_1}H$) then with high probability we will have piles of particles in I_{m-1} and in I_{m-1} at time $(n + 1)T$, and the event that guarantees this is measurable with respect to the graphical representation in $\mathcal{R}_{m,n}$. If we think of drawing arrows from (m, n) to $(m + 1, n + 1)$ and to $(m - 1, n + 1)$ whenever the good event in $\mathcal{R}_{m,n}$ occurs then the connection with oriented percolation should be clear.

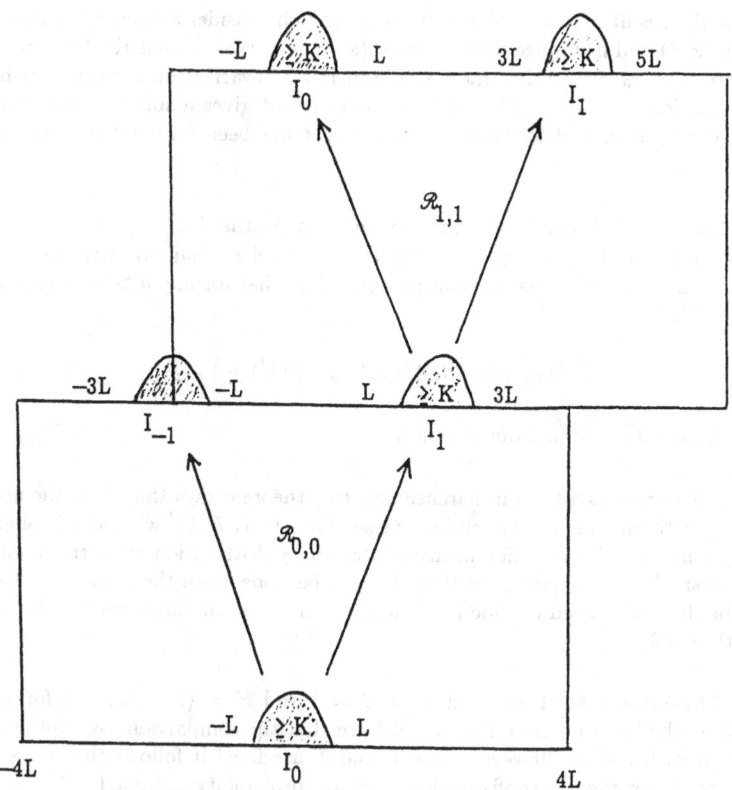

Figure 4.2

To formulate our theorem we let $X_n = \{m : (m,n) \in \mathcal{L}_0, \xi_{nT} \in \sigma_{m2Le_1} H\}$. Intuitively, $m \in X_n$ if there is a pile of particles in I_m at time nT

Theorem 4.3. If the comparison assumptions hold then we can define random variables $\omega(x,n)$ so that X_n dominates an M dependent oriented percolation process with initial configuration $W_0 = X_0$ and density at least $1 - \gamma$, i.e., $X_n \supset W_n$ for all n.

Again the details are hidden away in the appendix so that they can be digested after the reader has seen that this is a useful result.

Our first indication that Theorems 4.1–4.3 are useful is a simple proof of a general result about the existence of stationary distributions, which contains as a special case a number of earlier results. To formulate our result we will consider a fixed set of increasing birth rates $c_1(x,\xi)$ and introduce death rates $c_0(x,\xi) \equiv \epsilon$. We say that the birth rates are *robust* if there is an $\epsilon_0 > 0$ so that there is a translation invariant stationary distribution with a positive density of 1's for $\epsilon < \epsilon_0$. Our next result gives a sufficient condition for robustness. It may look a little strange at first but it has been formulated to be easy to prove and to apply.

Theorem 4.4. Let $\bar\xi_t^{L,\rho}$ denote the process with no deaths, i.e., $\epsilon = 0$, starting from $\bar\xi_0^{L,\rho}(x) = 1$ for $x \in [-L,L]^d$, $= 0$ otherwise and modified so that no births are allowed outside $[-\rho L, \rho L]^d$. Suppose that we can pick $\rho \geq 3$ so that for any $\delta > 0$ we can pick L and $T < \infty$ so that

$$P(\bar\xi_T^{L,\rho}(x) = 1 \text{ for all } x \in [-3L, 3L]^d) \geq 1 - \delta$$

Then the birth rates are robust (and fertile).

Ignoring the undefined term in parentheses, this theorem says that if, in the absence of deaths, the birth mechanism can triple the size of a cube $[-L,L]^d$ with high probability, then there is a nontrivial translation invariant stationary distribution when the death rate $c_0(x,\xi) \equiv \epsilon$ is small. The requirement that this can be done when the model is "modified so that no births are allowed outside $[-\rho L, \rho L]^d$" is a technical condition that is usually satisfied with $\rho = 3$.

Proof of Theorem 4.4: If we let $K = \rho$, $J = 1$ and $H = \{\xi : \xi(x) = 1 \text{ for all } x \in [-L,L]^d\}$. then the hypotheses of Theorem 4.4 are that the comparison assumptions hold for the system with $\epsilon = 0$. However, once L and T are fixed it follows that for $\epsilon \leq \epsilon_0$, the good event G_ξ for the one configuration in H has probability at least $1 - 2\delta$, since the probability a death occurs at some site in the space time box $[-3L, 3L]^d \times [0,T]$ is less than δ when ϵ_0 is sufficiently small.

To construct our stationary distribution, we consider the process ξ_t^1 starting from $\xi_0^1(x) = 1$ for all x. In this case $X_0 = 2\mathbf{Z}$ so using Theorems 4.3 and 4.2 with $p = 1$, it follows that if $\epsilon \leq \epsilon_0$ then

$$\liminf_{n\to\infty} P(\xi_{nT}^1(0) = 1) \geq 19/20$$

Using (2.7) now it follows that there is a nontrivial stationary distribution. □

We will now give three examples to shows that is easy to check the conditions of Theorem 4.4.

Corollary 4.5. If we fix $\lambda = 1$ in the contact process with neighborhood $\mathcal{N} = \{x : \|x\|_p \leq r\}$ where $r \geq 1$ then there is a nontrivial stationary distribution when the death rate $\delta < \delta_0$.

PROOF: Take $\rho = 3$ and $L = 1$. Since 1's can never flip to 0 it is easy to see that

$$\lim_{T\to\infty} P(\bar{\xi}_T^{L,\rho}(x) = 1 \text{ for all } x \in [-3L, 3L]^d) = 1$$

so the hypotheses of Theorem 4.4 are satisfied. □

Example 4.1. One Dimensional Counting Rules. Suppose $d = 1$, $\mathcal{N} = \{z : |z| \le k\}$, and let

$$n_1(x,\xi) = |\{z \in \mathcal{N} : \xi(x+z) = 1\}|$$

be the number of neighbors of x that are 1. We call a birth rate $c_1(x,\xi)$ a *counting rule* if it only depends on the number of 1's in the neighborhood, i.e., $c_1(x,\xi) = b(n_1(x,\xi))$ Clearly a counting rule birth rate is increasing if and only if $j \to b(j)$ is nondecreasing. Let $j_0 = \min\{j : b_j > 0\}$ and call j_0 the *order* of the birth rate. The next result is due to Mityugin.

Corollary 4.6. When $d = 1$ and $\mathcal{N} = \{j : |j| \le k\}$, increasing counting rule birth rates are robust if and only if their order $j_0 \le k$.

PROOF: If $j_0 > k$ then a string of at least $k+1$ consecutive 0's can never flip back to 1 even if all the other sites are 1. If $c_0(x,\xi) \equiv \epsilon > 0$ then such a string will eventually be created and grow to cover the whole line, so there cannot be a nontrivial stationary distribution.

If $j_0 \le k$, we take $\rho = 3$ and choose L so that $2L + 1 \ge k$. When $\epsilon = 0$ the 1's never flip back to 0. The 0 at $L+1$ has k neighbors that are 1 and hence flips to 1 at rate $b(k) \ge b(j_0) > 0$. Once the 0 at $L+1$ flips to 1, the 0 at $L+2$ will flip to 1 at rate $b(k)$, so

$$\lim_{T\to\infty} P(\bar{\xi}_T^{L,\rho}(x) = 1 \text{ for all } x \in [-3L, 3L]) = 1$$

and the hypotheses of Theorem 4.4 are satisfied. □

Things get more interesting in two dimensions.

Example 4.2. Two Dimensional Threshold Birth Rates. Suppose $d = 2$ and $\mathcal{N} = \{z : \|z\|_\infty = 1\}$, i.e., in addition to the four nearest neighbors we use the four diagonally adjacent points:

$$\mathcal{N} = \left\{ \begin{matrix} (-1,1) & (0,1) & (1,1) \\ (-1,0) & & (1,0) \\ (-1,-1) & (0,-1) & (1,-1) \end{matrix} \right\}$$

This is sometimes called the *Moore neighborhood* in honor of one of the pioneers in the field of cellular automata. Let $n_1(x,\xi) = |\{j \in \mathcal{N} : \xi(x) = 1\}$ be the number of neighbors in state 1 and let

$$c_1(x,\xi) = \begin{cases} 1 & \text{if } n_1(x,\xi) \ge \theta \\ 0 & \text{if } n_1(x,\xi) < \theta \end{cases}$$

This is called a threshold θ since the birth rate 1 if there are at least θ 1's in the neighborhood then the birth rate is 1, and otherwise it is 0. From Theorem 4.4 we get easily that

Corollary 4.7. Two dimensional threshold birth rates for the Moore neighborhood in two dimension are robust if $\theta \le 3$.

PROOF: Take $\rho = 3$, $L = 1$, and draw a picture.

$$\begin{matrix} 4 & 3 & 2 & 3 & 4 \\ 3 & 1 & 1 & 1 & 3 \\ 2 & 1 & 1 & 1 & 2 \\ 3 & 1 & 1 & 1 & 3 \\ 4 & 3 & 2 & 3 & 4 \end{matrix}$$

We start with the 3×3 square of 1's occupied by 1's. If $\theta \le 3$ then the four sites marked with 2's have birth rate 1 and will eventually become occupied. Once they do, the eight sites marked 3 have three occupied neighbors and will become occupied. Finally the four sites marked 4 will become occupied. At this point we have shown how the process can fill up $[-2,2]^2$. Repeating the argument, it is easy to see that

$$\lim_{T\to\infty} P(\bar{\xi}_T^{L,\rho}(x) = 1 \text{ for all } x \in [-3,3]^2) = 1$$

the hypothesis of Theorem 4.4 is satisfied and the result follows. □

In the last argument it was important that we used the Moore neighborhood, instead of the usual nearest neighbors $\{z : |z| = 1\}$. If we use the nearest neighbors then, no matter how big L, is if we start with $[-L,L]^2$ occupied nothing happens since any site outside $[-L,L]^2$ has at most one occupied neighbor.

$$\begin{matrix} 0 & 0 & 0 & 0 & 0 \\ 0 & 0 & x & 0 & 0 \\ 1 & 1 & 1 & 1 & 1 \end{matrix}$$

Since births are impossible outside any rectangle containing the 1's in the initial configuration, it is clear that the threshold two birth rate for the nearest neighbors *dies out* whenever the death rate is $c_0(x,\xi) \equiv \epsilon > 0$. That is, if there are only finitely many 1's in ξ_0, then

$$P(\xi_t \not\equiv 0) \to 0 \quad \text{as } t \to \infty$$

Here $\xi_t \equiv 0$ is short for $\xi_t(x) = 0$ for all x. Note that the all 0's state is absorbing so $t \to P(\xi_t \not\equiv 0)$ is decreasing. The opposite of dies out is *survives*. That is, if L is large enough and we start with 1's on $[-L,L]^d$ then

$$\lim_{t\to\infty} P(\xi_t \not\equiv 0) > 0 \quad \text{as } t \to \infty$$

We say that a birth rate is *fertile* if it survives when $c_0(x,\xi) = \epsilon$ and $\epsilon < \epsilon_0$. As the parenthetical phrase in Theorem 4.4 indicates, our sufficient conditions for robustness are also sufficient for fertility.

Having two notions of what it means for birth rates to be large enough, fertility and robustness, it is natural ask what is the relationship between these two notions:

1. Results of Bezuidenhout and Gray imply that increasing birth rates that are fertile are also robust, but the two notions are not equivalent.

2. As we have shown the two dimensional threshold two system using the nearest neighbors is not fertile. However, Bramson and Gray (1991) have shown that it is robust. Intuitively the process cannot grow outside of a rectangle but it is good at filling in holes that develop so it can have a nontrivial stationary distribution when ϵ is small.

In the case of the Moore neighborhood in two dimensions, it is easy to see that the threshold 4 system is not fertile but techniques of Bramson and Gray (1991) can be used to show that it is robust. The threshold 5 system has finite configurations of 0's that cannot be filled in

$$\begin{matrix} & 0 & 0 & \\ 0 & 0 & 0 & 0 \\ 0 & 0 & 0 & 0 \\ & 0 & 0 & \end{matrix}$$

so an easy argument shows that it is not robust. An interesting open problem is to look at the neighborhoods $\mathcal{N} = \{z : \|z\|_p \le r\}$ (or even just take $p = \infty$) and find the largest thresholds for which the threshold θ birth rule on that neighborhood is robust (resp. fertile).

Further results. There are many other results proving the existence of phase transitions for processes with state space $\{0,1\}^S$. Gray and Griffeath (1982) proved a "stability theorem for attractive nearest neighbor spin systems on **Z**" by the contour method, a result which was reproved by the methods of this section by Bramson and Durrett (1988). Gray (1987) proved results for the one dimensional majority vote model. Chen (1992) used ideas from bootstrap percolation to study a model with sexual reproduction. In general the numerical bounds on critical values from this method are terrible but Durrett (1992c) has shown that in some cases you can get good bounds.

Bramson and Neuhauser (1993) studied perturbations of one dimensional cellular automata. Their results are exciting because they apply to a number of examples that are not attractive. An important special case is that if one considers the addition mod 2 automaton:

$$\eta_{n+1}(x) = (\eta_n(x-1) + \eta_n(x+1)) \pmod 2$$

and adds spontaneous deaths at a small rate ϵ then there is a stationary distribution close to product measure with density 1/2. Figure 4.3 shows the cellular automaton starting from a single 1 at 0, which generates a discrete version of the Sierpinski gasket. Figure 4.4 shows what happens when we introduce spontaneous deaths at rate $\epsilon = 0.01$. Note that there are many more occupied sites in the model with extra deaths.

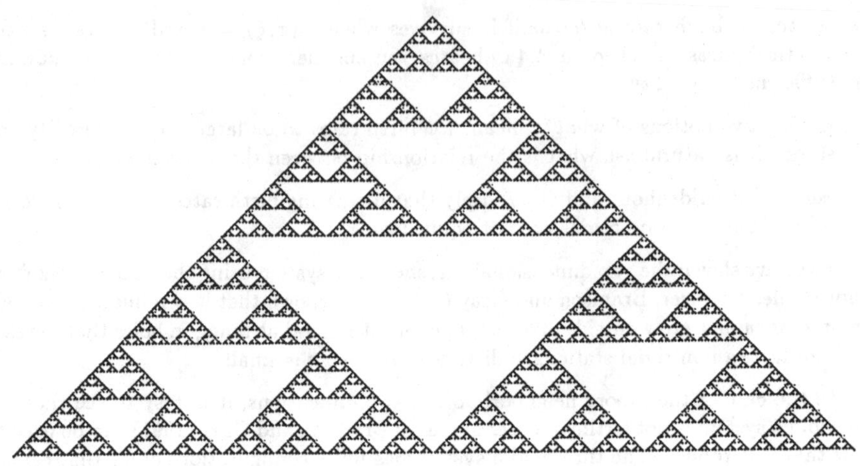

Figure 4.3. Pascal's triangle mod 2

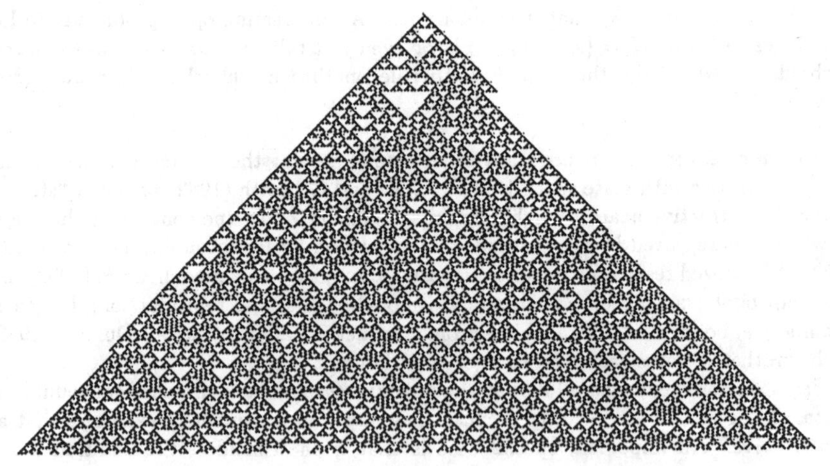

Figure 4.4. Plus spontaneous deaths with probability 0.01

5. Threshold Models

We begin by recalling a definition given in Section 1.

Example 5.1. The threshold voter model. The state space is $\{0,1\}^S$ and the flip rates are

$$c_i(x,\xi) = \begin{cases} 1 & \text{if } n_i(x,\xi) \geq \theta \\ 0 & \text{if } n_i(x,\xi) < \theta \end{cases}$$

Here, as usual, $n_i(x,\xi) = |\{y \in \mathcal{N} : \xi(x+y) = i\}|$ is the number of neighbors of type i and we assume $\mathcal{N} = \{y : \|y\|_p \leq r\}$ for some $1 \leq p \leq \infty$ and $r \geq 1$.

Our first goal is to show that the behavior of the threshold 1 voter model is much different from that of the basic voter model. We begin with one case in which the behavior is the same.

Theorem 5.1. Suppose $d = 1$ and $\mathcal{N} = \{-1,1\}$. Then the threshold 1 voter model clusters starting from any translation invariant initial state ξ_0. That is, for any $x \neq y$ we have $P(\xi_t(x) \neq \xi_t(y)) \to 0$.

PROOF: To motivate the proof, take a look at Figure 5.1 which shows a simulation of the system on $\{0,1,\ldots,719\}$ with periodic boundary conditions (i.e., 0 and 719 are neighbors). The initial configuration at the top of the page is product measure with density 1/2. As we go down the page from time 0 at the top to time 690 at the bottom, it should be clear that intervals of sites with the same opinion can be destroyed but cannot be created. Thus the number of intervals per unit distance will go to 0, i.e., the system clusters.

To turn the last paragraph into a proof, we define a process on $1/2 + \mathbf{Z}$ so that

$$\zeta_t(x) = |\xi_t(x-1/2) - \xi_t(x+1/2)|$$

In words, there is a 1 at x if and only if $\xi_t(x-1/2) \neq \xi_t(x+1/2)$. ζ is called the *boundary process of* ξ since the 1's mark the boundaries between clusters of the voters with the same opinion. To see how ζ evolves consider the following picture

ξ	1		1		0		1		1		0		0
ζ		0		1		1		0		1		0	
x	1	1.5	2	2.5	3	3.5	4	4.5	5	5.5	6	6.5	7

Isolated 1's in ζ like the one at 5.5 perform random walks: the 1 at 5 flips to 0 at rate one and when this occurs the boundary jumps from 5.5 to 4.5; similarly, the 0 at 6 flips to 1 at rate one and when this occurs the boundary jumps from 5.5 to 6.5. When a 1 is adjacent to another 1 (like those at 2.5 and 3.5) they annihilate at rate 1, since when the 0 at 3 flips to 1 the two boundaries disappear.

Let $u(t) = P(\zeta_t(x) = 1)$, which is independent of x since we have supposed that ξ_0 is translation invariant. Since 1's can be destroyed in ζ but cannot be created, it should not be surprising that $u(t) \to 0$ as $t \to \infty$. To prove this, we note that

$$\frac{du}{dt} = -P(\zeta_t(x) = 1, \zeta_t(x-1) = 1) - P(\zeta_t(x) = 1, \zeta_t(x+1) = 1) \tag{5.1}$$

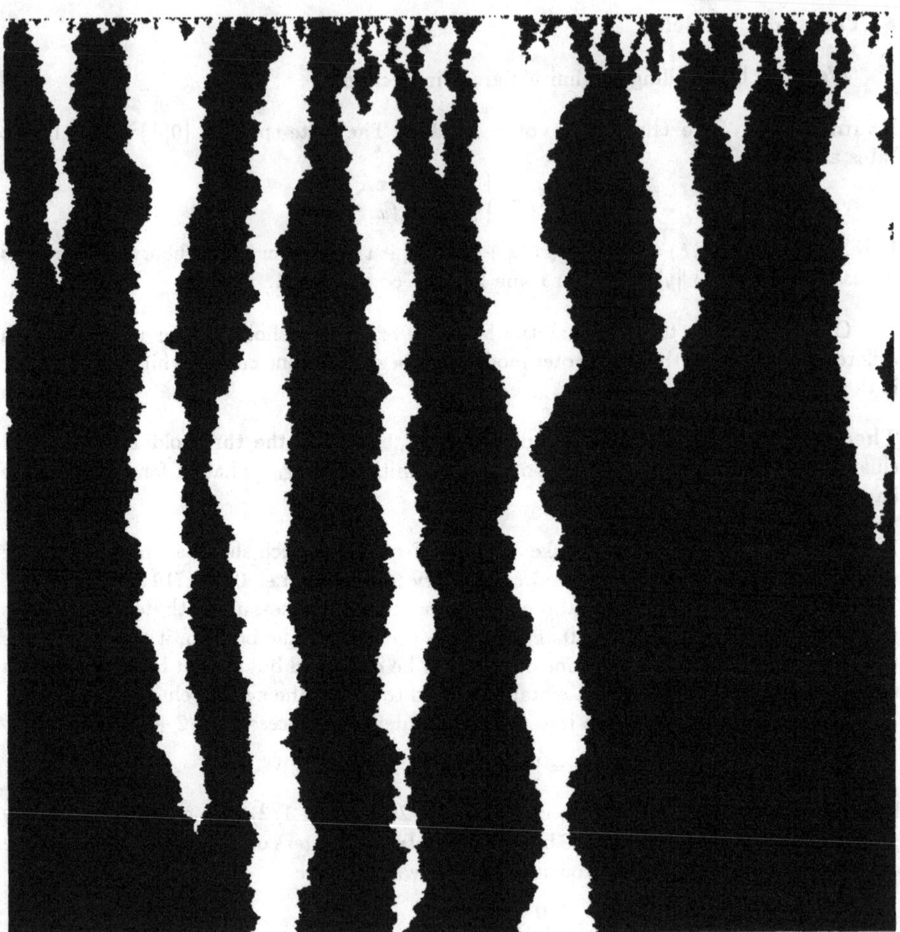

Figure 5.1. One dimensional nearest neighbor threshold voter model.

This can be proved by using $\frac{d}{dt}T_t f = T_t L f$ or more intuitively by noting that the right hand side gives the two ways that a 1 at x can be destroyed. The terms that involve a 1 moving to x or moving away from x cancel.

Translation invariance implies that the right hand side of (5.1) is

$$-2P(\zeta_t(0.5) = 1, \zeta_t(-0.5) = 1) \equiv -v(t)$$

The first step in proving $u(t) \to 0$ is to show that if $t \geq 1$

$$v(t) \geq g(u(t-1)) \text{ where } g(x) > 0 \text{ when } x > 0 \tag{5.2}$$

To do this we note that if $u(s) \geq 1/L$ where L is an integer then

$$P(\zeta_s \text{ has at least two 1's in } (-L, L]) \geq \frac{1}{2L-1} \tag{5.3}$$

for otherwise we get a contradiction

$$2 \leq 2Lu(s) = E \sum_{x \in (-L,L]} \zeta_s(x) < 1 \cdot \frac{2L-2}{2L-1} + 2L \cdot \frac{1}{2L-1} = \frac{4L-2}{2L-1} = 2$$

Now if we have an initial configuration in which there are at least two 1's in $(-L, L]$ there is a probability $\geq \epsilon_L > 0$ that no particles will enter $(-L, L]$ before time 1, the two particles closest to 0 will move to 0.5 and -0.5, and none of the other particles in $(-L, L]$ will move. Combining this observation with (5.3) proves (5.2). To complete the proof of Theorem 5.1 now, we observe that $u(t)$ is decreasing so $u(t) \to u(\infty) \geq 0$ as $\to \infty$. If $u(\infty) > 0$ then for all t we have

$$\frac{du}{dt} = -v(t) \leq -g(u(\infty)) < 0$$

so integrating we find $u(t) \to -\infty$ a contradiction. □

Remark. The argument above applies to any one dimensional nearest neighbor system in which $c_i(x, \xi) = f(n_i(x, \xi))$ with $f(0) = 0$, the so-called *nonlinear voter models*. In the case of the basic voter model, i.e., $f(2) = 2f(1)$ the boundary process is an *annihilating random walk*. That is, particles perform independent random walks until they hit at which time the two particles annihilate. Theorem 3.1 shows that for the basic voter model clustering occurs for any initial configuration. Theorem 4 in Cox and Durrett shows that for the threshold voter model clustering occurs for any initial configuration. We

Conjecture 5.1. In any one dimensional nearest neighbor nonlinear voter model clustering occurs for any initial configuration.

Our next goal is to show that coexistence is possible in the threshold 1 voter model even in one dimension. To do this we will use some ideas from Liggett (1993) to compare with

Example 5.2. The threshold contact process. The state space is $\{0,1\}^S$ and the flip rates are

$$c_1(x,\xi) = \begin{cases} \lambda & \text{if } n_1(x,\xi) \geq \theta \\ 0 & \text{if } n_1(x,\xi) < \theta \end{cases}$$
$$c_0(x,\xi) = 1$$

Here $c_1(x,\xi)$ is the same as in the threshold voter model but we have set $c_0(x,\xi) \equiv 1$.

(5.4) **Lemma.** If the threshold θ contact process with $\lambda = 1$ has a nontrivial stationary distribution then so does the threshold θ voter model.

PROOF: To construct the stationary distribution we will start the threshold voter model ξ from $\nu_{1/2}$, product measure with density 1/2, and compare with the threshold contact process ζ to show that clustering does not occur.

The first step in doing this is to show that the upper invariant measure π for the threshold voter model with $\lambda = 1$ is stochastically smaller than $\nu_{1/2}$ To do this we compare the threshold contact process ζ with the "independent flips process" η_t in which $c_i(x,\eta) \equiv 1$, i.e., each site flips at rate 1 independently of the others. Since sites in η flip to 1 at rate one independent of what is around them, if we start ζ and η with $\zeta_0 = \eta_0$ having distribution π and construct the two processes using the recipe in Section 2 then $\zeta_t(x) \leq \eta_t(x)$ for all t and x. This is true since 1's flip to 0 at rate 1 in both processes while 0's flip to 1 at rate 1 always in η, but at rate 1 in ζ only if there are enough 1 neighbors. On the graphical representation then we find that each flip preserves the inequality and the result can be proved like (2.5).

Now since the sites in η flip independently it is easy to see that as $t \to \infty$ η_t converges to $\nu_{1/2}$. The inequality $\zeta_t(x) \leq \eta_t(x)$ and the fact that ζ_t always has distribution π imply that π is stochastically smaller than $\nu_{1/2}$. To prove this we observe that if f is increasing and depends on only finitely many coordinates then $Ef(\zeta_t) \leq Ef(\eta_t)$ and since any such f is bounded and continuous letting $t \to \infty$ gives

$$\int f(\xi)d\pi \leq \int f(\xi)d\pi$$

checking the definition we gave in (2.11).

Now the result of Holley in the remark (2.12) implies that we can define ξ_0 with distribution $\nu_{1/2}$ and ζ_0 with distribution π, so that that $\xi_0(x) \geq \zeta_0(x)$ for all x. Since sites in ζ flip to 0 at rate one, while those in ξ only flip to 0 at rate one when there are enough 0 neighbors, and the rates of flipping to 1 are the same, if we construct the two processes using the recipe in Section 2 then $\xi_t(x) \geq \zeta_t(x)$ for all x and t. To construct a stationary distribution for ξ, let μ_t be the distribution of ξ_t, form the Cesaro average

$$\bar{\mu}_T = \frac{1}{T}\int_0^T \mu_t dt$$

and let $\bar{\mu}_\infty$ be the limit of a weakly convergent subsequence. It follows from (2.13) that $\bar{\mu}_\infty$ is a stationary distribution. To see that it concentrates on configurations with infinitely many 1's we note that the inequality $\xi_t(x) \geq \zeta_t(x)$ implies that $\bar{\mu}_\infty$ is larger than the upper invariant measure π, which is spatially ergodic by (2.15) and hence concentrates on configurations with infinitely many 1's. To see that $\bar{\mu}_\infty$ concentrates on configurations with infinitely many 0's, note that the initial distribution $\nu_{1/2}$ and the threshold voter model are symmetric under the interchange of 0's and 1's, so the limit measure $\bar{\mu}_\infty$ is as well. □

Liggett (1993) has shown

Theorem 5.2. When $d = 1$ and $\mathcal{N} = \{-2, -1, 1, 2\}$ or $d = 2$ and $\mathcal{N} = \{y : \|y\|_1 = 1\}$ the threshold 1 contact process with $\lambda = 1$ has a nontrivial stationary distribution.

Since enlarging the neighborhood $\mathcal{N}$ makes it easier for the threshold 1 contact process to have a nontrivial stationary distribution, it follows from (5.4) and Theorem 5.2 that

Theorem 5.3. Suppose $\mathcal{N} = \{z : \|z\|_p \leq r\}$ with $1 \leq p \leq \infty$ and $r \geq 1$. With the exception of the one dimensional nearest neighbor case, the threshold one voter model always has a nontrivial stationary distribution.

By another comparison argument Liggett shows that to prove Theorem 5.3 it is enough to consider the case $d = 1$ and $\mathcal{N} = \{-2, -1, 1, 2\}$ - map $\mathbf{Z}^2$ to $\mathbf{Z}$ by $(x, y) \to x + 2y$ and notice that the image of the two dimensional threshold contact process dominates the one dimensional one. A simulation of the case $d = 1$ and $\mathcal{N} = \{-2, -1, 1, 2\}$ given in Figure 5.2, which parallels the one for the nearest neighbor case in Figure 5.1, makes it clear that Theorem 5.3 is true. However, the proof of Theorem 5.2 (which implies 5.3) requires a tricky generalization of the result Holley and Liggett (1978) that the one dimensional nearest neighbor contact process has $\lambda_c \leq 2$. Therefore we content ourselves to prove less (and more).

Theorem 5.4. Suppose $\mathcal{N} = \{y : \|y\|_p \leq r\}$ with $r \geq 1$. For any threshold θ if $r \geq r_0(d, \theta)$ then there is a nontrivial stationary distribution for threshold θ contact process with $\lambda = 1$ and hence also for the threshold θ voter model.

PROOF: We will use the comparison theorem from Section 4. To do this, it is convenient to suppose that ξ has been constructed from a percolation substructure with rate 1 Poisson processes $\{T_n^x, n \geq 1\}$ at which times we draw arrows from $y + x$ to x for all $y \in \mathcal{N}$, and rate 1 Poisson processes $\{U_n^x, n \geq 1\}$ at which times we write a δ at x.

Exercise. This shows that the threshold contact process can be constructed from a percolation substructure defined in Section 3. What is the dual process?

Suppose $r = (2d + 2)L$. To check the comparison assumptions, let H be the configurations that have at least θ 1's in $[-L, L]^d$. Let $\gamma > 0$. If T is small enough then the probability that $U_1^x > T$ for all of our θ 1's, is $e^{-\theta T} > 1 - \gamma/5$. Now since $r = (2d + 2)L$, the neighborhood of each site in $I_1 = [L, 3L] \times [-L, L]^{d-1}$ contains all the sites in $[-L, L]^d$

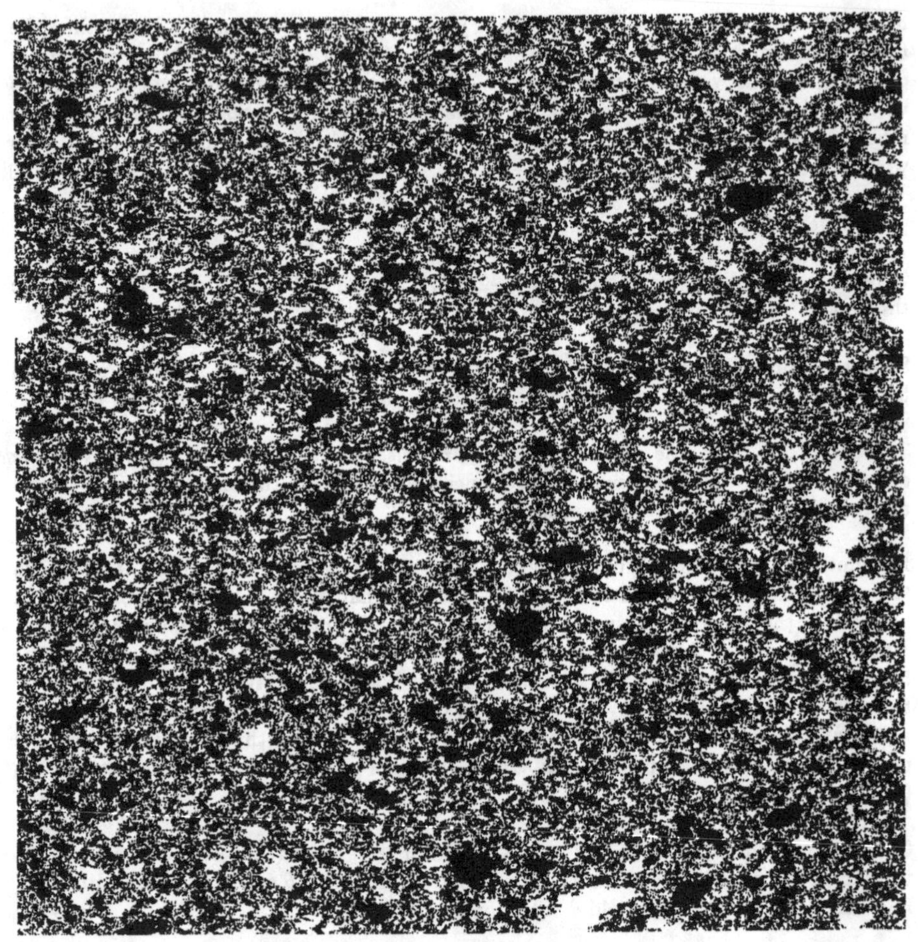

Figure 5.2. One dimensional threshold voter model, range two.

(distances are largest for the L^1 norm and for the points $(3L, L, \dots, L)$ and $(-L, L, \dots, L)$). Now as long as there are at least θ 1's in $[-L, L]^d$, each site in $[L, 3L]$ will flip to 1 at rate 1. If r and hence L is sufficiently large then with probability at least $1-\gamma/5$ at least θ sites will flip to 1 by time T. A similar remark applies to the sites in $I_{-1} = [-3L, -L] \times [-L, L]^{d-1}$, and our first estimate implies that in each case the probability one of our θ 1's flips back to 0 by time T is $\leq \gamma/5$.

The results in the last paragraph show that if we start with θ 1's in $I_0 = [-L, L]^d$ then with probability at least $1-\gamma$ there will be at least θ 1's in I_1 and in I_{-1} at time T. Our good event is measurable with respect to the graphical representation in $[-3L, 3L]^d$ so we have checked the comparison assumptions of Section 4 with $k_0 = 3$ and $j_0 = 1$. If we start the threshold contact process with all sites occupied then Theorem 4.3 implies our process dominates an oriented percolation starting with all sites wet, so Theorem 4.2 shows

$$\liminf_{n\to\infty} P(0 \in X_{2n}) \geq 19/20$$

Now $0 \in X_{2n}$ means that there are at least θ 1's in $[-L, L]^d$ at time $2nT$ and ξ_{2nT} is translation invariant so it follows that

$$\liminf_{n\to\infty} P(\xi_{2nT}(0) = 1) = \liminf_{n\to\infty} \frac{1}{(2L+1)^d} \sum_{x\in[-L,L]^d} P(\xi_{2nT}(x) = 1)$$
$$\geq \frac{1}{(2L+1)^d} \cdot \theta \cdot \frac{19}{20} > 0$$

To pass from this result to the whole sequence we notice that since a 1 survives for t units of time with probability e^{-t}, $P(\xi_{2nT+t}(0) = 1) \geq e^{-t} P(\xi_{2nT}(0) = 1)$. Combined with the last result this implies

$$\liminf_{T\to\infty} \frac{1}{T} \int_0^T P(\xi_s(0) = 1)\, ds > 0$$

and it follows from (2.13) that there is a nontrivial stationary distribution. □

The last result shows that if the threshold is small compared to the number of neighbors then *coexistence* occurs in the threshold voter model, i.e. there is a stationary distribution that concentrates on configurations with infinitely 1's and infinitely many 0's. Our next result due to Durrett and Steif (1993) shows that if the threshold is too large the system *fixates*, i.e., with probability one each site changes its state only finitely many times.

Theorem 5.5. Suppose $\mathcal{N} = \{y : \|y\|_p \leq r\}$. If $\theta > (|\mathcal{N}| - 1)/2$ then the system fixates.

The borderline case in this result, $\theta = (|\mathcal{N}|+1)/2$ ($|\mathcal{N}|$ is always odd), is called *the majority vote process*, since you change your mind if you are in the minority in your neighborhood.

PROOF: Our proof is based on an idea of Grannan and Swindle. Let $\delta_{x,y}(t)$ be 1 if $\xi_t(x) \neq \xi_t(y)$, 0 otherwise, and define the *energy* at time t to be

$$\mathcal{E}_t = \sum_{x,y:y-x\in\mathcal{N}} e^{-\epsilon\|x+y\|_2} \delta_{x,y}(t)$$

where $\epsilon > 0$ is to be chosen later. Since $0 \le \mathcal{E}_0 < \infty$, we can prove Theorem 5.5 by showing

(5.5) If $\theta > (|\mathcal{N}| - 1)/2$ and ϵ is small then a flip at x decreases the energy by at least $\gamma(x) > 0$.

To prove (5.5) we note that if $\alpha = |\{y \in x+\mathcal{N} : \xi_t(y) \neq \xi_t(x)\}|$ and $N = \sup\{\|x\|_2 : x \in \mathcal{N}\}$ then the drop in energy due to a flip at x is at least

$$e^{-2\epsilon\|x\|_2}\left[e^{-\epsilon N}\alpha - e^{\epsilon N}(|\mathcal{N}| - 1 - \alpha)\right] \tag{5.6}$$

since (i) the site x now agrees with the α sites it used to disagree with and now disagrees with the other $|\mathcal{N}| - 1 - \alpha$ neighbors and (ii) even in the worst case all the points in $\{y \in x + \mathcal{N} : \xi_t(y) \neq \xi_t(x)\}$ have $\|x + y\|_2 \le 2\|x\|_2 + N$ and the other points $y \in x + \mathcal{N}$ have $\|x+y\|_2 \ge 2\|x\|_2 - N$. In order for a flip to occur we must have $\alpha \ge \theta > (|\mathcal{N}| - 1)/2$ and hence $|\mathcal{N}| - 1 - \alpha < \alpha$. Since the last two number are integers smaller than $|\mathcal{N}|$, (5.5) follows from (5.6). □

Refinements of Theorem 5.4. Before we stated Theorem 5.5, we said "if the threshold is small compared to the number of neighbors" then the threshold contact process with $\lambda = 1$ has a nontrivial stationary distribution (and hence there is coexistence in the threshold voter model). What we would like to concentrate on now is:

How large can θ be when the range is r?

The comparison theorem involves obnoxiously small constants (when $M = 1$ Theorems 4.1 and 4.2 require $\gamma \le 6^{-100}$). So we cannot hope to get a nontrivial result for $r = 10$, or even $r = 10{,}000$, but it is not unreasonable to look at how θ behaves asymptotically with r. The results were are about to give foreshadow the developments in the next section, but are not needed for them, or for any subsequent section, and can be skipped without loss.

Here and until the end of the section we suppose $\mathcal{N} = \{z : \|z\|_p \le r\}$, let $N = |\mathcal{N}|$, and we investigate what happens for fixed p as $r \to \infty$ First let's see what we get when we follow the proof of Theorem 5.4.

(5.7) There is a $c_p > 0$ so that if $\theta \le c_p\sqrt{N}$ and if r (and hence N) is large then the threshold θ contact process with $\lambda = 1$ has a nontrivial translation invariant stationary distribution.

PROOF: Taking $T = \gamma/5\theta$ gives $e^{-\theta T} = e^{-\gamma/5} \ge 1 - \gamma/5$. Having fixed the time, the number of sites in $[L, 3L] \times [-L, L]^{d-1}$ that flip to 1 by time T has a binomial distribution with parameters $n = (2L+1)^d$ and $p = 1 - e^{-T} \ge \gamma/6\theta$ when θ is large. If we let Z be the number of sites in $[L, 3L] \times [-L, L]^{d-1}$ that flip to 1 by time T then Z has mean $\ge (2L+1)^d\gamma/6\theta$ and variance $\le (2L+1)^d\gamma/6\theta$ so if we set $(2L+1)^d\gamma/6\theta = 2\theta$ (sticklers for details should take the smallest integer L so that $\ge$ holds) Chebyshev's inequality implies that

$$P(Z \le \theta) \le \frac{(2L+1)^d\gamma/6\theta}{\theta^2} \le \frac{2}{\theta} \to 0$$

as $\theta \to \infty$. Now $\theta^2 = (2L+1)^d\gamma/12 \geq c_pN$ since $r = (2d+2)L$ and the result follows. □

By choosing a more intelligent block event we can get

(5.8) There is a $c_p > 0$ so that if $\theta \leq c_pN$ and if r (and hence N) is large then the threshold θ contact process with $\lambda = 1$ has a nontrivial translation invariant stationary distribution.

PROOF: Let $\theta = (2L+1)^d/5$ and let H be the configurations that have at least $(2L+1)^d/4$ 1's in $[-L,L]^d$. If we pick $r = (2d+2)L$ then $\theta \geq c_pN$ for all r and as long as there are at least θ 1's in $[-L,L]^d$ the number of 1's in $[-L,L]^d$ (or in $[L,3L] \times [-L,L]^{d-1}$), behaves like a Markov chain that jumps $k \to k+1$ at rate $(2L+1)^d - k$ and $k \to k-1$ at rate k. Now when $k \leq (2L+1)^d/3$ this chain jumps at rate $(2L+1)^d$ moving up with probability at least 2/3 and down with probability at most 1/3. A comparison with asymmetric simple random walk shows

(i) with high probability it will take a long time (i.e., at least $e^{c(2L+1)^d}$ for some $c > 0$) for the total number of 1's in $[-L,L]^d$ to go below θ

(ii) we can pick a large time T (that is independent of L) so that if L is large then with high probability the number of 1's in $[L,3L] \times [-L,L]^{d-1}$ and in $[-3L,-L] \times [-L,L]^{d-1}$ at time T will be at least $(2L+1)^d/4$

We leave it to the reader to fill in the missing details since we know how to prove a sharp result:

(5.9) Let $c < 1/4$. If $\theta \leq cN$ and if r (and hence N) is large then the threshold θ contact process with $\lambda = 1$ has a nontrivial translation invariant stationary distribution.
Let $c > 1/4$. If $\theta \geq cN$ and if r (and hence N) is large then the threshold θ contact process with $\lambda = 1$ has only the trivial stationary distribution.

The proof of the first conclusion is closely related to that of Theorem 6.1. For details and the proof of the converse see Durrett (1992).

6. Cyclic Models

As already suggested by our remarks on refinements in the last section, we can considerably close the gap between Theorems 5.4 and 5.5 if we look at systems with large range. The proof of our main result, Theorem 6.1, is no harder for a class of models that includes a multicolor version of the threshold voter model, so we formulate the result in that generality.

Example 6.1. Cyclic Color Model. The states of each site are $\{0, 1, \dots, \kappa - 1\}$ and the flip rates are

$$c_i(x, \xi) = \begin{cases} 1 & \text{if } \xi(x) = i - 1 \text{ and } n_i(x, \xi) \geq \theta \\ 0 & \text{otherwise} \end{cases}$$

Here and throughout this section, arithemtic is done modulo κ so $0 - 1 = \kappa - 1$. When $\kappa = 2$ the last definition reduces to the threshold voter model. The dynamics here were invented by David Griffeath as a generalization of the voter model. The cyclic color model is closely related to the hypercycle of evolutionary biology. See Eigen and Schuster (1979) and Boerlijst and Hogeweg (1991).

Our main result also applies to two other examples

Example 6.2. Greenberg Hastings Model. The states of each site are $\{0, 1, \dots, \kappa - 1\}$ and the flip rates are

$$\begin{aligned} c_1(x, \xi) &= 1 \qquad \text{if } \xi(x) = 0 \text{ and } n_1(x, \xi) \geq \theta \\ c_i(x, \xi) &= 1 \qquad \text{if } \xi(x) = i - 1 \end{aligned}$$

In words, we need an above threshold number of 1's to make the transition from $0 \to 1$ but then the rest of the transitions happen at rate 1. When $\kappa = 2$ this reduces to the threshold contact process with $\lambda = 1$.

Example 6.3. Host Parasitoid Interactions. Insect parasitoids lay their eggs on or in the bodies of other arthropods, and the parasitoid larvae kill their host as they feed on it. Hassell, Comins, and May (1991) introduced a cellular automaton model for this system. The corresponding particle system model has nine states $\{0, 1, \dots 8\}$ and makes transitions as follows:

$$\begin{aligned} c_1(x, \xi) &= 1 \qquad \text{if } \xi(x) = 0 \text{ and } n_1(x, \xi) \geq \theta \\ c_4(x, \xi) &= 1 \qquad \text{if } \xi(x) = 3 \text{ and } n_5(x, \xi) \geq \theta \\ c_i(x, \xi) &= 1 \qquad \text{if } i \neq 1, 4 \text{ and } \xi(x) = i - 1 \end{aligned}$$

As they explain on page 256, the first transition corresponds to colonization of empty sites (state 0) by the host, the second to a mature parasitoid (state 5) colonizing a mature host (state 3), and the others to the aging and/or death of host and parasitoid.

To indicate what common features of the last three models are needed to apply Theorem 6.1, we say that ξ is a *cyclic model* if the states of each site are $\{0, 1, \dots, \kappa - 1\}$ and makes transitions as follows:

$$c_i(x, \xi) = 1 \qquad \text{if } \xi(x) = i \text{ and } n_{g(i)}(x, \xi) \geq \theta_i$$

Here $g(i) \in \{0, 1, \ldots, \kappa - 1\}$ and we set $\theta_i = 0$ if the transition happens at rate 1 independent of the states of the neighbors. Let $\theta = \max_i \theta_i$.

Theorem 6.1. Let $\epsilon > 0$ and suppose $\theta \leq (1-\epsilon)|\mathcal{N}|/2\kappa$. If $r \geq R_\epsilon$ then there is a stationary distribution close to the uniform product measure.

Recall that we suppose $\mathcal{N} = \{y : \|y\|_p \leq r\}$ and that the uniform product measure is the one in which the coordinates are independent and have $P(\xi(x) = i) = 1/\kappa$. When $\kappa = 2$ this says that for thresholds $a|\mathcal{N}|$ with $a < 1/4$ there is coexistence for large r. (This result was stated in (5.9).) In contrast Theorem 5.4 says that when $a \geq 1/2$ the system fixates for any r. We

Conjecture 6.1. When $\theta = a|\mathcal{N}|$ in the threshold voter model and $1/4 < a < 1/2$, clustering occurs for large r.

We will explain our reasons after we give the proof. Theorem 6.1 concentrates on the behavior for large range. For results about the one dimensional cyclic color model, see Bramson and Griffeath (1987) (1989), or for a treatment of the corresponding cellular automaton, see Fisch (1990a), (1990b), (1991).

PROOF IN $d = 1$: Let $a = \theta/|\mathcal{N}|$. By assumption $a \leq (1-\epsilon)/2\kappa$. Pick $\beta \in (0, \epsilon/4]$ so that $B = 1/\beta$ is an integer, pick $\rho < \sigma < 1/\kappa$ so that $(1-2\beta)\rho \geq (1-\epsilon)/\kappa$, then pick r large enough so that

$$(1-\beta)\rho \cdot \frac{2r}{2r+1} \geq (1-\epsilon)/\kappa$$

Let $K = \beta r$ and note that $BK = r$. For each $m \in \mathbf{Z}$, we call $[mK, (m+1)K)$ a *house*. We say that a house is *good* at time 0 if it contains at least σK sites in each of the states $0, 1, \ldots, \kappa-1$. We say that the interval $[-r, r)$ is *good* at time 0 if all the houses it contains are good. This will be our event H when we apply Theorem 4.3.

We have chosen our constants so that as long as each house in $[-r, r)$ is *reasonable* i.e., contains at least ρK sites of each color, each site in $[-r-K, r+K)$ will see at least θ sites of each color. To check this, note that the worst case occurs when $x \in [r, r+K)$, but even in this case all the sites in $[K, r)$ are in its neighborhood and if all of the houses in $[K, r)$ are reasonable then the number of sites of a given color in x's neighborhood will be at least

$$\begin{aligned} \rho(r-K) = \rho r(1-\beta) &= \frac{\rho(1-\beta)}{2} \cdot \frac{2r}{2r+1} \cdot 2r+1 \\ &\geq \frac{(1-\epsilon)}{2\kappa} \cdot (2r+1) \geq \theta \end{aligned}$$

So as long as each house in $[-r, r)$ stays reasonable, the sites in $[-r-K, r+K)$ flip from i to $i+1$ at rate 1 (here $(\kappa-1)+1 = 0$) and hence behave like independent Markov chains. These "single site" Markov chains are irreducible on $\{0, 1, \ldots, \kappa-1\}$ and hence converge to the equilibirum distribution, which assigns probability $1/\kappa$ to each state. Let $p_t(i, j)$ be the transition probability of the single site Markov chain, let $\sigma' \in (\sigma, 1/\kappa)$ and pick S so that $p_S(0,i) \geq \sigma'$ for all i. Let $T = 2BS$. By using a simple large deviations result (see (6.2) below) it is easy to show that with high probability

(a) All the houses in $[-r, r)$ stay reasonable until time T.

(b) The houses $[r + (j-1)K, r + jK)$ and $[-r - jK, -r - (j-1)K)$ will be good at time jS and stay reasonable to time T.

(c) All the houses in $[r, 3r)$ and $[-3r, -r)$ will be good at time T.

Figure 6.1 gives a picture of this expansion. The gray shaded area gives the space time region occupied by reasonable houses.

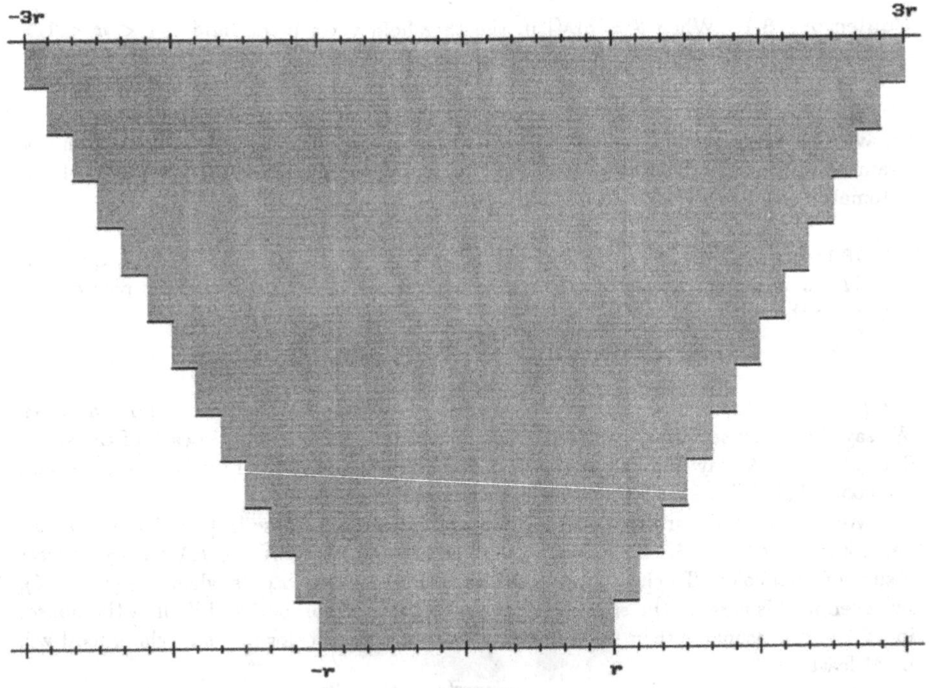

Figure 6.1

Since our good event is measurable with respect to the Poisson processes in $[-3r, 3r) \times [0, T)$ we have verified the comparison assumptions with $L = r$, $K = 3$, $J = 1$. If we start our cyclic system from uniform product measure then X_0 is a Bernoulli set with density $p > 0$. (p is close to 1 if L is large but we do not need that.) Applying Theorems 4.3 and 4.2 now it follows that

$$\liminf_{n \to \infty} P(0 \in X_n) \geq 19/20$$

Arguing as in the end of the proof of Theorem 5.4 it is easy to improve this conclusion to

$$\liminf_{n\to\infty} P(\text{ all } \kappa \text{ colors are in } [-r,r)) > 0$$

and it follows from (2.13) that there is anontrivial stationary distribution. By using an improvement of Theorem 4.2 given in the appendix (see Theorem A.3)

(6.1) **Lemma.** If $p > 0$ and $\gamma \le 6^{-4(2M+1)^2}$ then

$$\liminf_{n\to\infty} P(\{-2K,\ldots,2K\} \cap W_{2n}^p \neq \emptyset) \ge 1 - \epsilon_K$$

where $\epsilon_K \to 0$ as $K \to \infty$

we can show that the stationary distribution we constructed concentrates on configurations in which there are infinitely many sites in each state. ((6.1) shows directly that with probability one each state appears somewhere in the configuration, but the distribution is stationary so if there were only finitely many sites in some state we would have positive probability of having 0 in that state a contradiction.)

By the arguments in the last paragraph it is enough to show that (a), (b), and (c) hold. The first step is proving the large deviations estimate.

(6.2) **Lemma.** *Let* $X_1,\ldots X_n$ *be i.i.d. with* $P(X_i = 1) = p$, $P(X_i = 0) = 1 - p$. *Then*

$$P(X_1 + \ldots + X_n \le n(p - \epsilon)) \le \exp(-\epsilon^2 n/2).$$

Remark. This result and its proof are standard but we need to know that the right hand side does not depend on p.

Proof: If $\alpha > 0$ then

$$P(X_1 + \ldots + X_n \le n(p - \epsilon))\, e^{-\alpha n(p-\epsilon)} \le (pe^{-\alpha} + (1-p))^n$$

Taking log's, dividing by n, rearranging and then using $\log(1 + x) \le x$ we have

$$\begin{aligned}\frac{1}{n}\log P(X_1 + \ldots + X_n \le n(p-\epsilon)) &\le \alpha(p-\epsilon) + \log(1 + p(e^{-\alpha} - 1))\\ &\le \alpha(p-\epsilon) + p(e^{-\alpha} - 1) = -\alpha\epsilon + p(e^{-\alpha} - 1 + \alpha)\end{aligned}$$

Now $e^{-\alpha} - 1 + \alpha = \alpha^2/2! - \alpha^3/3! + \ldots \le \alpha^2/2$ for $0 < \alpha < 1$, so taking $\alpha = \epsilon$ and using $p \le 1$ gives

$$P(X_1 + \ldots + X_n \le n(p - \epsilon)) \le \exp(-\epsilon^2 n/2)$$

and completes the proof of (6.2). □

Let Z_t be a copy of the single site Markov chain, let $p_t(i,j) = P_i(Z_t = j)$, and observe that $p_t(i,j) = p_t(0, j - i)$. Until the first time some house in $[-r,r)$ becomes unreasonable, the sites in each house in $[-r,r)$ behave like independent copies of the single site Markov chain so we consider a collection of $K = r\beta$ independent copies of Z_t and let v_i be the number of "sites" in state i at time 0. The expected number of sites in state j at time t is $w_j(t) = \sum_i v_i p_t(i,j)$. To prove (a) we apply (6.2) with $n = v_i \geq \sigma K$ to the sites that start in state i to see that with probability at least $1 - \exp(-\epsilon^2 \sigma K/2)$, at least $v_i(p_t(i,j) - \epsilon)$ of the sites that start in state i will be in state j at time t. Taking $\epsilon = (\sigma - \rho)$ and summing over i gives

$$\sum_i v_i(p_t(i,j) - \epsilon) \geq \sigma K \sum_i p_t(i,j) - K\epsilon \geq (\sigma - \epsilon)K = \rho K$$

since $\sum v_i = K$ and $\sum_i p_t(i,j) = \sum_i p_t(0,j-i) = 1$. So with probability at least $1 - \kappa \exp(-\epsilon^2 \sigma K/2)$, at least ρK sites will be in state j at time t.

The last bound is for a fixed time but it is easy to extend it to cover the interval $[0,T]$. Let $\delta = \epsilon^2 \sigma/2$, let $J = \exp(\delta K/2)$, and $t_k = k/J$ for $1 \leq k \leq JT$. The probability that the number of sites in state i is less than ρK at some time t_k is at most

$$\kappa JT \exp(-\epsilon^2 \sigma K/2) = \kappa T \exp(-\delta K/2)$$

The probability that two sites flip in some interval (t_{k-1}, t_k) is at most

$$JT \binom{K}{2} J^{-2} \leq K^3 T \exp(-\delta K/2).$$

When we never have two flips in any interval, the state at each $t \in (t_{k-1}, t_k)$ agrees with the state at one of the two endpoints. Combining the last two estimates we have that the probability a collection of K independent single site chains becomes unreasonable before time T

$$\leq (\kappa + K^3) T \exp(-\delta K/2)$$

Since the sites in $[-r,r)$ behave like independent single site chains until some house becomes unreasonable, the probability of the event in (a) is at least

$$1 - 2B(\kappa + K^3) T \exp(-\delta K/2)$$

The proof that the house $[r, r+K)$ will be good at time S is similar but simpler. If all the houses in $[-r,r)$ stay reasonable until time S then each site in $[r, r+K)$ always sees an above threshold number of sites of each color and flips to the next color at rate 1. We again consider a collection of K independent single site chains but this time starting from an arbitrary initial configuration. The choice of S guarantees that $p_S(i,j) \geq \sigma'$ so applying (6.2) to K i.i.d. random variables with $p = \sigma'$ we conclude that the fraction of sites in state j is at least σK with probability at least $1 - \kappa \exp(-(\sigma' - \sigma)^2 K/2)$. Once we know that with high probability $[r, r+K)$ is good at time S and all the houses in $[-r,r)$ are reasonable at all times in $[0,T]$, we can repeat the proof of (a) to conclude that the house $[r, r+K)$ stays reasonable at all times in $[S,T]$. This verifies (b) when $j = 1$ but by continuing in the same way we can prove the result for $2 \leq j \leq 2B$. Now (b) implies that all the houses in $[-3r+K, 3r-K)$ are reasonable at time $T-S$ we can repeat the proof that the house $[r, r+K)$ is good at time S to conclude that all the houses in $[-3r, 3r)$ are good at time T and the proof is complete. □

PROOF IN $d>1$: Let $B_p(x,r) = \{y : \|x-y\|_p \le r\}$. The key to the proof is the following fact, which basically says that large balls are almost flat.

(6.3) Lemma. *Suppose $\lambda < 1/2$. There are constants R_0, δ, and M_0, so that if $M \ge M_0$ and $R \ge R_0$ then for $x \in B_2(0,(R+\delta)M)$.*

$$|B_2(0,RM) \cap B_p(x,M)| \ge \lambda |B_p(x,M)|$$

PROOF: In one dimension we can take $R_0 = 1$ and $\delta = 1-2\lambda$. Turning to dimensions $d>1$, let $Q = \{x \in R^d : \|x\|_p \le 1\}$ and let q be its volume. To prove the result it is convenient to scale space by $1/M$ and translate so that x/M sits at the origin. Any $d-1$ dimensional hyperplane through the origin divides Q into two pieces with volume $q/2$. For $i=1,2,3$ let $\lambda < \lambda_3 < \lambda_2 < \lambda_1 < 1/2$. By continuity, there is a $\delta > 0$ so that if a hyperplane passes within a distance δ of the origin then it divides Q into two pieces each of which has volume at least $q\lambda_1$. Another application of continuity shows that if R_0 is large and $D = B_2(y,r)$ with $r \ge R_0$ and $B_2(y,r) \cap B_2(0,\Delta) \ne \emptyset$ then the volume of $D \cap Q$ is at least $q\lambda_2$.

The last step is to argue that if M is large then the lattice behaves like the "continuum limit" considered above. Pick $\epsilon > 0$ so that if $D = B_2(y,r)$ is as above then the volume of $B_2(y, r-\epsilon) \cap (1-\epsilon)Q$ is always larger than $q\lambda_3$. Then pick M_0 so that $1/M_0 < \epsilon$ and if $M \ge M_0$ then $|B_p(0,M)|/qM^d < \lambda_3/\lambda$. Let $\mathcal{X} = (Z^d/M) \cap D \cap Q$. The first part of the choice of M_0 implies that if $M \ge M_0$ then

$$B_2(y, r-\epsilon) \cap Q(1-\epsilon) \subset \cup_{x \in \mathcal{X}}\, x + \left[\frac{-1}{2M}, \frac{1}{2M}\right]^d$$

so $M^{-d}|\mathcal{X}| \ge q\lambda_3 \ge \lambda |B_p(0,M)| M^{-d}$, by the second part of the choice of M_0 and the proof is complete. □

To use this lemma we pick $\lambda < 1/2$ and $\rho < 1/\kappa$ so that $\lambda\rho > a$, use (6.3) to pick R_0, Δ, M_0, and then pick $M_1 \ge M_0$ so that

$$\lambda\rho K^d |B_p(0,M_1)| > a|B_p(0, K(M_1+d))| \quad \text{holds for large } K \tag{6.4}$$

Let $\sigma \in (\rho, 1/\kappa)$ and suppose that the range of interaction is $r = K(M_1+d)$. For $z \in \mathbf{Z}^d$ let

$$I_z = [z_1 K, (z_1+1)K) \times \cdots \times [z_d K, (z_d+1)K)$$

and call I_z a *house*. We say that a house is *good* at time 0 if it contains at least σK^d sites of each color. We say that ξ_0 is good if all the houses I_z, $z \in B_2(0, R_0 M_1)$ are good. This will be our event H when we apply Theorem 4.3.

We have set things up so that as long as each house in $B_2(0, R_0M_1)$ is *reasonable*, i.e. contains at least ρK^d sites of each color, each site in each house in $B_2(0,(R_0+\delta)M_1)$ sees at least θ sites of each color. To check this note if $z \in B_2(0,(R_0+\delta)M_1)$ then all the sites in any house I_w with $w \in B_2(0,R_0M_1) \cap B_p(z, M_1)$ are within p-norm distance $r = (M_1+d)K$ of each site in I_z. (To see note that $\|z-w\|_p \le M_1$ so $\|zK - wK\|_p \le M_1K$ and if we use 1 to denote a vector of 1's then $\|zK-(w+1)K\|_p \le (M_1+d)K$, with $p=1$ being the worst case.) By (6.3)

$$|B_2(0,R_0M_1)\cap B_p(x,M_1)| \ge \lambda |B_p(x,M_1)|$$

Multiplying the last inequality by ρK^d and using the choice of M_1 and K in (6.4) that

$$\begin{aligned}\rho K^d|B_2(0,R_0M_1)\cap B_p(x,M_1)| &\ge \lambda\rho K^d |B_p(x,M)| \\ &\ge a|B_p(x,K(M_1+d)| = \theta\end{aligned}$$

Pick B so that $B\delta > 2R_0$ and hence

$$B_p(x,(R_0+B\delta)M_1) \supset B_p(x, 3R_0M_1)$$

Let $\sigma' \in (\sigma, 1/\kappa)$, choose S so that $p_S(0,i) \ge \sigma'$ for all i, and let $T = BS$. Let $D_j = B_2(0,(R_0+j\delta))$ (D is for disk) and $A_j = D_j - D_{j-1}$ (A is for annulus). By repeating the one dimensional proof we can show that with high probability

(a) All the houses in D_0 stay reasonable until time T.

(b) The houses in A_j will be good at time jS and stay reasonable to time T.

(c) All the houses in $D_B \supset B_2(0,3R_0)$ will be good at time T.

and the desired result follows from an application of Theorems 4.3 and 4.2 as before. □

We will now give the promised explanation of the conjecture for the case $\kappa = 2$. First consider the situation in $d=1$ and for ease of exposition call the two states "yellow" and "blue". As our proof shows if we have a sufficiently large interval of sites in which two colors occur with approximately equal frequency then the distribution of colors in this region will quickly converge to a product measure with density 1/2 and the region will expand, no matter what it encounters outside. For the region to expand we need $\theta = a|\mathcal{N}|$ with $a < 1/4$ for if $a > 1/4$ and all sites in $[r, 2r)$ are yellow then the random region cannot expand since the site at r will have about $r/2 < a(2r+1)$ blue sites in its neighborhood. Applying the same reasoning to yellow sites in $x \in [br, r)$, who have about $(2-b)r/2$ blue sites in their neighborhood, we see that if $(2-b)/2 < 2a$, i.e. $b > 2-4a$ then the yellow sites in $[br, r)$ will not flip to blue but since $a < 1/2$ the blue sites will flip to yellow at rate one.

Similar reasoning applies to the system in $d > 1$ with $1/4 < a < 1/2$ and shows that a large enough ball of yellow sites will expand through a random region. The trouble with turning this into a proof is that we cannot guarantee that the blob will always find itself in competion with a random region. Indeed in a deterministic version of the threshold voter model in $d = 1$ (see Durrett and Steif (1993)) this naive picture is not correct since there are "blockades" that in some circumstances will stop the advance of blobs. However, we believe that this will not happen in random systems or in $d > 1$. In support of this claim, we note that Andjel, Mountford, and Liggett (1992) have shown that clustering occurs in $d = 1$ when $\mathcal{N} = \{-k, \dots, k\}$ and $\theta = k$. The important special property of this example is that if an interval of 1's (or 0's) is long enough only the site on either end can flip.

7. Long Range Limits

In the last section, we saw that the cyclic color model and Greenberg Hastings models simplified considerably when the range of interaction was large. In this section we show that the contact process also simplifies in this way.

Example 7.1. The basic contact process. As usual the neighborhood is $\mathcal{N} = \{x : \|x\|_p \le r\}$. We will write the contact process as a set valued process with the state = the set of sites occupied by particles and formulate the dynamics as follows:

(i) Each particle dies at rate 1, and gives birth at rate β.

(ii) A particle born at x is sent to a site y chosen at random from $x + \mathcal{N}$.

(iii) If y is vacant, it becomes occupied. If y is already occupied the birth has no effect.

If r is large and the contact process starts from a single occupied site then at least until the number of particles is a significant fraction of $|\mathcal{N}|$, the contact process will behave like a *branching random walk*, i.e., the process that obeys (i) and (ii) but allows any number of particles per site.

The total number of particles at time t in a branching random walk is a *branching process* – a Markov chain Z_t in which transitions from k to $k+1$ occur at rate $k\beta$ and transitions k to $k-1$ occur at rate k. Let $T_y = \inf\{t : Z_t = y\}$ and use P_x to denote the law of the branching process with $Z_0 = x$. Well known properties of the exponential distribution imply that

$$P_k(T_{k+1} < T_{k-1}) = \frac{\beta}{\beta+1} \qquad \text{for } k > 0$$

so Z_t is a time change of an asymmetric random walk S_n that, when $k > 0$, makes transitions $k \to k+1$ with probability $\beta/(\beta+1)$ and $k \to k-1$ with probability $1/(\beta+1)$ and has 0 as an absorbing state, i.e., once $S_n = 0$ we will have $S_m = 0$ for all $m > n$. Using this observation and well known formulas for simple random walk it follows that

$$P_1(T_0 < \infty) = \begin{cases} 1 & \text{if } \beta \le 1 \\ 1/\beta & \text{if } \beta > 1 \end{cases}$$

so the critical value of β for the survival of the branching process is 1.

The main result in this section is that as the range $r \to \infty$ the critical value for survival of the contact process converges to that of the branching process. Let $\tau^0 = \inf\{t : \xi_t^0 = \emptyset\}$ where ξ_t^0 denotes the contact process starting from a single particle at the origin, i.e., $\xi_0^0 = \{0\}$. Let $\beta_c = \inf\{\beta : P(\tau^0 = \infty) > 0\}$.

Theorem 7.1. As $r \to \infty$, $\beta_c \to 1$ and if $\beta > 1$

$$P(x \in \xi_\infty^1) \to \frac{\beta-1}{\beta}$$

Remark. Schonmann and Vares (1986) have shown that if we consider the basic contact process in d dimensions with $\mathcal{N} = \{x : \|x\|_1 = 1\}$ and we let $\beta = 2d\lambda$ then the conclusions of Theorem 7.1 and (7.18) below hold.

PROOF: To begin we note that we can construct the contact process from a branching random walk by suppressing births onto occupied sites. So we can define the contact process and the branching random walk on the same space so that the branching random walk always has more particles than the contaact process, and it follows that $\beta_c \geq 1$ for all r. To prove the rest of the result we note that taking $A = \mathbf{Z}^d$ and $B = \{0\}$ in the duality equation (5.3) gives

$$P(\xi_t^1 \cap \{0\} \neq \emptyset) = P(\xi_t^0 \cap \mathbf{Z}^d \neq \emptyset) = P(\tau^0 > t)$$

Letting $t \to \infty$ we have

(7.1) $$P(0 \in \xi_\infty^1) = P(\tau^0 = \infty)$$

So to prove Theorem 8.1 it suffices to show that

(7.2) If $\beta > 1$ then $P(\tau^0 = \infty) \to (\beta - 1)/\beta$

for this implies that $\limsup_{r\to\infty} \beta_c \leq 1$. To prove (8.2) we scale space by dividing by r and consider the contact process on $\mathbf{Z}^d/r$ to facilitate taking the limit $r \to \infty$. Our approach will be to use the comparison theorem, so we let $I_k = k2Le_1 + [-L, L]^d$ and consider a modification of the contact process $\bar{\xi}_t$ in which births are not allowed outside $(-4L, 4L)^d$. The two key ingredients in the proof are

(7.3) Let $\delta > 0$. If we pick L large, set $T = L^2$, and pick K large then for $r \geq r_0$, $\bar{\xi}_T$ will have at least K particles in I_1 and in I_{-1} with probability at least $1 - \delta$ whenever $\bar{\xi}_0$ has at least K particles in I_0

(7.4) Consider the process starting from $\xi_0^0 = \{0\}$. If we pick S large then for $r \geq r_1 \geq r_0$, ξ_S^0 will have at least K particles in I_0 with with probability at least $((\beta - 1)/\beta) - \delta$

Once this is done (7.2) follows by using Theorem 4.3 to compare

$$X_n = \{m : |\xi_{S+nT}^0 \cap I_m| \geq K\}$$

with a one-dependent oriented percolation with density $\geq 1 - \delta$ and Theorem 4.1 to conclude that the cluster containing $(0, 0)$ in the percolation model will be infinite with probability at least $1 - 55\delta^{1/9}$. For these two facts imply that

$$P(\tau^0 = \infty) \geq \frac{\beta - 1}{\beta} - \delta - 55\delta^{1/9}$$

PROOF OF (7.3): The starting point is the observation that if we let $r \to \infty$ then the contact process on $\mathbf{Z}^d/r$ converges to a branching random walk η_t in which

(i) Each particle dies at rate 1, and gives birth at rate β.

(ii) A particle born at x is sent to a point y chosen at random from $\{y : \|y-x\|_p \le 1\}$.

This should be intutively clear since if we start with one particle at 0, fix T and let $r \to \infty$ then the probability of a *collision* (birth onto an occupied site) by time T goes to 0 as $r \to \infty$, and the displacements of the individual particles converge to a uniform distribution on $\{y : \|y\|_p \le 1\}$.

We will prove the convergence of the contact process on $\mathbf{Z}^d/r$ to the branching random walk later (see the "continuity argument" below). We have introduced this result now to motivate the first step of the proof, which is to prove the analogue of (7.3) for the branching random walk η_t, which is given in (7.12) below. Let η_t^x denote the branching random walk starting from $\eta_0^x = \{x\}$. To leave room for the limit $r \to \infty$ we consider $\bar{\eta}_t^x$ a modification of η_t^x in which particles that land outside $(-4L+1, 4L-1)^d$ are killed. Let $m(t,x,A) = E|\bar{\eta}_t^x \cap A|$ be the mean number of particles in A at time t for the modified branching random walk starting with a single particle at x. We claim that

$$(7.5) \qquad m(t,x,A) = e^{(\beta-1)t} P(\bar{W}_t^x \in A)$$

where $\bar{W}_t^x$ is a random walk that starts at x, jumps at rate β, has jumps that are uniform on $\{y : \|y\|_p \le 1\}$, and is killed when it lands outside $(-4L+1, 4L-1)^d$. To check this claim note that both sides of (7.5) satisfy the same differential equation: if $A \subset (-4L+1, 4L-1)^d$ then

$$\frac{dm(t,x,A)}{dt} = -m(t,x,A) + \int m(t,x,dy)\nu(A-y)$$

where $A - y = \{x - y : x \in A\}$ and ν is the uniform probability measure on $\{y : \|y\|_p \le 1\}$.

Let $I_1' = 2Le_1 + [-L+1, L-1]^d$, i.e. I_1 shrunk by a little bit. Donsker's theorem implies that if $T = L^2$ and $x/L \to \theta \in [-1,1]^d$

$$(7.6) \qquad P(\bar{W}_T^x \in I_1') \to \psi(\theta)$$

where $\psi(\theta) = P_\theta(B_t \in [-4,4]^d \text{ for } t \le 1,\ B_1 \in 2e_1 + [-1,1]^d)$ and B_t is a constant multiple of d-dimensional Brownian motion. $\psi(\theta) > 0$ and is continuous, so a simple argument (suppose not and extract a convergent subsequence) shows

$$(7.7) \qquad \liminf_{L\to\infty} \left[\inf_{x \in [-L,L]^d} P(\bar{W}_T^x \in I_1') \right] \ge \inf_{\theta \in [-1,1]^d} \psi(\theta) > 0.$$

It follows from (7.5)–(7.7) that we can pick L large enough so that

$$(7.8) \qquad \inf_{x \in [-L,L]^d} E|\bar{\eta}_T^x \cap I_1'| \ge 2.$$

Let $\bar{\eta}_t^A$ denote the modified branching random walk with $\bar{\eta}_0^A = A$. (7.8) implies

$$(7.9) \qquad E|\bar{\eta}_T^A \cap I_1'| \ge 2|A|$$

while an obvious comparison and a well known fact about branching processes (see Athreya and Ney (1972) for this and other facts about branching processes we will use) implies

$$\text{(7.10)} \qquad \mathrm{var}(\bar\eta_T^x \cap I_1') \le E|\bar\eta_T^x \cap I_1'|^2 \le E|\eta_T^x|^2 = C_T < \infty$$

Combining the last two conclusions and using Chebyshev's inequality it follows that if $A \subset [-L, L]^d$ has $|A| = K$ then

$$\text{(7.11)} \qquad P(|\bar\eta_T^A \cap I_1'| < K) \le \frac{\mathrm{var}(|\bar Z_T^A \cap I_1'|)}{(2|A| - K)^2} \le \frac{K \sup_x \mathrm{var}(|\bar Z_T^x \cap I_1'|)}{K^2} \le \frac{C_T}{K}$$

From the last result it follows that

(7.12) If $\delta > 0$ and K is large then for any $A \subset [-L, L]^d$ with $|A| = K$.

$$P(|\bar\eta_T^A \cap I_1'| < K) \le \delta/10$$

Continuity Argument. (7.12) shows that if $A \subset I_0$ has $|A| = K$ then with high probability $\bar\eta_T^A$ will have at least K particles in I_{-1} and in I_1. The next step is to prove the corresponding result for the contact process. To avoid some technicalities we will give the details only for the case in which $\mathcal{N} = \{z : \|z\|_\infty \le r\}$ and then indicate the extension to $p < \infty$ in a remark after the proof.

Let $\bar\xi_t^A$ be a modification of the contact process with $\bar\xi_0^A = A$ in which births outside $(-4L, 4L)^d$ are not allowed. We begin by observing that the number of births up to time t in the contact process, V_t, is dominated by a branching process $\bar V_t$ in which births occur at rate β and deaths occur at rate 0. If $|A| = K$ then $E\bar V_t = Ke^{\beta t} < \infty$, so our comparison and Chebyshev's inequality imply

$$\text{(7.13)} \qquad P(V_T > r^{1/3}) \le P(\bar V_T > r^{1/3}) \le \frac{Ke^{\beta T}}{r^{1/3}} \to 0$$

since T is fixed and $r \to \infty$.

Let $G_1 = \{V_t \le r^{1/3}\}$. Here G is for good event and the subscript indicates it is the first of several we will consider. When G_1 occurs, the probability of having a birth land on an occupied site (a "collision") is

$$\text{(7.14)} \qquad \le r^{1/3} \frac{r^{1/3}}{(2r+1)^d} \to 0$$

since there are at most $r^{1/3}$ births and even if all the particles are in $\{x : \|x\|_\infty \le 1\}$ (on $\mathbf{Z}^d/r$) each birth has probability at most $r^{1/3}/(2r+1)^d$ of landing on an occupied site. Let G_2 be the event that there are no collisions by time t.

To deal with the spatial location of particles, we will create a coupling of the displacements of the particles in the branching random walk to those of particles in the contact process. To couple the displacements we observe that if U is uniform on $\{y : \|y\|_\infty \le 1\}$

and $\pi_r(x)$ is the closest point in $\mathbf{Z}/r^d$ to x (with some convention for breaking ties) then $U^r = \pi_r(U(1+1/2r))$ is uniform on $\mathcal{N}/r$.

Now if the U_i are the displacements of particles in the branching random walk, we will use the U_i^r for the displacements in the contact process. When our good events G_1 and G_2 occur, we have G_3 = all of the points in the contact process ξ_s^A are within $r^{1/3}/r$ (in $\| \|_\infty$) of their counterparts in the branching process η_s^A. Passing to the truncated processes and noting that the branching particles are required to stay in $(-4L+1, 4L-1)^d$ for $0 \le s \le T$, while the contact process particles are required to stay in $(-4L, 4L)^d$, it follows that on G_3 we have $|\bar\xi_T^A \cap I_1| \ge |\bar\eta_T^A \cap I_1'|$ Combining the last observation with (7.12) gives (7.3). □

Remark. If $p < \infty$ then $U^r = \pi_r((1+1/2r)U)$ is not uniform on $\mathcal{N}/r$ but is within C/r of uniform in the total variation norm. In the last paragraph of the proof we then have $P(\|U_i - U_i^r\|_\infty > 1/r) \le C/r$, which since there are at most $r^{1/3}$ transitions on G_1, is good enough for the proof.

PROOF OF (7.4): By the continuity argument it is enough to show that we can pick S so that η_S^0 will have at least K particles in I_0 with probability at least $((\beta-1)/\beta) - \delta/2$. However, this follows from

(7.16) If Ω_∞ is the event that the branching process does not die out, then for any $L > 0$ and $K < \infty$,

$$P(|\eta_t^0 \cap [-L, L]^d| < K, \Omega_\infty) \to 0$$

Indeed as Asmussen and Kaplan (1976) have shown (see Theorem 2 on p. 5)

(7.17) There is a constant $\sigma > 0$ so that

$$\sqrt{t}e^{-(\beta-1)t}|\eta_t^0 \cap [-L, L]^d \to W \cdot \frac{(2L+1)^d}{(2\pi\sigma^2)^{d/2}}$$

where $W = \lim_{t\to\infty} e^{-(\beta-1)t}|\eta_t^0| > 0$ a.s. on Ω_∞

This completes the proof of (7.4) and hence of (7.2). □

The argument just used on the long range contact process can also be applied to

Example 7.2. Succesional dynamics. We suppose that the set of states at each site are 0 = grass, 1 = a bush, 2 = a tree and formulate the dynamics as

$$\begin{aligned} c_0(x,\xi) &= \delta_{\xi(x)} \\ c_1(x,\xi) &= \lambda_1 n_1(x) \quad \text{if } c_i(x) = 0 \\ c_2(x,\xi) &= \lambda_2 n_2(x) \quad \text{if } c_i(x) \le 1 \end{aligned}$$

The title of this example and its formulation are based on the observation that if an area of land is cleared by a fire, then regowth will occur in three stages: first grass appears then small bushes and finally trees, with each species growing up through and replacing the previous one. With this in mind, we allow each type to give birth onto sites occupied by lower numbered types.

Theorem 7.2. Let $\beta_i = \lambda_i|\mathcal{N}|$. Suppose that $\beta_2 > \delta_2$ and

$$(\star) \qquad \beta_1 \cdot \frac{\delta_2}{\beta_2} > \delta_1 + \beta_2 \cdot \frac{\beta_2 - \delta_2}{\beta_2}$$

If r is large then there is a nontrivial translation invariant stationary distribution in which all three types have positive density.

SKETCH OF PROOF: The fact that the 2's do not feel the presence of the 1's implies that the set of sites occupied by 2's is a contact process. To construct a stationary distribution we start with the 2's in their upper invariant measure and we put 1's at all sites not occupied by 1's to get a process ξ_t^{12}. This is the analogue of starting an attractive system from all 1's and a result of Durrett and Moller imples that as $t \to \infty$, $\xi_t^{12} \Rightarrow \xi_\infty^{12}$ a translation invariant stationary distribution.

To prove that ξ_∞^{12} is nontrivial we will prove an analogue of (7.3). The first step is to prove the following result about the long range contact process (which is here considered as a subset of $\mathbf{Z}^d$)

(7.18) If $\beta > 1$ and $x \neq y$ then as $r \to \infty$

$$P(x, y \in \xi_\infty^1) \to \left(\frac{\beta - 1}{\beta}\right)^2$$

In words, the equilibrium distribution converges to a product measure as $r \to \infty$. Of course, the last conclusion only says that the sites are asymptotically pairwise independent, but the argument can easily be generalized to a finite number of x's.

PROOF: By duality (see the proof of (7.1))

$$P(x, y \in \xi_\infty^1) = P(\tau^x = \infty, \tau^y = \infty)$$

Our comparison of the contact process with a branching process at the beginning of the proof of Theorem 8.1 shows that $P(\tau^x = \infty) \leq (\beta - 1)/\beta$ for all r. If we pick K and L as in (7.3) and then pick S large as in (7.4) then for $r \geq r_1$ we have

$$\begin{aligned} \frac{\beta - 1}{\beta} + \delta &\geq P(\tau^x > S) \\ &\geq P(|\xi_S^x \cap [-L, L]| > K) \geq \frac{\beta - 1}{\beta} - \delta \end{aligned}$$

Our choice of K and L and the comparison with oriented percolation shows that

$$P(|\xi_S^x \cap [-L, L]| > K, \tau^x < \infty) \leq 55\delta^{1/9}$$

Combining the last two estimates shows

$$|P(\tau^x = \infty) - P(\tau^x > S)| \le \delta + 55\delta^{1/9}$$

With this in hand the desired result follows easily since continuity argument shows that for any fixed S as $r \to \infty$

$$P(\tau^x > S, \tau^y > S) \to P(\eta_S^0 \neq \emptyset)^2 \qquad \square$$

Turning now to the heart of the proof we will again scale space by dividing by r and consider the contact process on $\mathbf{Z}^d/r$ to facilitate taking the limit. The approach we will take is a combination of that of Durrett and Swindle (1991) and Durrett and Schinazi (1993). We will concentrate on explaining the main ideas and refer the reader to those papers for the details. Pick $\rho > (\beta_2 - \delta_2)/\beta_2$ so that

$$(\star\star) \qquad \beta_1(1-\rho) > \delta_1 + \beta_2\rho$$

By dividing space into cubes of side δr then using (7.18) and the weak law one can prove that with high probability all sites in our space time box have at most $\rho|\mathcal{N}|$ neighbors in state 2. (Recall that the set of 2's at any time is distributed according to the upper invariant measure.) This means that a single 1 will have births that land on an occupied site at rate $\ge \beta_1(1-\rho)$ while it dies at rate δ_1 and is smothered by a 2 at rate $\le \beta_2\rho$.

The inequality $(\star\star)$ implies that a single particle gives birth faster than it dies. If we start with a fixed number of 1's then in the limit $r \to \infty$ the 1's dominate a supercritical branching random walk. If this fixed number K is large and L and $T = L^2$ are chosen appropriately then for large r a truncated version of the process which is not allowed to give birth outside $(-4L, 4L)^d$ will with high probability have at least K particles in I_1 and in I_{-1} whenver the initial configuration has at least K particles in I_0.

The last result is an analogue of (8.3) but there is one problem. The event that $\xi_t(x) = 2$, which is the same as the survival of the dual contact process of 2's starting from (x, t), does not have a finite range of dependence. To avoid this problem we adopt the more liberal viewpoint that x is occupied by a 2 at time t if the dual process escapes from a certain space-time box. If the box is large enough the liberalization of the definition does not increase the density of 2's by enough to violate $(\star\star)$, we can verify the comparison assumptions of Theorem 4.3 and the desired result follows from Theorem 8.2.

8. Rapid Stirring Limits

The point of this section is that if we take a fixed interacting particle system, scale space by ϵ and "stir" the particles at rate ϵ^{-2} then as $\epsilon \to 0$ the particle system converges to the solution of a reaction diffusion equation. To be precise, we consider processes $\xi_t^\epsilon : \epsilon \mathbf{Z}^d \to \{0, 1, \ldots, \kappa - 1\}$ that evolve as follows

(i) there are *translation invariant finite range flip rates*

$$c_i(x, \xi) = h_i(\xi(x), \xi(x + \epsilon y_1), \ldots, \xi(x + \epsilon y_N))$$

(ii) *rapid stirring:* for each $x, y \in \epsilon \mathbf{Z}^d$ with $\|x - y\|_1 = \epsilon$ we exchange the values at x and y at rate ϵ^{-2}. That is, we change the configuration from ξ to $\xi^{x,y}$ where

$$\xi^{x,y}(y) = \xi(x) \qquad \xi^{x,y}(x) = \xi(y) \qquad \xi^{x,y}(z) = \xi(z) \quad z \neq x, y$$

The reader should note that in (i) changing ϵ scales the lattice but does not change the interaction between the sites. In (ii) we superimpose stirring in such a way that the individual values will be moving according to Brownian motions (run at rate 2) in the limit. The motivation for modifying the system in this way comes from the following *mean field limit theorem* of De Masi, Ferrari, and Lebowitz (1986). The derivation of such "hydrodynamic limits" has become a major enterprise (see e.g., Spohn (1991) or DeMasi and Presutti (1992)) but this particular result is rather easy to establish.

Theorem 8.1. Suppose $\xi_0^\epsilon(x)$ are independent and let $u_i^\epsilon(t, x) = P(\xi_t^\epsilon(x) = i)$. If $u_i^\epsilon(0, x) = g_i(x)$ is continuous then as $\epsilon \to 0$, $u_i^\epsilon(t, x) \to u_i(t, x)$ the bounded solution of

$$\partial u_i / \partial t = \Delta u_i + f_i(u) \qquad u_i(0, x) = g_i(x) \tag{8.1}$$

where

$$f_i(u) = < c_i(0, \xi) 1_{(\xi(0) \neq i)} >_u - \sum_{j \neq i} < c_j(0, \xi) 1_{(\xi(0) = i)} >_u \tag{8.2}$$

and $< \phi(\xi) >_u$ denotes the expected value of $\phi(\xi)$ under the product measure in which state j has density u_j, i.e., when $\xi(x)$ are i.i.d. with $P(\xi(x) = j) = u_j$.

Theorem 8.1 is easy to understand. The stirring mechanism (i.e., (ii)) has product measures as its stationary distributions. See Griffeath (1979), Section II.10. When ϵ is small, stirring operates at a fast rate and keeps the system close to a product measure. The rate of change of the densities can then be computed assuming adjacent sites are independent. To help explain the somewhat ugly formula in (8.2) we will now consider two concrete examples.

Example 8.1. The basic contact process. In this case $c_0(x, \xi) = 1$ and $c_1(x, \xi) = \lambda n_1(x, \xi)$ where $n_i(x, \xi) = |\{y \in \mathcal{N} : \xi(x + y) = i\}$ is the number of neighbors in state i.

We claim that when $|\mathcal{N}| = N$ the equation in (9.1) becomes (we do not need an equation for $u_0 = 1 - u_1$)

$$\partial u_1/\partial t = \Delta u_1 - u_1 + N\lambda(1-u_1)u_1$$

To see the second term on the right hand side the equation, we note that particles die at rate 1 independent of the state of neighbors. For the third, we note that if we assume all sites are independent then the probability x is vacant and $y \in x + \mathcal{N}$ is occupied is $(1-u_1)u_1$. Each such pair produces a new particle at rate λ and there are N such pairs, so the total rate at which new particles are created (assuming that adjacent sites are independent) is $N\lambda(1-u_1)u_1$.

The equation in the last example is just the mean field equation for the contact process that we have seen several times before. To see something new we look at

Example 8.2. The threshold one voter model. In this case

$$c_i(x,\xi) = 1 \quad \text{if } n_i(x,\xi) \ge 1$$

and if we assume $|\mathcal{N}| = N$ then the limiting equation is (again we do not need an equation for $u_0 = 1 - u_1$)

$$\partial u_1/\partial t = \Delta u_1 - u_1(1-u_1^N) + (1-u_1)(1-(1-u_1)^N)$$

To see this note that if all sites are independent then the probability x is occupied and at least one neighbor is vacant is $u_1(1-u_1^N)$ and this is the rate at which 1's are destroyed. Intercahnging the roles of vacant and occupied in the last sentence gives the third term.

Having explained the formula in (8.2) we turn now to a result that extends Theorem 8.1 by showing that the particle system itself, not just its expected values are close to the p.d.e. To motivate the statement we note that the states of the sites in the model become independent in the limit $\epsilon \to 0$ and the number of sites per unit volume becomes large so it should not be surprising that in the limit $\xi_t^\epsilon(x)$ becomes deterministic.

Theorem 8.2. Let $\phi(x)$ be a smooth function with compact support. As $\epsilon \to 0$

$$\epsilon^d \sum_{y \in \epsilon \mathbf{Z}^d} \phi(y) 1_{(\xi_t^\epsilon(y)=i)} \to \int \phi(y) u_i(t,y)\,dy$$

in probability.

Although the indicator function of a bounded open set G is not continuous, this should be thought of as saying that

$$\epsilon^d \sum_{y \in \epsilon \mathbf{Z}^d \cap G} 1_{(\xi_t^\epsilon(y)=i)} \to \int_G u_i(t,y)\,dy$$

or more intuitively that the fraction of sites near y that are in state i converges to $u_i(t,y)$. The result for an open set G is also true, but is a little more difficult to prove precisely because 1_G is not continuous.

Theorem 8.2 provides a link between the particle system with fast stirring that we will exploit in the next lecture to prove the existence of stationary distributions for a predator prey model with fast stirring. Once Theorem 8.1 is established, the proof of Theorem 8.2 is easy: compute second moments and use Chebyshev's inequality. So we will concentrate on the proof of Theorem 8.1. The ideas behind the proof are simple: we will give an explicit construction of the process that allows us to define a dual processes by asking the question: "What is the state of x at time t?" and working backwards in time. The answer to this question can be determined by looking at the states of the sites in the "dual process" $I_\epsilon^{x,t}(s)$ at time $t-s$. The particles in $I_\epsilon^{x,t}(s)$ move according to stirring at a fast rate and give birth to new particles at rate

$$c^* = \sup_\xi \sum_i c_i(x,\xi)$$

We will show that for small ϵ the dual process is almost a branching random walk and converges to a branching Brownian motion as $\epsilon \to 0$. The proof of the last result leads easily to the conclusion that two dual processes $I_\epsilon^{x,t}(s)$ and $I_\epsilon^{y,t}(s)$ are asymptotically independent which gives the asymptotic independence of the sites in the parrticle systems. The convergence of the dual process to branching Brownian motion leads in a straightforward way to the convergence of the $u_i^\epsilon(t,x)$ to limits $u_i(t,x)$ and the asymptotic independence of adjacent sites implies that the $u_i(t,x)$ satisfy the limiting equations.

a. The dual process. The first step in the proof is to construct the process from a number of Poisson processes, all of which are assumed to be independent. The construction is similar in spirit to the one in Section 3 but it is convenient to do the details in a slightly different way. For each $x \in \epsilon\mathbf{Z}^d$, let $\{T_n^x, n \geq 1\}$, be a Poisson process with rate c^* and let $\{U_n^x, n \geq 1\}$ be a sequence of independent random variables that are uniform on $(0,1)$. At time T_n^x we compute the flip rates $r_i = c_i(x, \xi(T_n^x))$ and use U_n^x to determine what (if any) flip should occur at x at time T_n^x. To be precise we let $p_i = \sum_{j\leq i} r_j/c^*$ for $i = 0,\ldots,\kappa-1$ with $p_{-1} = 0$ and flip to i if $U_n^x \in (p_{i-1}, p_i)$. If $U_n^x \in (p_{\kappa-1}, 1)$ no flip occurs. To move the particles around, we let $\{S_n^{x,y}, n \geq 1\}$ be Poisson processes with rate ϵ^{-2} when $x, y \in \epsilon\mathbf{Z}^d$ with $\|x-y\|_1 = \epsilon$, and we declare that at time $S_n^{x,y}$ the values at x and y are exchanged.

The dual process $I_\epsilon^{x,t}(s)$ is naturally defined only for $0 \leq s \leq t$ but for a number of reasons, it is convenient to assume that the Poisson processes and uniform random variables in the construction are defined for negative times and define $I_\epsilon^{x,t}(s)$ for all $s \geq 0$. Let $\mathcal{N} = \{\epsilon y_1, \ldots, \epsilon y_N\}$ be the set of neighbors of 0. The dual process makes transitions as follows:

If $y \in I_\epsilon^{x,t}(s)$ and $T_n^y = t-s$ then we add all the points of $y + \mathcal{N}$ to $I_\epsilon^{x,t}(s)$.

If $y \in I_\epsilon^{x,t}(s)$ and $S_n^{y,z} = t-s$ then we move the particle at y to z.

For a picture of (a rather unlikely sample path for) the dual when $d = 1$ and $\mathcal{N} = \{-1,0,1\}$ see Figure 8.1

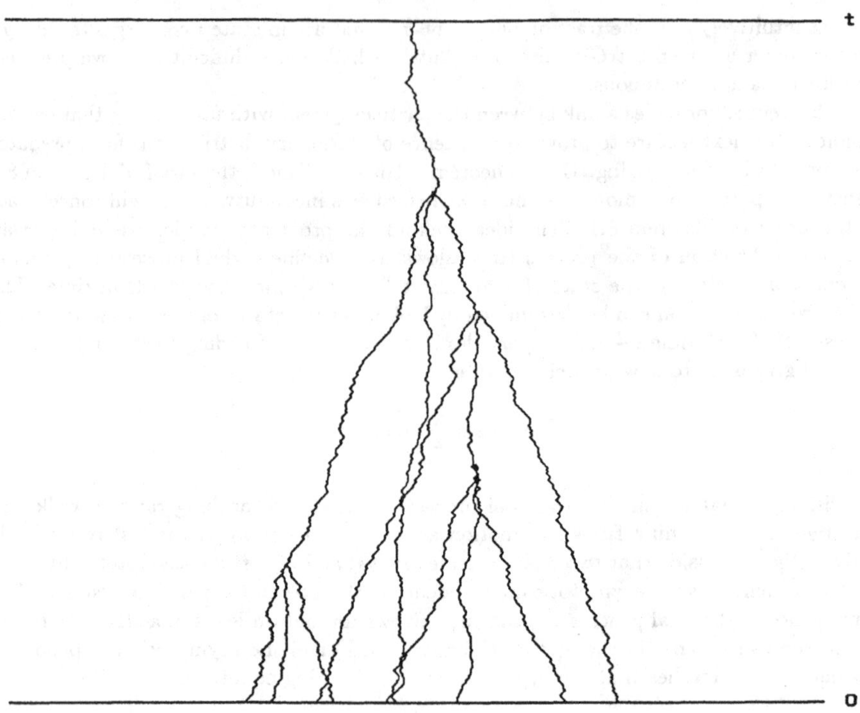

Figure 8.1

It is easy to see that we can compute the state of x at time t by knowing the states of the y in $I_\epsilon^{x,t}(s)$ at time $t-s$. We start with the values in $I_\epsilon^{x,t}(s)$ at time $t-s$ and work up to time t. At S arrivals we perform the indicated stirrings. When an arrival T_n^y occurs at a point of the dual, we look at the value of the process on $y+\mathcal{N}$, compute the flip rates r_i, and use U_n^x to determine what (if any) flip should occur.

To prepare for the proof of the convergence of $u_i^\epsilon(t,x)$ we will now give a more detailed description of $I_\epsilon^{x,t}(s)$. Let $X_\epsilon^0(0)=x$, let R_ϵ^1 be the smallest value of s so that we have a T arrival at $X_\epsilon^0(s)$ at time $t-s$, and set $X_\epsilon^i(s)=\epsilon y_i + X_\epsilon^0(s)$ for $1\le i\le N$. Finally, we set $\mu_\epsilon^1=0$ to indicate that 0 is the mother of the N particles created at time R_ϵ^1. Passing now to the inductive step of the definition, suppose that we have defined the process up to time R_ϵ^m with $m\ge 1$. The $mN+1$ existing particles move as dictated by stirring until R_ϵ^{m+1}, the first time $s>R_\epsilon^m$ that a T arrival occurs at the location of one of our moving particles $X_\epsilon^k(s)$ and then we set $X_\epsilon^{mN+i}(s)=\epsilon y_i + X_\epsilon^k(s)$ for $1\le i\le N$, and $\mu_\epsilon^{m+1}=k$. The new particles may be created at the locations of existing particles. If so we say that a *collision* occurs and call the new particle *fictitious*. We will prove later that the probability of a collision tends to 0 as $\epsilon\to 0$, but for proving the convergence of $u_i^\epsilon(t,x)$, it is convenient to allow the fictitious particles to move and give birth like other particles, so for each $m \geq 1$ we define an independent copy of the graphical representation which we use for the births and movement of the mth particle if it is fictitious. By definition all the offspring of fictitious particles are also fictitious.

b. The dual process is almost a branching random walk. The point of introducing fictitious particles is that $\mathcal{K}_t = mN + 1$ for $t \in [R^\epsilon_m, R^\epsilon_{m+1})$ defines a branching process in which each particle gives birth to N additional particles at rate c^*. Our next goal is to show that if ϵ is small then $I^{x,t}_\epsilon(s)$ is almost a branching random walk in which particles jump to a randomly chosen neighbor at rate $2d\epsilon^{-2}$ and give birth as above. To do this we will couple X^k_ϵ to independent random walks Y^k_ϵ that start at the same location at time β_k = the birth time of X^k_ϵ, and jump to a randomly chosen neighbor at rate $2d\epsilon^{-2}$.

We say X^k_ϵ is *crowded* at time s if for some $j \neq k$ $\|X^j_\epsilon(s) - X^k_\epsilon(s)\|_1 \leq \epsilon$. When X^k_ϵ is not crowded, we define the displacements of Y^k_ϵ to be equal to those of X^k_ϵ. When X^k_ϵ is crowded we use independent Poisson processes to determine the jumps of Y^k_ϵ. To estimate the difference between X^k_ϵ and Y^k_ϵ, we need to estimate the amount of time X^k_ϵ is crowded. Let $j \neq k$, $V^\epsilon_s = X^k_\epsilon(s) - X^j_\epsilon(s)$ and W^ϵ_s be a random walk that jumps to a randomly chosen neighbor at rate $4d\epsilon^{-2}$. (Notice that V^ϵ_s is the difference of two random walks and hence jumps at rate $4d\epsilon^2$. The transition probabilities of V^ϵ_s differ slightly from those of W^ϵ_s when $\|x\|_1 = \epsilon$. Here y denotes any point $\neq -x$ with $\|y\|_1 = \epsilon$.

jumps from x to	rate in V	rate in W
$-x$	ϵ^{-2}	0
0	0	$2\epsilon^{-2}$
$x+y$	$2\epsilon^{-2}$	$2\epsilon^{-2}$

From the last table it should be clear that $\|W^\epsilon_s\|_1$ is stochastically smaller than $\|V^\epsilon_s\|_1$, i.e., the two random variables can be constructed on the same space so that $\|W^\epsilon_s\|_1 \leq \|V^\epsilon_s\|_1$ for all s. To check this note that all the transition of V and W can be coupled except those in the first two lines of the table, but there $\|W\|_1$ jumps from 1 to 0 at rate ϵ^{-2} while $\|V\|_1$ jumps from 1 to 1 at rate $\epsilon^2/2$.

From the last comparison of $\|V\|_1$ and $\|W\|_1$ it follows that for any integer $M \geq 1$, $v^{M\epsilon}_t = |\{s \leq t : \|V^\epsilon_s\|_1 \leq M\epsilon\}|$ is stochastically smaller than $w^{M\epsilon}_t = |\{s \leq t : \|W^\epsilon_s\|_1 \leq M\epsilon\}|$. Well known asymptotic results for random walks imply that when $t\epsilon^{-2} \geq 2$

$$(8.3) \qquad Ew^{M\epsilon}_t \leq \begin{cases} CM^d\epsilon^2 & d \geq 3 \\ CM^2\epsilon^2\log(t\epsilon^{-2}) & d = 2 \\ CM\epsilon t^{1/2} & d = 1 \end{cases}$$

To see this note that $w^{M\epsilon}_t$ has the same distribution as $\epsilon^2 w^M_{t\epsilon^{-2}}$ and the last line is equal to $CM\epsilon^2(t\epsilon^{-2})^{1/2}$.

Let $\chi^k_\epsilon(t)$ be the amount of time X^k_ϵ is crowded in $[0,t]$. It is easy to see that

$$(8.4) \qquad E(\chi^k_\epsilon(t)|\mathcal{K}_t = K) \leq KEw^\epsilon_t$$

$$(8.5) \qquad E\mathcal{K}_t = \exp(\nu t) \text{ where } \nu = c^*N$$

$$(8.6) \qquad E(\chi^k_\epsilon(t)) \leq \exp(\nu t)Ew^\epsilon_t$$

To estimate the difference between $X_\epsilon^k(s)$ and $Y_\epsilon^k(s)$ we observe that if $\chi_\epsilon^k(t) = \tau$ then the number of "independent jumps" in the ith coordinate of Y_ϵ^k that occur in $[0,t]$ has a Poisson distribution with mean $\epsilon^{-2}\tau$. Let $\Delta_Y^i(s)$ be the net effect of the independent jumps on coordinate i up to time s. Recalling that changes in the ith coordinate of Y_ϵ^k have mean 0 and variance ϵ^2, it follows that $E\Delta_Y^i(s) = 0$ and

$$(8.7) \qquad E(\Delta_Y^i(s)^2) = E\chi_\epsilon^k(s)$$

Since $\Delta_Y^i(s)$ is a martingale, Kolmogorov's inequality implies

$$(8.8) \qquad E\left(\max_{0\le s\le t} \Delta_Y^i(s)^2\right) \le 4E(\Delta_Y^i(t)^2)$$

Using Markov's inequality (i.e., if $X \ge 0$ then $P(X > x) \le EX^r/x^r$) then (8.8), (8.7), (8.6), and (8.3) (noting that the worst case is $d = 1$) gives

$$(8.9) \qquad P\left(\max_{0\le s\le t} |\Delta_Y^i(s)| \ge \epsilon^{.3}\right) \le \epsilon^{-.6} E\left(\max_{0\le s\le t} \Delta_Y^i(s)^2\right) \le C\epsilon^{.4} t^{1/2} \exp(\nu t)$$

Here and in what follows C dentores a constant whose value is unimportant and that will change from line to line. The arguments leading to the last inequality also apply to $\Delta_X^i(t)$, the net effect of jumps in $[0,t]$ while X_ϵ^k is crowded, so

$$(8.10) \qquad P\left(\max_{0\le s\le t} \|X_\epsilon^k(s) - Y_\epsilon^k(s)\|_\infty \ge 2\epsilon^{.3}\right) \le C\epsilon^{.4} t^{1/2} \exp(\nu t)$$

The estimate in (8.10) shows that the X_ϵ^k are close to independent random walks. To see that with high probability no collisions occur, we pick M large enough so that $\|x\|_1 \le M$ for all $x \in \mathcal{N}$ and repeat the derivation of (8.6) with ϵ replaced by $M\epsilon$ to conclude that the expected number of births from X_ϵ^k while there is some other X_ϵ^j in $X_\epsilon^k + \mathcal{N}$ is smaller than

$$(8.11) \qquad C\epsilon t^{1/2} \exp(\nu t)$$

(8.5) and Markov's inequality imply that

$$(8.12) \qquad P(\mathcal{K}_t > K) \le K^{-1} \exp(\nu t)$$

When $\mathcal{K}_t \le K$, (8.11) implies that the expected number of collisions is smaller than

$$(8.13) \qquad KC\epsilon t^{1/2} \exp(\nu t)$$

Combining the last two results and setting $K = \epsilon^{-.5}$ shows that the probability of a collision is smaller than

$$(8.14) \qquad C\epsilon^{.5}(1+t)^{1/2} \exp(\nu t)$$

Having shown that collisions are unlikely we no longer have to worry about the labels μ_m^ϵ that tell us the mother of the N particles created at time R_m^ϵ since this will be clear from the evolution of the dual. A more significant consequence of the results in this subsection is that dual processes for different sites are asymptotically independent. To argue this, we say the two duals *collide* if a particle in one dual gives birth when crowded by a particle in the other one. The arguments leading to (8.14) show that with high probability two duals do not collide, and (8.10) implies that the movements of all the particles can be coupled to independent random walks.

c. Convergence of $u_i^\epsilon(t,x)$. The next step is to show that as $\epsilon \to 0$ the branching random walk Y converges to a branching Brownian motion Z. To do this we use Skorokhod's trick to embed the ith component of the kth walk, $Y_s^{k,i}$ in a a Brownian motion $Z_s^{k,i}$. Using some standard estimates (see Durrett and Neuhauser for details) it follows that

$$(8.15)\qquad P\left(\max_{0\le s\le t} \|Y_\epsilon^k(s) - Z^k(s)\|_\infty > 4\epsilon^{.3} \text{ for some } k \le K\right) \le KC\epsilon^{.32}(1+t)$$

To compute the state of x at time t, we need not only the dual process $I_\epsilon^{x,t}(s)$, $s \le t$ but also the labels μ_n^ϵ and the uniform random variables U_n^x. However, the uniform random variables are independent of the dual process and, as we pointed out in a remark after (8.14), the μ_n^ϵ are only needed when a collision occurs.

As we will now explain, the results in the last paragraph make it easy to show that $u_a^\epsilon(t,x) \to u_a(t,x)$ as $\epsilon \to 0$. Here and in what follows we will use a and b to denote possible states of the sites to ease the burden on the middle of the alphabet. The first step is to describe $u_a(t,x)$. Let Z_s be a branching Brownian motion starting with a single particle at x and let $\mathcal{K}_t$ be the number of particles at time t. For $0 \le k < \mathcal{K}_t$, we let $\zeta_0(k)$ be independent and $= a$ with probability $\phi_a(Z_t^k)$. Once the ζ_0 are defined, we work up the space time set $\{Z_{t-s}^k\} \times \{s\}$. The values of $\zeta_s(k)$, the state of Z_{t-s}^k at time s, stay constant as long as only stirring occurs. When $N+1$ branches Z_{t-s}^i, $Z_{t-s}^{kN+1}, \ldots Z_{t-s}^{(k+1)N}$ come together (corresponding to a birth in the dual), we compute the flip rate at Z_{t-s}^i assuming it is in state $\zeta_s(i)$ and its neighbors are in states $\zeta_s(kN+j), 1 \le j \le N$. We generate an independent random variable uniform on $(0,1)$ to determine what (if any) flip should occur at Z_{t-s}^i. After we decide if we should change $\zeta_s(i)$, we can ignore $\zeta_s(kN+j)$ for $1 \le j \le N$. When we reach time t we will only be looking at the value at $\zeta_t(0)$. We call this value, the *result of the computation* and let $u_a(t,x) = P(\zeta_t(0) = a)$.

The description in the last paragraph is much like the one given earlier for the dual with one exception: the uniform random variables come from an auxiliary i.i.d. sequence instead of being read off the graphical representation. When there are no collisions in the dual, then the family structure of the influence set and the branching Brownian motion are the same. In this case if the inputs $\zeta_0(k)$ and the uniform random variables used are the same, the two computations have the same result. We have supposed that the initial functions $\phi_b(x)$ are continuous so (2.19) implies that as $\epsilon \to 0$,

$$\max_k |\phi_b(X_\epsilon^k(t)) - \phi_b(Z^k(t))| \to 0$$

where the maximum is taken over particles alive at time t. The last observation implies that we can with high probability arrange for all the inputs to be the same and it follows that $u_a^\epsilon(t,x) \to u_a(t,x)$. The last proof extends trivially to show that if $x_\epsilon \to x$ then $u_a^\epsilon(t,x_\epsilon) \to u_a(t,x)$. At the end of subsection b, we observed that the influence sets from different points are asymptotically independent. Combining that observation with the proofs in this subsection implies that if $x_\epsilon \to x$ then

$$P(\xi_t^\epsilon(x_\epsilon + \epsilon y_j) = c_j, 0 \le j \le N) \to \prod_{j=0}^{N} u_{c_j}(t,x) \tag{8.16}$$

We are interested in statements that allow $x_\epsilon \to x$ since this form of the conclusion implies that convergence occurs uniformly on compact sets.

d. The limit satisfies the p.d.e. The first step is to write the limiting equation in integral form.

(8.17) **Lemma.** Suppose $f_a, 0 \le a < \kappa$ are continuous and $g_a, 0 \le a < \kappa$ are bounded and continuous. The following statements are equivalent:
(i) The functions $u_a(t,x)$ are a classical solution of

$$\frac{\partial u_a}{\partial t} = \Delta u_a - f_a(u) \qquad u_a(0,x) = g_a(x)$$

i.e., the indicated derivatives exist and are continuous.
(ii) The functions $u_a(t,x)$ are bounded and satisfy

$$u_a(t,x) = \int p_t(x,y) g_a(y)dy + \int_0^t ds \int p_s(x,y) f_a(u(t-s,y))dy$$

where $p_t(x,y)$ is the transition probability for Brownian motion run at rate 2.

Proof: (i) implies that $Z_s^a \equiv u_a(t-s,B_s) - \int_0^s f_a(u(t-r,B_r))dr$ is a bounded martingale, so $Z_0^a = EZ_t^a$ and (ii) follows from Fubini's theorem. To prove the converse, we begin by observing that if (ii) holds then $u_a(t,x)$ has the necessary derivatives and Z_s^a is a martingale, so (i) follows from Itô's formula. □

To get (ii) we will use the integration by parts formula. Let S_t^ϵ be the semigroup for the stirred particle system and T_t^ϵ be the semigroup for pure stirring. The integration by parts formula implies that for nice functions ψ we have

$$S_t^\epsilon \psi(\xi) = T_t^\epsilon \psi(\xi) + \int_0^t ds\, S_{t-s}^\epsilon L T_s^\epsilon \psi(\xi) \tag{8.18}$$

where L is the generator for the particle system with no stirring. We use (8.18) with $\psi_{x,a}(\xi) = 1$ if $\xi(x) = a$ and 0 otherwise. Now for this choice of ψ

$$T_s^\epsilon \psi_{x,a}(\xi) = \sum_y p_s^\epsilon(x,y) \psi_{y,a}(\xi) \tag{8.19}$$

where $p_s^\epsilon(x,y)$ is the transition probability of a random walk that jumps from y to z at rate $\epsilon^{-2}/2$ if $\|y-z\|_1=\epsilon$. Now if $c_b(y,\xi)=h_b(\xi(y+\epsilon y_0),\dots,\xi(y+\epsilon y_N))$ then

$$L\psi_{y,a} = -\sum_b h_{b_0}(a,b_1,\dots,b_N)\psi_{y,a}\prod_{j=1}^N \psi_{y+\epsilon y_j,b_j} + \sum_b h_a(b_0,b_1,\dots,b_N)\psi_{y,b_0}\prod_{j=1}^N \psi_{y+\epsilon y_j,b_j} \tag{8.20}$$

where the sums are over $b_0,\dots,b_N\in\{0,1,\dots,\kappa-1\}$. Substituting (8.19) and (8.20) into (8.18) gives

$$\begin{aligned}(8.21)\quad P(\xi_t^\epsilon(x)=a) &= \sum_y p_t^\epsilon(x,y)g_a(y)\\ &+\int_0^t ds\sum_y p_s^\epsilon(x,y)E\Bigg\{-\sum_b h_{b_0}(a,b_1,\dots,b_N)\psi_{y,a}(\xi_{t-s}^\epsilon)\prod_{j=1}^N\psi_{y+\epsilon y_j,b_j}(\xi_{t-s}^\epsilon)\\ &\qquad+\sum_b h_a(b_0,b_1,\dots,b_N)\psi_{y,b_0}(\xi_{t-s}^\epsilon)\prod_{j=1}^N\psi_{y+\epsilon y_j,b_j}(\xi_{t-s}^\epsilon)\Bigg\}\end{aligned}$$

The local central limit theorem implies

$$\sum_y |\epsilon^d p_s(x,y)-p_s^\epsilon(x,y)|\to 0 \tag{8.22}$$

as $\epsilon\to 0$. As we observed at the end of subsection c,

$$E\psi_{y,c_0}(\xi_{t-s}^\epsilon)\prod_{j=1}^N\psi_{y+\epsilon y_j,c_j}(\xi_{t-s}^\epsilon)\to\prod_{j=0}^N u_{c_j}(t-s,y)$$

and this convergence occurs uniformly on compact sets. Using (8.21), (8.22), and the dominated convergence theorem, gives

$$\begin{aligned}(8.23)\quad u_a(t,x) &= \int p_t(x,y)g_a(y)\,dy\\ &+\int_0^t ds\int dy\, p_s(x,y)\Bigg\{-\sum_b h_{b_0}(a,b_1,\dots,b_N)u_a(t-s,y)\prod_{j=1}^N u_{b_j}(t-s,y)\\ &\qquad+\sum_b h_a(b_0,b_1,\dots,b_N)u_{b_0}(t-s,y)\prod_{j=1}^N u_{b_j}(t-s,y)\Bigg\}\end{aligned}$$

The term in braces is

$$-\sum_{b\neq a} < c_b(0,\xi)1_{\{\xi(0)=a\}} >_{u(t-s,y)} + < c_a(0,\xi) >_{u(t-s,y)} = f_a(u(t-s,y)) \tag{9.24}$$

Combining this with (8.17) gives the conclusion of Theorem 8.1.

9. Predator Prey Systems

In this section we will show that if you "know enough" about the limiting p.d.e. in Theorem 8.1 then you can prove results about the existence of stationary distributions for the system with fast stirring. For our approach, what you need to know about the p.d.e. is the following:

($\star$) There are constants $A_i < a_i < b_i < B_i$, L, and T so that if $u_i(0,x) \in (A_i, B_i)$ when $x \in [-L, L]^d$ then $u_i(x,T) \in (a_i, b_i)$ when $x \in [-3L, 3L]^d$.

Theorem 9.1. If ($\star$) holds then there is a nontrivial translation invariant stationary distribution for the process with fast stirring.

As the reader can probably guess, ($\star$) and Theorem 9.2 combine to produce a block event that turns one "pile of particles" into two and has high probability when ϵ is small and then the result follows from our comparison theorem. The details are somewhat technical so we refer the reader to Section 3 of Durrett and Neuhauser (1993) and turn to the problem of checking that ($\star$) holds in one particular example. For other applications of this technique see Durrett and Neuhauser (1993) or Durrett and Swindle (1993).

Example 9.1. Predator Prey Systems. The state at time t is $\xi_t^\epsilon : \epsilon \mathbf{Z}^d \to \{0,1,2\}$. We think of 0 as vacant, 1 and 2 as occupied by a fish and shark respectively. As usual, $n_i(x,\xi)$ is the number of neighbors of x (i.e., y with $\|y-x\|_1 = \epsilon$) that are in state i. The system changes states at the following rates:

$$\begin{aligned}
c_1(x,\xi) &= \beta_1 n_1(x,\xi)/2d & \text{if } \xi(x) &= 0\\
c_0(x,\xi) &= \delta_1 & \text{if } \xi(x) &= 1\\
c_2(x,\xi) &= \beta_2 n_2(x,\xi)/2d & \text{if } \xi(x) &= 1\\
c_0(x,\xi) &= \delta_2 + \gamma n_2(x,\xi)/2d & \text{if } \xi(x) &= 2
\end{aligned}$$

The first two rates say that fish repopulate vacant sites at a rate proportional to the number of fish at neighboring sites and die at rate δ_1. That is, in the absence of sharks, the fish are a contact process. The third rate says that sharks reproduce when they eat fish. This transition is a little strange from a biological point of view, but it has the desirable property that sharks will die out when the density of fish is too small. The last rate says that sharks die at rate δ_2 when they are isolated and the rate increases linearly with crowding. Finally, the sharks and fish swim around: for each pair of neighbors x and y stirring occurs at rate ϵ^{-2}, i.e., the values at x and y are exchanged. Applying Theorem 8.1 gives

Theorem 9.2. *Suppose that $\xi_0^\epsilon(x)$, $x \in \epsilon\mathbf{Z}^d$ are independent and $u_i^\epsilon(t,x) = P(\xi_t^\epsilon(x) = i)$ for $i = 1,2$. If $u_i^\epsilon(0,x) = \phi_i(x)$, which is continuous, then as $\epsilon \to 0$, $u_i^\epsilon(t,x) \to u_i(t,x)$ the bounded solution of*

$$\begin{aligned}
\frac{\partial u_1}{\partial t} &= \Delta u_1 + \beta_1 u_1(1-u_1-u_2) - \beta_2 u_1 u_2 - \delta_1 u_1\\
\frac{\partial u_2}{\partial t} &= \Delta u_2 + \beta_2 u_1 u_2 - u_2(\delta_2 + \gamma u_2)
\end{aligned} \tag{9.1}$$

with $u_i(0,x) = \phi_i(x)$.

As in the two examples in Section 8, the reaction terms are computed by assuming that adjacent sites are independent. To get $\beta_1 u_1(1-u_1-u_2)$ for example we note that if x is vacant and neighbor y is occupied by a fish, an event of probability $(1-u_1-u_2)u_1$ when sites are independent, births from y to x occur at rate $\beta_1/2d$ and there are $2d$ such pairs.

When the initial functions $\phi_i(x)$ are constant, $u_i(t,x) = v_i(t)$ and the v_i's satisfy

$$\text{(9.2)} \qquad \begin{aligned} \frac{dv_1}{dt} &= v_1((\beta_1-\delta_1) - \beta_1 v_1 - (\beta_1+\beta_2)v_2) \\ \frac{dv_2}{dt} &= v_2(-\delta_2 + \beta_2 v_1 - \gamma v_2) \end{aligned}$$

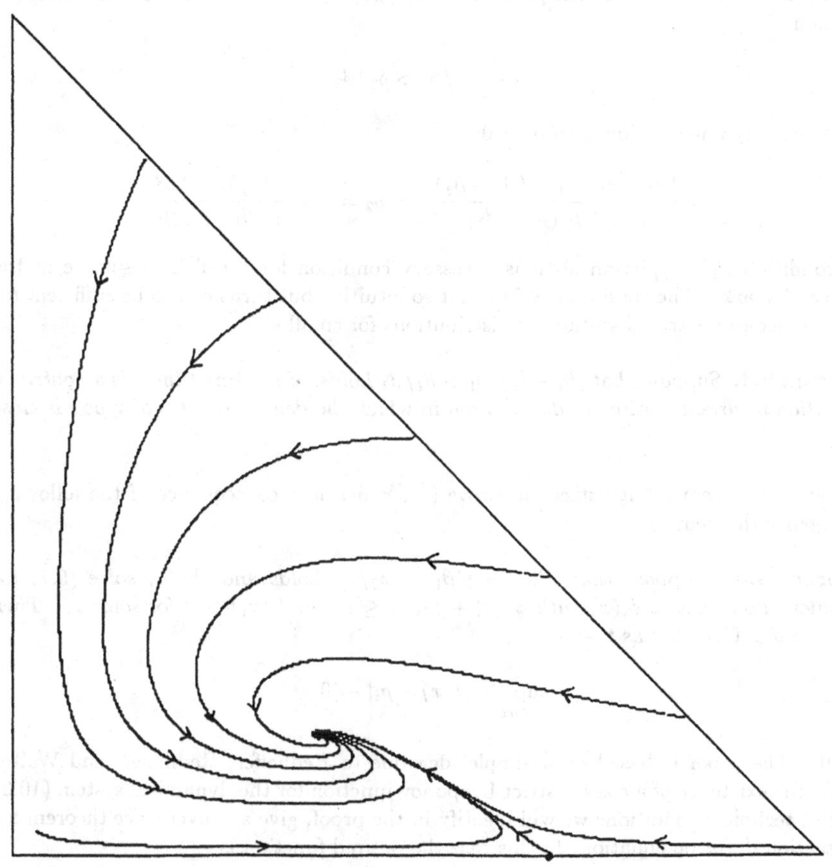

Figure 9.1

Here we have re-arranged the right hand side to show that the system is the standard predator–prey equations for species with limited growth. (See e.g., page 263 of Hirsch and Smale (1974).) Before we plunge into the details of analyzing (9.2), the reader should look at Figure 9.1, which gives some solutions of (9.2) with

$$\beta_1 = 3, \delta_1 = 1, \quad \beta_2 = 3, \delta_2 = 1, \quad \gamma = 1$$

In this case, as we will prove later, there is a fixed point at (8/21, 3/21) that is globally attracting.

The first step in understanding (9.1) is to look at (9.2) and ask: "What are the fixed points, i.e., solutions of the form $v_i(t) = \rho_i$?" It is easy to solve for the ρ_i. There is always the trivial solution $\rho_1 = \rho_2 = 0$. In the absence of sharks the fish are a contact process. So if $\beta_1 > \delta_1$ there is a solution $\rho_1 = (\beta_1 - \delta_1)/\beta_1$, $\rho_2 = 0$. If we impose the stronger condition

$$(\beta_1 - \delta_1)/\beta_1 > \delta_2/\beta_2 \tag{9.3}$$

there is exactly one solution with $\rho_2 > 0$:

$$\rho_1 = \frac{(\beta_1 - \delta_1)\gamma + \delta_2(\beta_1 + \beta_2)}{\beta_1\gamma + \beta_2(\beta_1 + \beta_2)} \qquad \rho_2 = \frac{(\beta_1 - \delta_1)\beta_2 - \delta_2\beta_1}{\beta_1\gamma + (\beta_1 + \beta_2)\beta_2}$$

The condition $\beta_1 > \delta_1$ is an obvious necessary condition for the fish to survive in the absence of sharks. The condition (9.3) is not so intuitive but turns out to be sufficient for the existence of nontrivial stationary distributions for small ϵ.

Theorem 9.3. *Suppose that $(\beta_1 - \delta_1)/\beta_1 > \delta_2/\beta_2$ holds. If ϵ is small there is a nontrivial translation invariant stationary distribution in which the density of sites of type i is close to ρ_i.*

In view of Theorem 9.1 it suffices to prove ($\star$), which is a consequence of the following convergence theorem.

Theorem 9.4. *Suppose that $(\beta_1 - \delta_1)/\beta_1 > \delta_2/\beta_2$ holds and the u_i solve (9.1) for continuous nonnegative $\phi_i(x)$ with $\phi_1(x) + \phi_2(x) \le 1$ and $\phi_i(x_i) > 0$ for some x_i. Then there is a $\sigma > 0$ so that as $t \to \infty$,*

$$\sup_{\|x\| \le \sigma t} |u_i(t,x) - \rho_i| \to 0.$$

Proof: The proof is based on a simple idea due to Redheffer, Redlinger, and Walter (1988): the existence of a convex strict Lyapunov function for the dynamical system (10.2) plus two technical conditions we will identify in the proof, give a convergence theorem for the reaction diffusion equation. In this case the desired function is

$$H(v_1, v_2) = \beta_2(v_1 - \rho_1 \log v_1) + (\beta_1 + \beta_2)(v_2 - \rho_2 \log v_2)$$

Being the sum of four convex functions, H is clearly convex. The next step is to check that it is a strict Lyapunov function: if (v_1, v_2) is a solution of the dynamical system that does not start at the fixed point then $\partial H(v_1, v_2)/\partial t < 0$. Differentiating gives

$$\frac{\partial H}{\partial v_1} = \beta_2\left(1 - \frac{\rho_1}{v_1}\right) \qquad \frac{\partial H}{\partial v_2} = (\beta_1 + \beta_2)\left(1 - \frac{\rho_2}{v_2}\right)$$

So using the chain rule and (9.2)

$$\begin{aligned} H_t \equiv \frac{\partial H(v_1, v_2)}{\partial t} &= \beta_2(v_1 - \rho_1)\{(\beta_1 - \delta_1) - \beta_1 v_1 - (\beta_1 + \beta_2)v_2\} \\ &\quad + (\beta_1 + \beta_2)(v_2 - \rho_2)\{-\delta_2 + \beta_2 v_1 - \gamma v_2\} \end{aligned}$$

Using the next two identities to subtract 0 from each term in braces

$$\begin{aligned} 0 &= (\beta_1 - \delta_1) - \beta_1\rho_1 - (\beta_1 + \beta_2)\rho_2 \\ 0 &= -\delta_2 + \beta_2\rho_1 - \gamma\rho_2 \end{aligned}$$

gives

$$\begin{aligned} H_t &= \beta_2(v_1 - \rho_1)\{-\beta_1(v_1 - \rho_1) - (\beta_1 + \beta_2)(v_2 - \rho_2)\} \\ &\quad + (\beta_1 + \beta_2)(v_2 - \rho_2)\{\beta_2(v_1 - \rho_1) - \gamma(v_2 - \rho_2)\} \\ &= -\beta_1\beta_2(v_1 - \rho_1)^2 - \gamma(\beta_1 + \beta_2)(v_2 - \rho_2)^2 \le 0 \end{aligned} \tag{9.4}$$

with strict inequality for $(v_1, v_2) \neq (\rho_1, \rho_2)$. The importance of the last conclusion is that $H(v_1(t), v_2(t))$ is strictly decreasing in t and hence all trajectories that begin in $(0, \infty)^2$ must end at the minimum of H, (ρ_1, ρ_2). For later purposes we would like to note that the level curves $H_t = -r$ are concentric ellipses.

The above computations that show H is a Lyapunov function obviously depend on the special form of (9.2). To prepare for other applications at the end of this section, we would like the reader to check that in what follows only equations (9.6) and (9.10) depend on the special form of H.

Since composing the Lyapunov function with solutions of the dynamical system shows that they converge to the fixed point, it is natural to look at $h(t,x) = H(u_1(t,x), u_2(t,x)) - H(\rho_1, \rho_2) \ge 0$ when u is a solution of (9.1). (Here we have subtracted the value of H at its minimum to make the minimum value 0.) To show the generality of this computation and to simplify notation we will write (9.1) as

$$\frac{\partial u_i}{\partial t} = \Delta u_i + f_i(u).$$

Differentiating and using the previous equation gives

$$\begin{aligned} \frac{\partial h}{\partial t} &= \sum_i \frac{\partial H}{\partial u_i}\frac{\partial u_i}{\partial t} = \sum_i \frac{\partial H}{\partial u_i} \cdot (\Delta u_i + f_i(u)) \\ \frac{\partial^2 h}{\partial x_m^2} &= \sum_i \frac{\partial H}{\partial u_i}\frac{\partial^2 u_i}{\partial x_m^2} + \sum_{i,j} \frac{\partial^2 H}{\partial u_i \partial u_j}\frac{\partial u_i}{\partial x_m}\frac{\partial u_j}{\partial x_m} \end{aligned}$$

Here and in what follows the indices i and j are summed from 1 to 2. Summing the second equation from $m = 1$ to d gives

$$\Delta h = \sum_i \frac{\partial H}{\partial u_i}\Delta u_i + \sum_{m,i,j} \frac{\partial^2 H}{\partial u_i \partial u_j}\frac{\partial u_i}{\partial x_m}\frac{\partial u_j}{\partial x_m}$$

so using $H_t = \sum_i \frac{\partial H}{\partial u_i} f_i(u)$ gives

$$\frac{\partial h}{\partial t} = \Delta h + H_t - \sum_{m,i,j} \frac{\partial^2 H}{\partial u_i \partial u_j}\frac{\partial u_i}{\partial x_m}\frac{\partial u_j}{\partial x_m}$$

Since H is convex the last term (including the minus sign) is nonpositive and we have

$$\frac{\partial h}{\partial t} \le \Delta h + H_t \tag{9.5}$$

To prove Theorem 9.4, we will use (9.5) to conclude

$$\sup_{\|x\|\le \epsilon t} h(t,x) \to 0 \tag{9.6}$$

If we were on a bounded set with Neumann boundary conditions this would be easy, since in this case $\inf u_i(t,x) > 0$ at positive times and thus $h(t,x)$ is bounded. If we let x_t be a place where $m_t = \max_x h(t,x)$ is attained then $\Delta h(t,x_t) \le 0$ so

$$\frac{dm_t}{dt} \le H_t \le \sup\{H_t(v) : H(v) = m_t\} < 0$$

and an argument like the one in the proof of Theorem 5.1 shows (9.6).

To prove (9.6) on $\mathbf{R}^d$ we have to deal with the fact that $h(t,x)$ may be unbounded. To do this we first get bounds on how fast the H_t will push h to 0 and then get *a priori* bounds on $h(t,x)$ inside $\|x\| \le at$ that will allow us to drive h to 0. To get upper bounds on H_t (recall it is ≤ 0), we let

$$g(h) = \inf\{-H_t(v_1,v_2) : H(v_1,v_2) \ge h\}$$

We have defined g this way to make it clear that $h \to g(h)$ is increasing. To determine the behavior as $h \to 0$ we observe that at (ρ_1,ρ_2) $\partial H/\partial v_i = 0$ and

$$\frac{\partial^2 H}{\partial v_1^2} = \frac{\beta_2\rho_1}{v_1^2} \qquad \frac{\partial^2 H}{\partial v_2^2} = \frac{(\beta_1+\beta_2)\rho_2}{v_2^2} \qquad \frac{\partial^2 H}{\partial v_1 \partial v_2} = 0$$

So near (ρ_1,ρ_2)

$$H(v_1,v_2) - H(\rho_1,\rho_2) \approx \frac{\beta_2}{\rho_1}(v_1-\rho_1)^2 + \frac{(\beta_1+\beta_2)}{\rho_2}(v_2-\rho_2)^2$$

and it follows from (9.4) that $g(h) \sim Bh$. Since $g(h)$ is increasing we have

$$(9.7) \qquad g(h) \geq \alpha h/(1+h) \quad \text{for some } \alpha > 0$$

The next step in bounding $h(t,x)$ is to examine the behavior of

$$w' = \frac{-\alpha w}{1+w} \qquad w(0) = W$$

Since $w(t) \geq W - \alpha t$ the time to reach $\eta > 0$ is at least $(W-\eta)/\alpha$. To see this estimate is fairly sharp observe that while $w(t) \geq W^{1/2}-1$ we have $w' \leq -\alpha(1-W^{-1/2})$ so $w(t)$ reaches $W^{1/2}-1$ at time $\leq W/(\alpha(1-W^{-1/2}))$. When $w(t) \leq W^{1/2}-1$ we have $w' \leq -\alpha w/W^{1/2}$ so the time to go from $W^{1/2}-1$ to η is at most $W^{1/2}\alpha^{-1}(\log(W^{1/2}) - \log\eta)$. Adding the two estimates we see that

(9.8) *For $W \geq 4$ and $\eta < 1$ the time to reach η is smaller than*

$$2\alpha^{-1}W + CW^{1/2}(\log W - \log\eta).$$

To get *a priori* bounds on h note that our hypotheses imply that $u_i(1,x)$ is positive and continuous so there are constants μ_i so that $u_i(1,x) \geq \mu_i$ for all x with $\|x\|_2 \leq 1$, and we can without loss of generality assume that the last conclusion holds at time 0.

(9.9) **Lemma.** *There is a constant K so that if $\|x\|_2 \leq at$ and $t \geq 1$ then $h(t,x) \leq Kt$.*

Proof: We have supposed that $\phi_1(x)+\phi_2(x) \leq 1$ so the probabilistic interpretation implies $u_1(t,x) + u_2(t,x) \leq 1$ for all t and x and it follows that

$$\frac{\partial u_1}{\partial t} \geq \Delta u_1 - (\beta_1 + \beta_2)u_1$$
$$\frac{\partial u_2}{\partial t} \geq \Delta u_2 - (\delta_2 + \gamma)u_2$$

To see these inequalities it is convenient to write the right–hand side in the form in (9.2). (Recall that our main assumption (9.3) implies $\beta_1 > \delta_1$.) Let $c_1 = (\beta_1+\beta_2)$ and $c_2 = \delta_2+\gamma$. Recalling that solutions of

$$\frac{\partial u}{\partial t} = \Delta v - cv \qquad u(0,x) = \phi(x)$$

are given by

$$u(t,x) = e^{-ct}\int (4\pi t)^{-d/2} e^{-\|x-y\|_2^2/4t}\phi(y)\,dy$$

and using the maximum principle (see (9.11) at the end of this section) we have that when $\|x\|_2 \leq at$

$$\begin{aligned} u_i(t,x) &\geq e^{-c_i t}\mu_i \int_{\|y\|\leq 1} (4\pi t)^{-d/2} e^{-(\|x\|_2+1)^2/4t}\,dy \\ &\geq C_d\mu_i(4\pi t)^{-d/2}\exp(-(c_i + (a^2/4))t - a/2 - (1/4t)) \end{aligned}$$

where C_d is the volume of $\{y : \|y\|_2 \le 1\}$. Combining the last expression with the fact that

(9.10) $$h(x) \le C(1 - \log(\min_i x_i))$$

completes the proof of (9.9) □

Let $a > 0$ be chosen so that $3\alpha^{-1}aK < (1-a)$, i.e., so that if $\omega(t)$ solves

$$\omega' = -\alpha\omega/(1+\omega) \quad \text{with} \quad \omega(0) = Kat$$

then for any $\eta > 0$ when $t > T_\eta$, $\omega((1-a)t) < \eta$. We will prove Theorem 9.4 with $\sigma = a/2$. Let $D_r = \{y : \|y\|_2 < r\}$ and define $h_1^t(t, x)$ to be the solution of

$$\frac{\partial h}{\partial t} = \Delta h - \alpha h/(1+h) \quad \text{in } \mathcal{D}_t \equiv [at, t] \times D_{at}$$
$$h(s, x) = Ks \quad \text{if } s = at, \text{ or } x \in \partial D_{at}$$

Since $h(s,x) \le h_1^t(s,x)$ when $s = at$ or $x \in \partial D_{at}$, and $g(h) \ge \alpha h/(1+h)$ it follows from the maximum principle that $h(s,x) \le h_1^t(s,x)$ for $(s,x) \in \mathcal{D}_t$.

To bound $h_1^t(t,x)$ we will use $h_2^t(s,x) = \omega(s - at)$. Another use of the maximum principle shows $h_1^t(s,x) \ge h_2^t(s,x)$ in $\mathcal{D}_t$. The last inequality is the opposite of the one we want but we will turn it around by showing that the difference is small when $\|x\|_2 \le at/2$. Intuitively the difference is only due to paths $(t-s, B_s)$ that escape from the space time cylinder $[at,t] \times D_{at}$ on the side. Here B_s is a Brownian motion run at twice its usual speed. When the starting point $\|B_0\|_2 \le at/2$ this event has exponentially small probability and brings a "reward" $\le Kt$ so the difference $h_1^t - h_2^t$ goes to 0 exponentially fast as $t \to \infty$.

To begin to turn our intuition into a proof, we let $\bar{g}(x) = \alpha x/(1+x)$ and observe that Itô's formula implies that if $\tau = \inf\{s : B_s \notin D_{at}\}$ then

$$h_i^t(t - (s \wedge \tau), B_{s\wedge\tau}) - \int_0^{s\wedge\tau} \bar{g}(h_i^t(t-r, B_r))\, dr \qquad s \le (1-a)t$$

is a bounded martingale. Using the martingale property at time $s = (1-a)t$ gives

$$h_i^t(t,x) = E_x\left(h_i^t(at, B_{(1-a)t}) - \int_0^{(1-a)t} \bar{g}(h_i^t(t-r, B_r))\, dr\,; \tau > (1-a)t\right)$$
$$+ E_x\left(h_i^t(t-\tau, B_\tau) - \int_0^{\tau} \bar{g}(h_i^t(t-r, B_r))\, dr\,; \tau \le (1-a)t\right)$$

Since $h_1^t(at,x) = h_2^t(at,x)$, $0 \le h_2^t(t-r, B_r) \le h_1^t(t-r, B_r)$ when $\tau \ge r$, and $\bar{g}$ is increasing, it follows that on $\{\tau > (1-a)t\}$ we have

$$h_1^t(at, B_{(1-a)t}) - \int_0^{(1-a)t} \bar{g}(h_1^t(t-r, B_r))\, dr$$
$$\le h_2^t(at, B_{(1-a)t}) - \int_0^{(1-a)t} \bar{g}(h_2^t(t-r, B_r))\, dr$$

Subtracting the two expressions for $h_i^t(t,x)$ and recalling $h_i^t \geq 0$ and $\bar{g} \geq 0$ gives

$$\begin{aligned} h_1^t(t,x) - h_2^t(t,x) &\leq 0 + E_x\left(h_1^t(t-\tau, B_\tau) + \int_0^\tau \bar{g}(h_2^t(t-r,B_r))\,dr; \tau \leq (1-a)t\right) \\ &\leq (Kt + \alpha t)P_x(\tau \leq (1-a)t) \end{aligned}$$

since $h_1^t(s,x) = Ks$ when $x \in \partial D_{at}$ and $0 \leq \bar{g} \leq \alpha$. Standard large deviations estimates for Brownian motion imply that for $\|x\|_2 \leq at/2$, $P_x(\tau < (1-a)t) \leq C\exp(-\delta t)$, so

$$\sup_{\|x\| \leq at/2} |h_1^t(t,x) - h_2^t(t,x)| \to 0 \quad \text{as } t \to \infty$$

Since $h_1^t(t,x) \geq h(t,x)$ for $x \in D_{at}$ and $h_2^t(t,x) = \omega((1-a)t) < \eta$ for $t > T_\eta$, Theorem 9.4 follows. □

For completeness we give a proof of

(9.11) **Maximum Principle.** *Suppose $f_1(h) \geq f_2(h)$ and the h_i solve*

$$\frac{\partial h_i}{\partial t} = \Delta h_i - f_i(h_i) \quad \text{in } \mathcal{D}_t$$

with $h_1(s,x) \leq h_2(s,x)$ if $s = at$, or $x \in \partial D_{at}$ then $h_1(s,x) \leq h_2(s,x)$ in $\mathcal{D}_t$.

PROOF: This is easier to prove than to find in the library. Suppose first that $f_1(h) > f_2(h)$ and $h_1(s,x) < h_2(s,x)$ if $s = at$, or $x \in \partial D_{at}$. Let s_0 be the smallest value of s for which there is an x with $h_1(s,x) \geq h_2(s,x)$. Continuity of the h_i implies that we can find an x_0 so that $h_1(s_0,x_0) = h_2(s_0,x_0)$. The strict inequality between the h_i on the boundary implies $x_0 \in D_{at}$ and $s_0 > 0$. The definition of s_0 implies that $h_1(s_0,x) \leq h_2(s_0,x)$ for all x. Since $h_1(s_0,x_0) = h_2(s_0,x_0)$, we must have $\nabla h_1(s_0,x_0) = \nabla h_2(s_0,x_0)$ and $\Delta h_1(s_0,x_0) \leq \Delta h_2(s_0,x_0)$. Using the last fact and $f_1(h) > f_2(h)$ it follows that at (s_0,x_0)

$$\frac{\partial h_1}{\partial t} = \Delta h_1 - f_1(h_1) < \Delta h_2 - f_2(h_2) = \frac{\partial h_2}{\partial t}$$

However this implies that $h_1(s_0-\epsilon,x) > h_2(s_0-\epsilon,x_0)$ for small ϵ contradicting the definition of s_0, so we must have $h_1(s,x) < h_2(s,x)$ for all $(s,x) \in \mathcal{D}_{at}$. To prove the result in (2.4) now let $f_0(h) = f_1(h) + \epsilon$ and change the boundary values to $h_1(s,x) - \epsilon$. The new solution $h_0^\epsilon(s,x) < h_2(s,x)$ and converges pointwise to $h_1(s,x)$ as $\epsilon \to 0$. □

The main reason for interest in Theorem 9.4 is that it applies to systems of equations. However, as the next two examples suggest we also get interesting information when we apply it to a single equation.

Example 9.2. The basic contact process. If we let $\beta = \lambda N$ where N is the number of neighbors and write u for u_1 then the equation in Example 8.1 can be written as

$$\frac{\partial u}{\partial t} = \Delta u - u + \beta(1-u)u \qquad u(0,x) = \phi(x) \tag{9.12}$$

To find a Lyapunov function we let $\rho = (\beta - 1)/\beta$ and write the dyanmical system as

$$\frac{dv}{dt} = v(-1 + \beta(1 - v)) = \beta v(\rho - v)$$

Taking $H(v) = v - \rho \log v$ and noticing $h'(v) = 1 - (\rho/v)$ we have

$$\frac{dH(v(t))}{dt} = -\beta(v - \rho)^2$$

Clearly H satisfies (9.10). Since $H'(\rho) = 0$ and $H''(v) = \rho/v^2$, repeating the proof of (9.7) shows it is satisfied. Since H is convex we get a convergence result like Theorem 9.4

Theorem 9.5. *Suppose that $\beta > 1$ u solves (9.12) for continuous $0 \le \phi(x) \le 1$ with $\phi(x_0) > 0$ for some x_0. Then there is a $\sigma > 0$ so that as $t \to \infty$,*

$$\sup_{\|x\| \le \sigma t} \left| u(t, x) - \frac{\beta - 1}{\beta} \right| \to 0.$$

Much better convergence results than this are known for this equation (see Aronson and Weinberger (1978) for more general results and Bramson (1983) for more detailed information), but the last result shows that $(\star)$ holds and we have

Theorem 9.6 *Suppose $\beta > 1$. If ϵ is small then the contact process with strring at rate ϵ^{-2} has a translation invariant stationary distribution in which the density of 1's is close to $(\beta - 1)/\beta$.*

Example 9.3. The threshold voter model. In this case if N is the number of neighbors and we write u for u_1 then the limiting equation in Example 8.2 is

$$\text{(9.13)} \qquad \frac{\partial u}{\partial t} = \Delta u - u(1 - u^N) + (1 - u)(1 - (1 - u)^N) \qquad u(0, x) = \phi(x)$$

When $N = 1$ the last two terms on the right hand side cancel so we will suppose that $N \ge 2$. For our Lyapunov function we take $H(v) = -\log v - \log(1 - v)$, which has

$$H'(v) = -\frac{1}{v} + \frac{1}{1 - v} = \frac{2v - 1}{v(1 - v)}$$

$$\begin{aligned} \frac{dH(v(t))}{dt} &= \frac{2v - 1}{v(1 - v)} \left\{ -v(1 - v^N) + (1 - v)(1 - (1 - v)^N \right\} \\ &= (2v - 1) \left\{ -(1 + v + \cdots + v^{N-1}) + (1 + (1 - v) + \cdots + (1 - v)^{N-1} \right\} \\ &= -(2v - 1)^2 \left\{ 1 + \sum_{j=2}^{N-1} \frac{(1 - v)^j - v^j}{1 - 2v} \right\} \end{aligned}$$

where the sum is 0 if $N = 2$. Since $(1-v)^j - v^j$ and $1-2v$ are both positive on $v < 1/2$ and negative on $v > 1/2$ their quotient is always positive. To compute the value at $v = 1/2$ we note that L'Hopital's rule implies that

$$\lim_{v\to 1/2} \frac{(1-v)^j - v^j}{1-2v} = \lim_{v\to 1/2} \frac{-j(1-v)^{j-1} - jv^{j-1}}{-2v} = j2^{-(j-1)}$$

so the term in braces is bounded away from 0 and ∞.

Since $H'(1/2) = 0$ and $H''(1/2) > 0$ it is easy to see as before that (9.7) holds. The other condition (9.10) does not hold as stated since $H(1) = \infty$. However it is easy to see that under suitable assumptions (9.9) holds and we have

Theorem 9.7. *Suppose that $N \geq 2$ u solves (9.13) for continuous $0 \leq \phi(x) \leq 1$ with $\phi(x_0) > 0$ for some x_0 and $\phi(x_1) < 1$ for some x_1. Then there is a $\sigma > 0$ so that as $t \to \infty$,*

$$\sup_{\|x\|\leq\sigma t} |u(t,x) - 1/2| \to 0.$$

Again better convergence results than this are known for this equation (see Aronson and Weinberger (1978) and Fife and McLeod (1977)) but the last result shows that $(\star)$ holds and we have

Theorem 9.8. *Suppose $N \geq 2$. If ϵ is small then the contact process with strring at rate ϵ^{-2} has a translation invariant stationary distribution in which the density of 1's is equal to 1/2.*

We get "equal to 1/2" rather than just "close to 1/2" by starting from product measure with density 1/2 and using the symmetry of the dynamics under interchange of 0's and 1's. Comparing this with Theorems 5.1 and 5.3, the only surprise is that in the nearest neighbor case there is a stationary distribution with fast stirring. We conjecture that the presence of stirring at any positive rate, there is a nontrivial stationary distribution. In support of this conjecture, Figure 9.2 shows a simulation of the nearest neighbor case with stirring rate = 3.

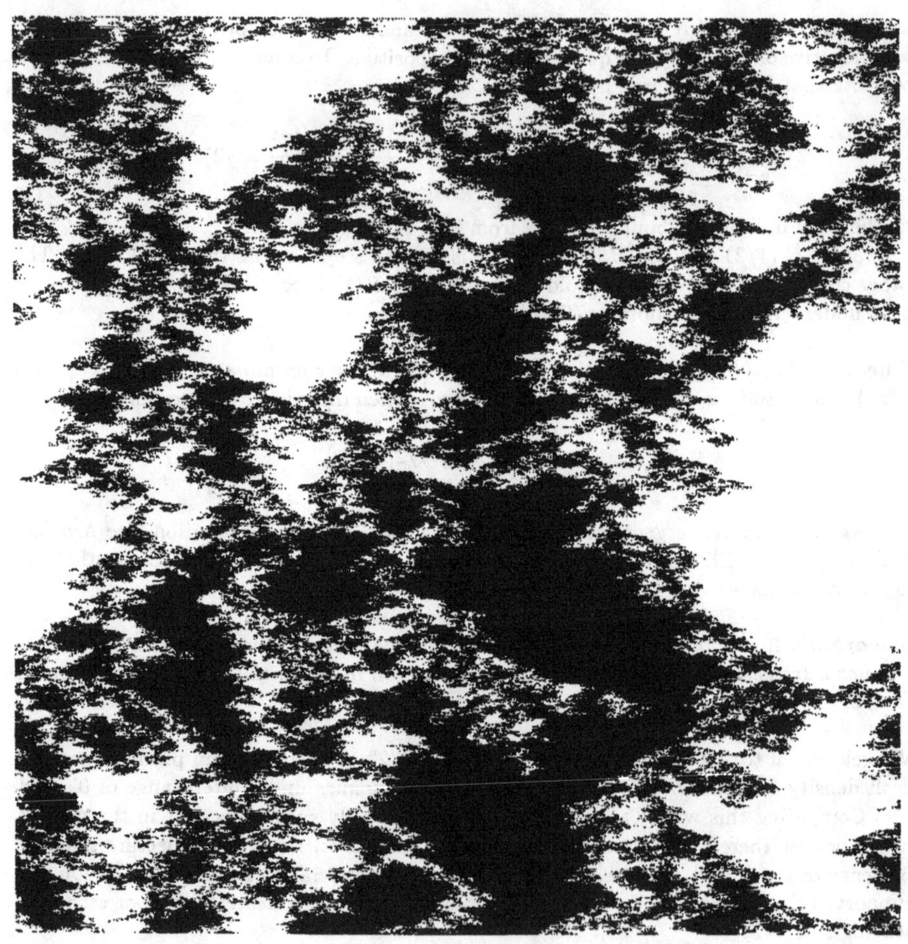

Figure 9.2. Threshold voter model, $d = 1$, $\mathcal{N} = \{-1, 1\}$, with stirring at rate 3

Appendix. Proofs of the Comparison Results

In this section we will prove Theorems 4.1, 4.2, and 4.3. The proofs are not beautiful but by now the reader has hopefully been convinced that they are useful. We begin by recalling the set-up and repeating some definitions that were more fully explained in Section 4. Let

$$\mathcal{L}_0 = \{(x,n) \in \mathbf{Z}^2 : x+n \text{ is even}, n \geq 0\}$$

and make $\mathcal{L}_0$ into a graph by drawing oriented edges from (x,n) to $(x+1,n+1)$ and from (x,n) to $(x-1,n+1)$. Given random variables $\omega(x,n)$ that indicate whether the sites are open (1) or closed (0), we say that (y,n) can be reached from (x,m) and write $(x,m) \to (y,n)$ if there is a sequence of points $x = x_m, \ldots, x_n = y$ so that $|x_k - x_{k-1}| = 1$ for $m < k \leq n$ and $\omega(x_k,k) = 1$ for $m \leq k \leq n$. We say that the $\omega(x,n)$ are "*M dependent with density at least* $1-\gamma$" if whenever (x_i,n_i), $1 \leq i \leq I$ is a sequence with $\|(x_i,m_i)-(x_j,m_j)\|_\infty > M$ if $i \neq j$ then

$$P(\omega(x_i,n_i) = 0 \text{ for } 1 \leq i \leq I) \leq \gamma^I \tag{A.1}$$

Let $\mathcal{C}_0 = \{(y,n) : (0,0) \to (y,n)\}$ be the set of all points in space-time that can be reached by a path from $(0,0)$. $\mathcal{C}_0$ is called the *cluster containing the origin.* When the cluster is infinite, i.e., $\{|\mathcal{C}_0| = \infty\}$ we say that *percolation occurs.* Our first result shows that if the density of open sites is high enough then percolation occurs.

Theorem A.1. If $\theta \leq 6^{-4(2M+1)^2}$ then $P(|\mathcal{C}_0| < \infty) \leq 55\,\theta^{1/(2M+1)^2} \leq 1/20$.

PROOF: The proof is by the contour method. Even though the argument is messy to write down, the idea is simple: if $|\mathcal{C}_0| < \infty$ then there is a "contour" of closed sites that stops the percolation from occuring. As we will show, the probability of a specific contour of length n is $\leq g(\theta)^n$ where $g(\theta) \to 0$ as $\theta \to 0$ and the number of contours of length n is $\leq 3^n$ so by summing a geometric series we see that the existence of a contour is unlikely if θ is small.

Most of the work goes into defining the contour. Before starting on this we have to discard a trivial case: if $(0,0)$ is closed, an event with probability $\leq \gamma$, then $\mathcal{C}_0 = \emptyset$. For the rest of the proof we will concentrate on the case in which $(0,0)$ is open and hence $(0,0) \in \mathcal{C}_0$. Let $D = \{z \in \mathbf{R}^2 : \|z\|_1 \leq 1\}$, where D is for diamond. To turn the cluster $\mathcal{C}_0$ into a solid blob, we look at

$$\mathcal{D}_0 = \cup_{(m,n)\in\mathcal{C}_0} ((m,n)+D)$$

where $(m,n)+D = \{(m,n)+z : z \in D\}$ is the set D translated by (m,n). When $(0,0) \in \mathcal{C}_0$, the lowest point in $\mathcal{D}_0$ is $(0,-1)$. If $|\mathcal{C}_0| < \infty$, then the open set

$$G = \{\mathbf{R} \times (-1,\infty)\} - \mathcal{D}_0$$

has exactly one unbounded component U. We call $\Gamma = \partial U \cap \mathcal{D}_0$ the *contour* associated with $\mathcal{C}_0$, and orient it so that the segment $(0,-1) \to (1,0)$, which is always present, is oriented in the direction indicated. For an example see Figure A.1.

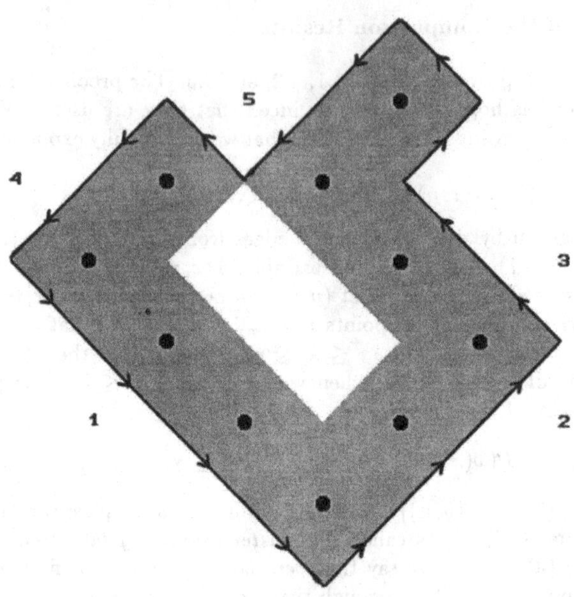

Figure A.1

The contour is made up of segments that are translates of the four sides of D

type	1	2	3	4
translate of	$(-1,0) \to (0,-1)$	$(0,-1) \to (1,0)$	$(1,0) \to (0,1)$	$(0,1) \to (-1,0)$

As we walk along the contour in the direction of the orientation, our left hand is always touching $\mathcal{D}_0$ and our right is always touching U. If we stand at the midpoint of one of the segments that make up Γ then the site in

$$\mathcal{L} = \{(m,n) \in \mathbf{Z}^2 : m+n \text{ is even}\}$$

closest to our right hand is called the *site associated with the segment.* A glance at Figure A.1 reveals that the sites associated with segments of types 3 and 4 must be closed but those associated with types 1 and 2 may be open or closed. Let n_i be the number of segments of type i in the contour. The segments of types 1 and 2 increase the x coordinate by 1, while those of types 3 and 4 decrease the x coordinate by 1. The contour ends where it begins so $n_3 + n_4 = n_1 + n_2$ and hence if the contour is composed of n segments we must have $n_3 + n_4 = n/2$. Now a closed site may be associated with one type 3 and one type 4 segement (see 5 on Figure A.1) but cannot be associated with more than one segment of each type, so if there are n segments in the contour there must be at least $n/4$ closed sites along it.

To count the number of contours of length n we note that the first segment is always $(0,-1) \to (1,0)$ and after that there are at most 3 choices at each stage (since we cannot retrace the step just made), so there are at most 3^{n-1} contours of length n. Suppose for the moment that the states of the sites are independent and open with probability $1-\gamma$. Noting that the length of the contour is ≥ 4, it follows that

$$P(0 < |\mathcal{C}_0| < \infty) \leq \sum_{n=4}^{\infty} 3^{n-1}\gamma^{n/4} = \frac{1}{3} \cdot \frac{(3\gamma^{1/4})^4}{1-3\gamma^{1/4}} = \frac{27\gamma}{1-3\gamma^{1/4}}$$

which is $< 1-\gamma$ if γ is small enough. (Recall $P(\mathcal{C}_0 = \emptyset) \leq \gamma$.) To extend the last result to the M dependent case, note that we can find a subset of the closed sites along the contour of size at least $n/4(2M+1)^2$ so that for each $z \neq w$ in this set $\|z-w\|_\infty > M$. (Pick any closed site to start, then throw out the $\leq (2M+1)^2 - 1$ closed sites in our set that are too close to the first one, pick another site, throw out the closed sites too close to it ...) Using (4.1) and noting our assumption on γ implies $3\gamma^{1/4(2M+1)^2} \leq 1/2$ we have

$$\begin{aligned} P(0 < |\mathcal{C}_0| < \infty) &\leq \sum_{n=4}^{\infty} 3^{n-1}\, \gamma^{n/4(2M+1)^2} \\ &= \frac{1}{3} \cdot \frac{(3\gamma^{1/4(2M+1)^2})^4}{1-3\gamma^{1/4(2M+1)^2}} \leq 54\,\gamma^{1/(2M+1)^2} \end{aligned}$$

Recalling now that $P(\mathcal{C}_0 = \emptyset) \leq \gamma \leq \gamma^{1/(2M+1)^2}$, we have proved Theorem 4.1. □

From the last proof it follows immediately that if we let $|\Gamma|$ denote the number of segments in the contour and assume $\gamma \leq 6^{-4(2M+1)^2}$ then

$$(A.2) \qquad P(L \leq |\Gamma| < \infty) \leq \sum_{n=L}^{\infty} 3^{n-1}\gamma^{n/4(2M+1)^2} = \frac{1}{3} \cdot \frac{(3\gamma^{1/4(2M+1)^2})^L}{1-3\gamma^{1/4(2M+1)^2}} \leq 2^{-L}$$

In order to prove the existence of stationary distributions we need results about M dependent oriented percolation starting from the initial configuration W_0^p in which the events $\{x \in W_0^p\}$, $x \in 2\mathbf{Z}$ are independent and have probability p. Let

$$W_n^p = \{y : (x,0) \to (y,n) \text{ for some } x \in W_0^p\}$$

Theorem A.2. *If* $p > 0$ *and* $\gamma \leq 6^{-4(2M+1)^2}$ *then*

$$\liminf_{n\to\infty} P(0 \in W_{2n}^p) \geq 1 - 55\,\theta^{1/(2M+1)^2} \geq 19/20$$

Proof: The first step is to look backwards in time to reduce the new problem to the old one solved in (A.2). This is the discrete time version of the duality considered in Section 3. To have the dual process defined for all time, it is convenient to introduce independent random variables $\omega(x,n)$ for $n < 0$ that have $P(\omega(x,n) = 1) = 1-\gamma$ and look at the percolation process on $\mathcal{L} = \{(x,n) \in \mathbf{Z}^2 : x+n \text{ is even}\}$. Later in the proof we will want to use the fact that $\gamma > 0$, so you should observe that the desired conclusion is trivial when $\gamma = 0$, i.e., all sites are open.

We say that (x,m) can be reached from (y,n) by a dual path (and write $(y,n) \to_* (x,m)$) if there is a sequence of points $x = x_m, \ldots, x_n = y$ so that $|x_k - x_{k-1}| = 1$ for $m < k \le n$ and $\omega(x_k,k) = 1$ for $m \le k \le n$. It should be clear from the definition that $(x,m) \to (y,n)$ if and only if $(y,n) \to_* (x,m)$, so

$$\{0 \in W_{2n}^p\} = \{(0,2n) \to_* (x,0) \text{ for some } x \in W_0^p\}$$

To estimate the right hand side it is convenient to introduce

$$\hat{W}_m^{2n} = \{x : (0,2n) \to_* (x, 2n-m)\}$$
$$\hat{\mathcal{C}}_{(0,2n)} = \{(x,t) : (0,2n) \to_* (x,t)\}$$

By conditioning on the value of $\hat{W}_{2n}^{2n}$, it is easy to see that

$$(A.3) \qquad P(0 \in W_{2n}^p) = 1 - E\left\{(1-p)^{|\hat{W}_{2n}^{2n}|}\right\}$$

so to complete the proof we want to show that if n is large and $\hat{W}_{2n}^{2n} \neq \emptyset$ then $|\hat{W}_{2n}^{2n}|$ is large with high probability. The process $\hat{W}_m^{2n}$ comes from random variables $\omega(x,n)$ that have property $(A.1)$, and the event on the left hand side of $(A.4)$ below implies that the contour associated with $\hat{\mathcal{C}}_{(0,2n)}$ has length at least $4n$, so (A.2) implies

$$(A.4) \qquad P(\hat{W}_{2n}^{2n} \neq \emptyset, |\hat{\mathcal{C}}_{(0,2n)}| < \infty) \le P(4n \le |\Gamma| < \infty) \le 2^{-4n}$$

Now the sites $(x,-1) \in \mathcal{L}$ are independent of those in $\mathcal{L}_0$ and are closed with probability γ so

$$(A.5) \qquad P\left(\hat{W}_{2n+1}^{2n} = \emptyset \,\middle|\, 0 < |\hat{W}_{2n}^{2n}| \le \sqrt{n}\right) \ge \theta^{2\sqrt{n}}$$

Combining $(A.4)$ and $(A.5)$ gives

$$(A.6) \qquad P(0 < |\hat{W}_{2n}^{2n}| \le \sqrt{n}) \le \frac{P(\hat{W}_{2n}^{2n} \neq \emptyset, |\hat{\mathcal{C}}_{(0,2n)}| < \infty)}{P\left(\hat{W}_{2n+1}^{2n} = \emptyset \,\middle|\, 0 < |\hat{W}_{2n}^{2n}| \le \sqrt{n}\right)} \le 2^{-4n}\gamma^{-2\sqrt{n}}$$

Using $(A.3)$ in the first step; then $(A.6)$ and $P(|\hat{W}_{2n}^{2n}| > 0) \ge P(|\hat{\mathcal{C}}_{(0,2n)}| = \infty)$ in the second; and finally, Theorem 4.1 in the third we have

$$\begin{aligned} P(0 \in W_{2n}^p) &\ge \left\{1 - (1-p)^{\sqrt{n}}\right\} P(|\hat{W}_{2n}^{2n}| \ge \sqrt{n}) \\ &\ge \left\{1 - (1-p)^{\sqrt{n}}\right\} \left(P(|\hat{\mathcal{C}}_{(0,2n)}| = \infty) - 2^{-4n}\gamma^{-\sqrt{n}}\right) \\ &\ge \left\{1 - (1-p)^{\sqrt{n}}\right\} \left(1 - 55\,\gamma^{1/(2M+1)^2} - 2^{-4n}\gamma^{-\sqrt{n}}\right) \end{aligned}$$

which proves the desired result. □

The arguments for the last two results can be extended easily to give the conclusion quoted in Section 6 as (6.1):

Theorem A.3. If $p > 0$ then

$$\liminf_{n\to\infty} P(\{-2K,\ldots,2K\} \cap W^p_{2n} \neq \emptyset) \geq 1 - \epsilon_K$$

where $\epsilon_K \to 0$ as $K \to \infty$.

PROOF: By the reasoning in the proof of Theorem A.2, we have $\{-2K,\ldots,2K\} \cap W^p_{2n} \neq \emptyset$ if and only if there is a path down from some (x,n) with $|x| \leq 2K$ to $(y,0)$ for some $y \in W^p_{2n}$. To estimate the probability that this occurs we suppose that all the sites $\{-2K+1,-2K+3,\ldots,2K-1\}$ are open at time $2n+1$, let

$$\hat{\mathcal{C}} = \{(x,t) : (y,2n+1) \to_* (x,t) \text{ for some } |y| \leq 2K-1\}$$

and turn the cluster $\hat{\mathcal{C}}$ into a solid blob by looking at

$$\hat{\mathcal{D}} = \cup_{(m,n)\in\hat{\mathcal{C}}} (m,n) + D$$

where $D = \{z \in \mathbf{R}^2 : \|z\|_1 \leq 1\}$. As in the proof of Theroem $A.1$ when $|\hat{\mathcal{C}}| < \infty$ we can define a contour associated with the cluster, and when the contour has length n there will be at least $n/4(2M+1)^d$ closed sites so that for each $z \neq w$ in this set $\|z-w\|_\infty > M$. Since this time the shortest contour has length $8K$ using $(A.2)$ gives

$$P(|\hat{\mathcal{C}}| < \infty) \leq 2^{-8K}$$

If we let

$$\hat{W}^{K,2n+1}_m = \{y : (x,2n+1) \to_* (y,2n-m) \text{ for some } |x| \leq 2K-1\}$$

then the argument in the proof of Theorem $A.2$ shows that

$$P(0 < |\hat{W}^{K,2n+1}_m| \leq \sqrt{n}) \leq 2^{-4n}\gamma^{-\sqrt{n}}$$

So repeating the last computation in the proof of Theorem A.2 proves the result with $\epsilon_K = 2^{-8K}$ □

Our last task is to prove Theorem 4.3. We begin by recalling the

Comparison Assumptions. We suppose given the following ingredients: a translation invariant finite range process $\xi_t : \mathbf{Z}^d \to \{0,1,\ldots \kappa - 1\}$ that is constructed from the graphical representation given in Section 2, an integer L, and a collection H of configurations determined by the values of ξ on $[-L,L]^d$ with the following property:

if $\xi \in H$ then there is an event G_ξ measurable with respect to the graphical representation in $[-k_0L, k_0L]^d \times [0, j_0T]$ and with $P(G_\xi) \geq (1-\theta)$ so that if $\xi_0 = \xi$ then on G_ξ, ξ_T lies in $\sigma_{2Le_1}H$ and in $\sigma_{-2Le_1}H$.

Here $(\sigma_y\xi)(x) = \xi(x+y)$ denote the translation (or shift) of ξ by y and $\sigma_y H = \{\sigma_y\xi : \xi \in H\}$. If we let $M = \max\{j_0, k_0\}$ then the space time regions

$$\mathcal{R}_{m,n} = (m2Le_1, nT) + \{[-k_0L, k_0L]^d \times [0, j_0T]\}$$

that correspond to points $(m,n), (m',n') \in \mathcal{L}$ with $\|(m,n)-(m',n')\|_\infty > M$ are disjoint.

Theorem A.4. If the comparison assumptions hold then we can define random variables $\omega(x,n)$ so that $X_n = \{m : (m,n) \in \mathcal{L}_0, \xi_{nT} \in \sigma_{m2Le_1}H\}$ dominates an M dependent oriented percolation process with initial configuration $W_0 = X_0$ and density at least $1-\gamma$, i.e., $X_n \supset W_n$ for all n.

PROOF: We will define the $\omega(x,n)$ in the oriented percolation by induction. We begin by setting $V_0 = X_0$ and defining a slightly enlarged version of the percolation process V_n consisting of all the y so that can be reached from some $(x_0, 0)$ with $x_0 \in V_0$ by a sequence $x_0, x_1, \ldots x_n = y$ so that $|x_k - x_{k-1}| = 1$ for $1 \leq k \leq n$ and $\omega(x_k, k) = 1$ for $0 \leq k < n$, i.e., the last point in the sequence does not have to be open. Since $V_n \supset W_n$ it suffices to show that $X_n \supset V_n$.

Let $n \geq 0$ and suppose that V_n and the $\omega(x,\ell)$ with $\ell < n$ have been defined so that $X_n \supset V_n$. To define the $\omega(m,n)$, and hence V_{n+1}, we consider two cases.

CASE 1. $m \in V_n \subset X_n$. We set $\omega(m,n) = 1$ if $G_{\sigma_{-m2Le_1}\xi_{nT}}$ occurs in the graphical representation translated by $-m2Le_1$ in space and $-nT$ in time, 0 otherwise. By assumption this event is determined by the Poisson points in $\mathcal{R}_{m,n}$, has probability at least $1-\gamma$, and guarantees that $(m+1), (m-1) \in X_{n+1}$.

CASE 2. $m \notin V_n$. In this case, the value of $\omega(m,n)$ is not important for the evolution of the percolation process so we set $\omega(m,n)$ equal to an independent random variable that is 1 with probability $1-\gamma$ and 0 with probability γ.

If $m \in V_{n+1}$ then either $m-1 \in V_n$ and $\omega(m-1,n) = 1$ or $m+1 \in V_n$ and $\omega(m+1,n) = 1$. In either case the observation in Case 1 implies that $m \in X_{n+1}$. The last conclusion and induction imply that $X_n \supset V_n$ for all n. The last detail to check is that the $\omega(m,n)$ satisfy $(A.1)$ and again we use induction. If $I = 1$ the conclusion is true, so suppose now that $k > 1$ and that the conclusion is true for $I = k-1$. Let (x_i, n_i) $1 \leq i \leq k$ be a seqeunce of points with $\|(x_i,n_i)-(x_j,n_j)\|_\infty > M$ if $i \neq j$ and suppose that the sequence has been indexed so that $n_k \geq n_j$ for all $j < k$. Let $\mathcal{F}$ be the information contained in the graphical representation up to time n_kT or in one of the space time boxes $\mathcal{R}_{m_i,n_i}$ with $i < k$. The comparison assumptions and the fact that $n_k \geq n_j$ for $j < k$ imply that

$$P(\omega(m_k, n_k) = 0|\mathcal{F}) \leq \gamma$$

Integrating the last inequality over $E_{k-1} = \{\omega(m_i, n_i) = 0 \text{ for } i \leq k-1\} \in \mathcal{F}$ which by induction has probability smaller than γ^{k-1} gives

$$\gamma^k \geq \int_{E_{k-1}} \gamma \, dP \geq \int_{E_{k-1}} P(\omega(m_k, n_k) = 0 | \mathcal{F}) \, dP$$
$$= P(E_{k-1} \cap \{\omega(m_k, n_k) = 0\}) = P(E_k)$$

which verifies $(A.1)$ and completes the proof. □

REFERENCES

Andjel, E.D. T.M. Liggett, and T. Mountford (1992) Clustering in one dimensional threshold voter models. *Stoch. Processes Appl.* **42**, 73–90

Aronson, D.G. and H.F. Weinberger (1978) Multidimensional diffusion equations arising in population genetics. *Advances in Math.* **30**, 33–76

Asmussen, S. and N. Kaplan (1976) Branching random walks, I. *Stoch. Processes Appl.* **4**, 1–13

Bezuidenhout, C. and L. Gray (1993) Critical attractive spin systems. *Ann. Probab.*, to appear

Bezuidenhout, C. and G. Grimmett (1990) The critical contact process dies out. *Ann. Probab.* **18**, 1462–1482

Bezuidenhout, C. and G. Grimmett (1991) Exponential decay for subcritical contact and percolation processes. *Ann. Probab.* **19**, 984–1009

Boerlijst, M.C. and P. Hogeweg (1991) Spiral wave structure in pre-biotic evolution: hypercycles stable against parasites. *Physica D* **48**, 17–28

Bramson, M. (1983) Convergence of solutions of the Kolmogorov equation to travelling waves. *Memoirs of the AMS*, **285**

Bramson, M. and R. Durrett (1988) A simple proof of the stability theorem of Gray and Griffeath. *Probab. Th. Rel. Fields* **80**, 293–298

Bramson, M., R. Durrett, and G. Swindle (1989) Statistical mechanics of Crabgrass. *Ann. Prob.* **17**, 444–481

Bramson, M. and L. Gray (1992) A useful renormalization argument. Pages ??? in *Random Walks, Brownian Motion, and Interacting Particle Systems*, edited by R. Durrett and H. Kesten, Birkhauser, Boston

Bramson, M. and D. Griffeath (1987) Survival of cyclic particle systems. Pages 21–30 in *Percolation Theory and Ergodic Theory of Infinite Particle Systems* edited by H. Kesten, IMA Vol. 8, Springer

Bramson, M. and D. Griffeath (1989) Flux and fixation in cyclic particle systems *Ann. Probab.* **17**, 26–45

Bramson, M. and C. Neuhauser (1993) Survival of one dimensional cellular automata. Preprint

Chen, H.N. (1992) On the stability of a population growth model with sexual reproduction in $\mathbf{Z}^2$. *Ann. Probab.* **20**, 232–285

Cox, J.T. (1988) Coalescing random walks and voter model consensus times on the torus in $\mathbf{Z}^d$. *Ann. Probab.*

Cox, J.T. and R. Durrett (1988) Limit theorems for the spread of epidemics and forest fires. *Stoch. Processes Appl.* **30**, 171–191

Cox, J.T. and R. Durrett (1992) Nonlinear voter models. Pages 189–202 in *Random Walks, Brownian Motion, and Interacting Particle Systems*, edited by R. Durrett and H. Kesten, Birkhauser, Boston

Cox, J.T. and D. Griffeath (1986) Diffusive clustering in the two dimensional voter model. *Ann. Probab.* **14**, 347–370

DeMasi, A., P. Ferrari, and J. Lebowitz (1986) Reaction diffusion equations for interacting particle systems. *J. Stat. Phys.* **44**, 589–644

DeMasi, A. and E. Presutti (1991) *Mathematical Methods for Hydrodynamic Limits.* Lecture Notes in Math **1501**, Springer, New York

Durrett, R. (1980) On the growth of one dimensional contact processes. *Ann. Probab.* **8**, 890–907

Durrett, R. (1984) Oriented percolation in two dimensions. *Ann. Probab.* **12**, 999–1040

Durrett, R. (1988) *Lecture Notes On Particle Systems And Percolation.* Wadsworth, Belmont, CA

Durrett, R. (1991a) Stochastic models of growth and competition. Pages 1049–1056 in *Proceedings of the International Congress of Mathematicians, Kyoto*, Springer, New York

Durrett, R. (1991b) The contact process, 1974–1989. Pages 1–18 in *Proceedings of the AMS Summer seminar on Random Media.* Lectures in Applied Math **27**, AMS, Providence, RI

Durrett, R. (1991c) Some new games for your computer. *Nonlinear Science Today* Vol. 1, No. 4, 1–7

Durrett, R. (1992a) Multicolor particle systems with large threshold and range. *J. Theoretical Prob.*, 5 (1992), 127–152

Durrett, R. (1992b) A new method for proving the existence of phase transitions. Pages 141-170 in *Spatial Stochastic Processes*, edited by K.S. Alexander and J.C. Watkins, Birkhauser, Boston

Durrett, R. (1992c) Stochastic growth models: bounds on critical values. *J. Appl. Prob.* **29**

Durrett, R. (1992d) *Probability: Theory and Examples.* Wadsworth, Belmont, CA

Durrett, R. (1993) Predator-prey systems. Pages 37–58 in *Asymptotic Problems in Probability Theory: Stochastic Models and Diffusions on Fractals*, edited by K.D. Elworthy and N. Ikeda, Pitman Research Notes in Math 283, Longman, Essex, England

Durrett, R. and D. Griffeath (1993) Asymptotic behavior of excitable cellular automata. Preprint

Durrett, R. and S. Levin (1993) Stochastic spatial models: A user's guide to ecological applications. *Phil. Trans. Roy. Soc. B*, to appear

Durrett, R. and A.M. Moller (1991) Complete convergence theorem for a competition model. *Probab. Th. Rel. Fields* **88**, 121–136

Durrett, R. and C. Neuhauser (1991) Epidemics with recovery in $d = 2$. *Ann. Applied Probab.* **1**, 189–206

Durrett, R. and C. Neuhauser (1993) Particle systems and reaction diffusion equations. *Ann. Probab.*, to appear

Durrett, R. and R. Schinazi (1993) Asymptotic critical value for a competition model. *Ann. Applied Probab.*, to appear

Durrett, R. and J. Steif (1993) Fixation results for threshold voter models. *Ann. Probab.*, to appear

Durrett, R. and G. Swindle (1991) Are there bushes in a forest? *Stoch. Proc. Appl.* **37**, 19–31

Durrett, R. and G. Swindle (1993) Coexistence results for catalysts. Preprint

Eigen, M. and P. Schuster (1979) *The Hypercycle: A Principle of Natural Self-Organization*, Springer, New York

Fife, P.C. and J.B. McLeod (1977) The approach of solutions of nonlinear diffusion equations to travelling front solutions. *Arch. Rat. Mech. Anal.* **65**, 335–361

Fisch, R. (1990a) The one dimensional cyclic cellular automaton: a system with deterministic dynamics which emulates a particle system with stochastic dynamics. *J. Theor. Prob.* **3**, 311–338

Fisch, R. (1990b) Cyclic cellular automata and related processes. *Physica D* **45**, 19–25

Fisch, R. (1992) Clustering in the one dimensional 3-color cyclic cellular automaton. *Ann. Probab.* **20**, 1528–1548

Fisch, R., J. Gravner, and D. Griffeath (1991) Threshold range scaling of excitable cellular automata. *Statistics and Computing* **1**, 23–39

Fisch, R., J. Gravner, and D. Griffeath (1992) Cyclic cellular automata in two dimensions. In *Spatial Stochastic Processes* edited by K. Alexander and J. Watkins, Birkhauser, Boston

Fisch, R., J. Gravner, and D. Griffeath (1993) Metastability in the Greenberg Hastings model. *Ann. Applied. Probab.*, to appear

Grannan, E. and G. Swindle (1991) Rigorous results on mathematical models of catalyst surfaces. *J. Stat. Phys.* **61**, 1085–1103

Gravner, J. and D. Griffeath (1993) Threshold growth dyanmics. *Transactions A.M.S.*, to appear

Gray, L. and D. Griffeath (1982) A stability criterion for attractive nearest neighbor spin systems on **Z**. *Ann. Probab.* **10**, 67–85

Gray, L. (1987) Behavior of processes with statistical mechanical properties. Pages 131–168 in *Percolation Theory and Ergodic Theory of Infinite Particle Systems* edited by H. Kesten, IMA Vol. 8, Springer

Griffeath, D. (1979) *Additive and Cancellative Interacting Particle Systems.* Lecture Notes in Math **724**, Springer

Harris, T.E. (1972) Nearest neighbor Markov interaction processes on multidimensional lattices. *Adv. in Math.* **9**, 66–89

Harris, T.E. (1976) On a class of set valued Markov processes. *Ann. Probab.* **4**, 175–194

Hassell, M.P., H.N. Comins, and R.M. May (1991) Spatial structure and chaos in insect population dynamics. *Nature* **353**, 255–258

Hirsch, M.W. and S. Smale (1974) *Differential Equations, Dynamical Systems, and Linear Algebra*, Academic Press, New York

Holley, R.A. (1972) Markovian interaction processes with finite range interactions. *Ann. Math. Stat.* **43**, 1961–1967

Holley, R.A. (1974) Remarks on the FKG inequalities. *Commun. Math. Phys.* **36**, 227–231

Holley, R.A., and T.M. Liggett (1975) Ergodic theorems for weakly interacting systems and the voter model. *Ann. Probab.* **3**, 643–663

Holley, R.A., and T.M. Liggett (1978) The survival of contact processes. *Ann. Probab.* **6**, 198–206

Kinzel, W. and J. Yeomans (1981) Directed percolation: a finite size renormalization approach. *J. Phys. A* **14**, L163–L168

Liggett, T.M. (1985) *Interacting Particle Systems.* Springer, New York

Liggett, T.M. (1993) Coexistence in threshold voter models. *Ann. Probab.*, to appear

Neuhauser, C. (1992) Ergodic theorems for the multitype contact process. *Probab. Theory Rel. Fields.* **91**, 467–506

Redheffer, R., R. Redlinger, and W. Walter (1988) A theorem of La Salle-Lyapunov type for parabolic systems. *SIAM J. Math. Anal.* **19**, 121–132

Schonmann, R.H. and M.E. Vares (1986) The survival of the large dimensional basic contact process. *Probab. Th. Rel. Fields* **72**, 387–393

Spohn, H. (1991) *Large Scale Dynamics of Interacting Particle Systems*, Springer, New York

Zhang, Yu (1992) A shape theorem for epidemics and forest fires with finite range interactions. Preprint